21世纪高等学校计算机专业实用规划教材

Java EE（SSM）企业应用实战

◎千锋教育高教产品研发部／编著

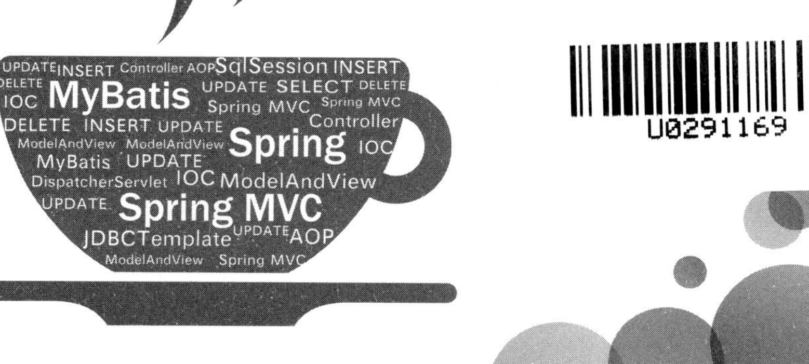

清華大学出版社
北京

内 容 简 介

本书全面介绍了Java EE中MyBatis、Spring和Spring MVC三大框架的基本知识和使用方法。书中对知识点的讲解由浅入深、通俗易懂，同时配备大量的操作案例，通过案例的演示帮助读者理解技术原理并提高实际操作能力。全书主要讲解了MyBatis、Spring、Spring MVC的相关知识，最后是一个项目案例，通过项目案例帮助读者掌握SSM框架整合的技术，让读者适应企业级开发的技术需要，为大型项目开发奠定基础。

本书适合作为高等院校计算机类相关课程的教材，同时也可作为编程人员的学习指南。

图书在版编目(CIP)数据

Java EE(SSM)企业应用实战/千锋教育高教产品研发部编著. —北京：清华大学出版社，2019(2024.2重印)
(21世纪高等学校计算机专业实用规划教材)
ISBN 978-7-302-53015-2

Ⅰ. ①J… Ⅱ. ①千… Ⅲ. ①JAVA语言－程序设计－高等学校－教材 Ⅳ. ①TP312.8

中国版本图书馆CIP数据核字(2019)第093946号

责任编辑：闫红梅
封面设计：胡耀文
责任校对：徐俊伟
责任印制：丛怀宇

出版发行：清华大学出版社
网　　址：https://www.tup.com.cn，https://www.wqxuetang.com
地　　址：北京清华大学学研大厦A座　　**邮　　编**：100084
社 总 机：010-83470000　　**邮　　购**：010-62786544
投稿与读者服务：010-62776969，c-service@tup.tsinghua.edu.cn
质量反馈：010-62772015，zhiliang@tup.tsinghua.edu.cn
课件下载：https://www.tup.com.cn，010-83470236
印 装 者：三河市龙大印装有限公司
经　　销：全国新华书店
开　　本：185mm×260mm　　**印　张**：24.75　　**字　　数**：612千字
版　　次：2019年8月第1版　　**印　　次**：2024年2月第10次印刷
印　　数：16501～18500
定　　价：69.00元

产品编号：078610-02

编 委 会

为什么要写这样一本书

当今世界是知识爆炸的世界，科学技术与信息技术急速地发展，新型技术层出不穷。但教科书却不能将这些知识内容及时编入，致使教科书的有些知识内容显得陈旧不实用。在初学者还不会编写代码的情况下，就开始讲解算法，会使初学者感到晦涩难懂，让初学者难以入门。

IT 行业的从业者不仅需要掌握理论知识，更需要具有技术过硬、综合能力强的实用技能。所以，高校毕业生求职面临的第一道门槛就是技能与经验的考验。学校往往只注重学生的理论知识，忽略了对学生的实践能力培养，因而导致学生无法将理论知识应用到实际工作中。

如何解决这一问题

为了解决这一问题，本书倡导的是快乐学习，实战就业。在语言描述上力求准确、通俗、易懂，在章节编排上力求循序渐进，在语法阐述时尽量避免术语和公式，从项目开发的实际需求入手，将理论知识与实际应用相结合。目标就是让初学者能够快速成长为初级程序员，并拥有一定的项目开发经验，从而在职场中拥有一个高起点。

千锋教育

前　言

在瞬息万变的IT时代，一群怀揣梦想的人创办了千锋教育，投身到IT培训行业。八年来，一批批有志青年加入千锋教育，为了梦想笃定前行。千锋教育秉承"用良心做教育"的理念，为培养"顶级IT精英"而付出一切努力。为什么会有这样的梦想？我们先来听一听用人企业和求职者的心声：

"现在符合企业需求的IT技术人才非常紧缺，这方面的优秀人才我们会像珍宝一样对待，可为什么至今没有合格的人才出现？"

"面试的时候，用人企业问能做什么，这个项目如何来实现，需要多长的时间，我们当时都蒙了，回答不上来。"

"这已经是面试过的第10家公司了，如果再不行的话，是不是要考虑转行了，难道大学四年都白学了？"

"这已经是参加面试的第N个求职者了，为什么都是计算机专业毕业，但当问到项目如何实现时，却怎么连思路都没有呢？"

……

这些心声并不是个别现象，而是现实社会中的普遍现象。高校的IT教育与企业的真实需求脱节，如果高校的相关课程仍然不进行更新的话，毕业生将面临难以就业的困境，很多用人单位表示，高校毕业生表象上知识丰富，但在学校所学的知识绝大多数在实际工作中用之甚少，甚至完全用不上。针对上述存在的问题，国务院也作出了关于加快发展现代职业教育的决定，千锋教育所做的事情就是配合高校达成产学合作。

千锋教育致力于打造IT职业教育全产业链人才服务平台，在全国拥有数十家分校，数百名讲师，坚持以教学为本的方针，全国采用面对面教学，传授企业实用技能。教学大纲紧跟企业需求，拥有全国一体化就业体系。千锋的价值观即"做真实的自己，用良心做教育"。

针对高校教师的服务

(1) 千锋教育基于8年来的教育培训经验，精心设计了"教材＋授课资源＋考试系统＋测试题＋辅助案例"的教学资源包，节约教师的备课时间，缓解教师的教学压力，显著提高教学质量。

(2) 本书配套代码视频，网址为 http://www.codingke.com/。

(3) 本书配备了千锋教育优秀讲师录制的教学视频，按本书知识结构体系部署到了教学辅助平台(扣丁学堂)上，这些教学视频既可以作为教学资源使用，也可以作为备课参考。

高校教师如需配套教学资源，请关注(扣丁学堂)师资服务平台，扫描下方二维码关注微信公众平台获取。

扣丁学堂

针对高校学生的服务

(1) 学 IT 有疑问,就找千问千知,它是一个有问必答的 IT 社区,平台上的专业答疑辅导老师承诺:工作时间 3 小时内答复读者在学习 IT 中遇到的专业问题。读者也可以通过扫描下方的二维码,关注千问千知微信公众平台,浏览其他学习者在学习中分享的问题和收获。

(2) 学习太枯燥,想了解其他学校的伙伴都是怎样学习的吗?可以加入"扣丁俱乐部"。"扣丁俱乐部"是千锋教育联合各大校园发起的公益计划,专门面向对 IT 感兴趣的大学生,为其提供免费的学习资源和问答服务,已有超过 30 多万名学习者从中获益。

就业难,难就业,千锋教育让就业不再难!

千问千知

关于本书

本书既可作为高等院校本、专科计算机相关专业的授课教材,也可作为计算机类培训教材,其中包含了千锋教育 Java EE(SSM 框架)全部的课程内容,是一本适合广大计算机编程爱好者的优秀读物。

抢红包

添加小千 QQ 号或微信号 2133320438,不仅可以获取本书配套源代码及习题答案,还可能获得小千随时发放的"助学金红包"。

致谢

千锋教育高教产品研发部将千锋 Java 学科多年积累的教学实战案例进行整合,通过反复精雕细琢,最终完成了本书。另外,多名院校老师也参与了本书的部分编写与指导工作,除此之外,千锋教育 500 多名学员也参与了本书的试读工作,他们站在初学者的角度对本书提供了许多宝贵的修改意见，在此一并表示衷心的感谢。

意见反馈

在本书的编写过程中，虽然力求完美，但不足之处在所难免，欢迎各界专家和读者朋友们给予宝贵意见，联系方式：huyaowen@1000phone.com。

千锋教育高教产品研发部

2018年12月25日于北京

目　录

第1章 MyBatis 基础

本章学习目标

- 了解 ORM 框架的概念
- 理解 MyBatis 的基本概念
- 理解 MyBatis 的功能架构和工作流程
- 掌握 MyBatis 的下载和安装
- 掌握 MyBatis 入门程序的编写

Java 程序依靠 JDBC 实现对数据库的操作，但是在大型企业项目中，由于程序与数据库交互次数较多以及读写数据量较大，仅仅使用 JDBC 操作数据库无法满足性能要求，同时，JDBC 的使用也会带来代码冗余、复用性低等问题，因此，企业级开发中一般使用 MyBatis 等 ORM 框架操作数据库。接下来，本章将对 MyBatis 框架涉及的基础知识进行详细讲解。

1.1 MyBatis 概述

1.1.1 传统 JDBC 的劣势

JDBC 是 Java 程序实现数据访问的基础，它提供了一套操作数据库的 API，一般通过加载驱动、获取连接、获取执行者对象、发送 SQL 语句等步骤实现数据库操作。但是，传统的 JDBC 编程存在一定的局限性，具体如下。

1）代码烦琐

使用 JDBC 编程时，代码量较大，尤其是当数据表字段较多时，代码显得烦琐、累赘并且使开发人员的工作量增加。

2）表关系维护复杂

数据表之间存在各种关系，包括一对一、一对多、多对多、级联等。如果采用 JDBC 编程的方式维护数据表之间的关系，过程较为复杂并且容易出错。

3）硬编码

当使用 JDBC 编程时，SQL 语句都是硬编码到 Java 程序中，如果改变 SQL 语句，那么需要重新编译 Java 代码，不利于系统后期的维护。

4）性能问题

在批量处理数据的时候，JDBC 编程存在效率低下的问题，此时，程序将向数据库发送大批量的同类 SQL 语句请求，浪费数据库资源，影响运行效率。

由于 JDBC 存在的缺陷，企业中通常使用 ORM 框架来完成数据库的操作。

1.1.2 ORM 简介

ORM 的全称是 Object Relational Mapping，即对象-关系映射。ORM 是一种规范，它是将简单 Java 对象（POJO）和数据库表记录进行映射，使数据库表中的记录和 POJO 一一对应，如图 1.1 所示。

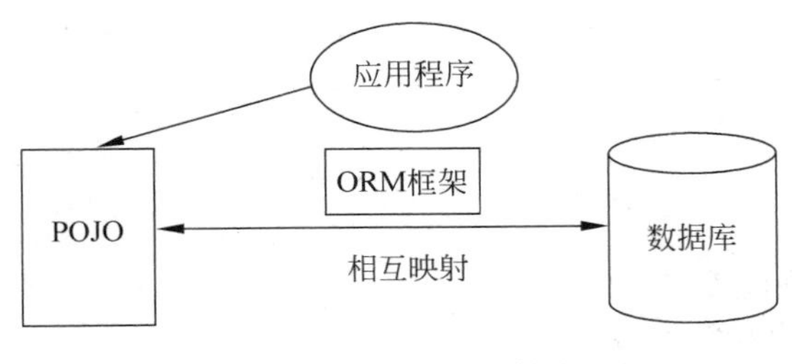

图 1.1 ORM 映射

在当今企业级应用的开发环境中，对象和关系数据是业务实体的两种表现形式。业务实体在内存中表现为对象，在数据库中表现为关系数据。当采用面向对象的方法开发程序时，一旦需要操作数据库，就必须回到关系数据库的访问方式，这给开发人员编程带来一系列的困扰。于是，ORM 应运而生。ORM 可以把关系数据库包装成为面向对象的模型，并成为应用程序和数据库交互的桥梁。

ORM 框架是对象-关系映射的系统化解决方案，当 ORM 框架完成映射后，开发人员既可以利用面向对象程序设计语言的简单易用性，又可以充分发挥关系数据库的优势。目前，在 Java 应用领域流行的 ORM 框架有 Hibernate、MyBatis 等，下面对这两种 ORM 框架做简要介绍。

1）Hibernate

Hibernate 是较为优秀的 ORM 框架之一，已被选为 JBoss 服务器的持久层解决方案。

Hibernate 建立在 POJO 和数据库表记录的直接映射关系之上，它通过 XML 映射文件（或注解）提供的规则实现关系映射，提供一种全表映射的模型，程序可以通过 POJO 直接操作数据库中的数据。Hibernate 封装性较高，开发人员通过 XML 映射文件（或注解）定义好映射规则后，Hibernate 会根据映射规则自动生成 SQL 语句并调用 JDBC 的 API 执行，这减少了开发人员编写 SQL 语句的烦琐，大大提升开发效率。但是，由于全表映射的特性，Hibernate 也存在一些局限，例如，无法根据不同的条件组装不同的 SQL，对多表关联和复杂 SQL 查询支持较差，不能有效支持存储过程和 SQL 语句优化等。随着互联网行业的发展，Hibernate 的局限性逐渐显现并影响其市场份额，目前，MyBatis 已成为互联网企业的首选。

2）MyBatis

MyBatis 是一种“半自动化”的 ORM 框架，和 Hibernate 不同，MyBatis 需要手动提供 POJO、SQL 语句并匹配映射关系，正因为此，它可以更加灵活地生成映射关系。MyBatis 充分允许开发人员利用数据库的各项功能，例如存储过程、视图、复杂查询等，具有高度灵活、可优化、易维护等优点。与 Hibernate 相比，使用 MyBatis 的编码量较大，但这并不影响它在一些复杂的和需要优化性能的项目中使用。

1.1.3 MyBatis 简介

MyBatis 的前身是 Apache 组织的一个开源项目 iBatis。2010 年，iBatis 由 Apache Software Foundation 迁移到了 Google Code，并且改名为 MyBatis。2013 年 11 月迁移到 Github，目前 MyBatis 由 Github 维护。

MyBatis 是一款优秀的 ORM 框架，它支持自定义 SQL、存储过程以及高级映射。MyBatis 避免了几乎所有的 JDBC 代码和手动设置参数以及获取结果集，它使用 XML 文件或注解进行配置和映射，将接口和 Java 的 POJO 映射成数据库中的记录。

MyBatis 作为 ORM 框架，其核心思想是剥离出程序中的大量 SQL 语句并将这些 SQL 语句配置到映射文件中。因此，使用 MyBatis 可以根据开发需要灵活编写 SQL 语句并指定映射规则，同时，程序也和 SQL 语句分离，实现在不修改程序代码的情况下变更 SQL 语句的功能，提升了程序的扩展性。

此外，MyBatis 支持动态列、动态表名、存储过程，同时提供了简易的日志、缓存和级联功能。

1.1.4 MyBatis 的功能架构

MyBatis 具有自身独特的功能架构，具体如图 1.2 所示。

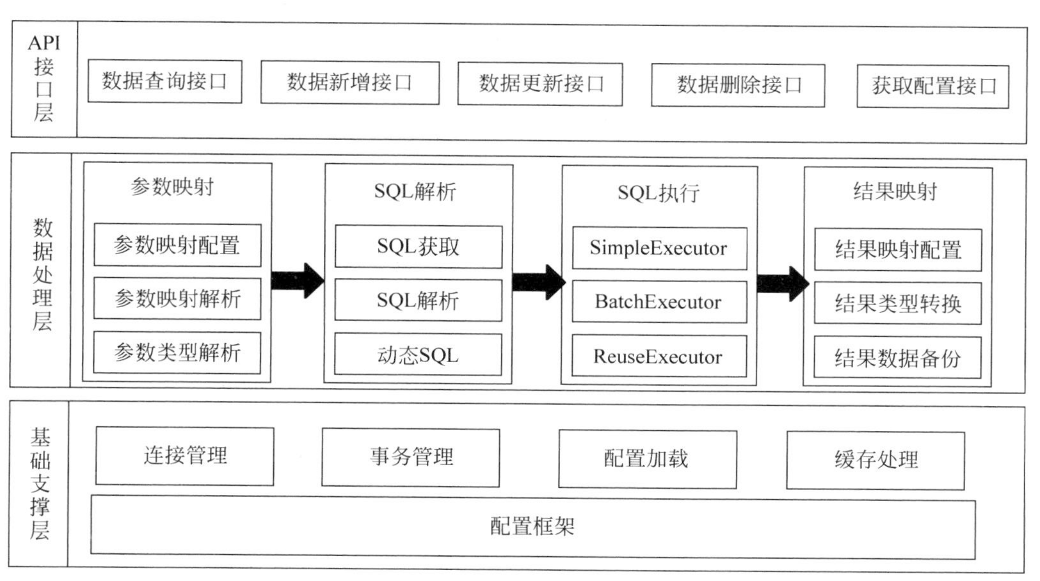

图 1.2 MyBatis 的功能架构

从图 1.2 中可以看出，MyBatis 的功能架构由三层组成，包括 API 接口层、数据处理层、基础支撑层。

API 接口层：提供给外部使用的接口 API，开发人员通过这些 API 来操纵数据库。API 接口层接收到调用请求时，会调用数据处理层来完成具体的数据处理。

数据处理层：负责具体的参数映射、SQL 解析、SQL 执行和结果映射等。它主要的功能是根据调用的请求完成一次数据库操作。

基础支撑层：负责最基础的功能支撑，包括连接管理、事务管理、配置加载、缓存处理等，MyBatis 将这些共用的功能抽取出来作为最基础的组件，为上层的数据处理层提供支持。

1.1.5 MyBatis 的工作流程

在理解了 MyBatis 的功能架构之后，接下来学习 MyBatis 的工作流程，如图 1.3 所示。

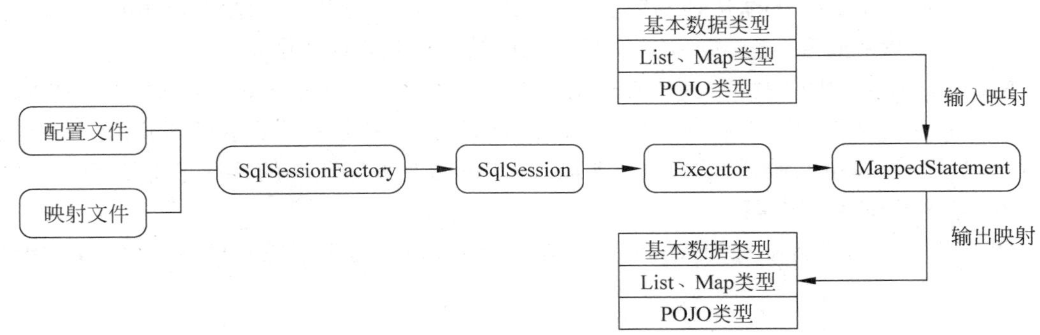

图 1.3　MyBatis 的工作流程

图 1.3 展示了 MyBatis 的工作流程，具体来说，可分为以下步骤。

(1) MyBatis 读取配置文件和映射文件。其中，配置文件设置了数据源、事务等信息；映射文件设置了 SQL 执行相关的信息。映射文件要引入到配置文件中才能被执行。

(2) MyBatis 根据配置信息和映射信息生成 SqlSessionFactory 对象，SqlSessionFactory 对象的重要功能是创建 MyBatis 的核心类对象 SqlSession。

(3) SqlSession 中封装了操作数据库的所有方法，开发者一般通过调用 SqlSession 完成数据库操作，但实际上，SqlSession 并没有直接操作数据库，它通过更底层的 Executor 执行器接口操作数据库。Executor 接口有两个实现类，一个是普通执行器，另外一个是缓存执行器。

(4) Executor 执行器将要处理的 SQL 信息封装到一个 MappedStatement 对象中。在执行 SQL 语句之前，Executor 执行器通过 MappedStatement 对象将输入的 Java 数据映射到 SQL 语句，在执行 SQL 语句之后，Executor 执行器通过 MappedStatement 对象将 SQL 语句的执行结果映射为 Java 数据，其中，作为输入参数和输出结果的映射类型可以为 Java 基本数据类型，也可为 List 类型、Map 类型或 POJO 类型。

以上是 MyBatis 实现一次数据库操作的基本流程，大家先对此做初步了解，以后的章节会有更加深入的讲解。

1.2 MyBatis 的重要 API

1.1.5 节介绍了 MyBatis 的工作流程，接下来对 MyBatis 的两个重要类 SqlSessionFactory 和 SqlSession 做重点讲解。

1. SqlSessionFactory

SqlSessionFactory 的首要功能是创建 SqlSession 对象，因此，每一个 MyBatis 应用程序

都以 SqlSessionFactory 为基础。SqlSessionFactory 存在于 MyBatis 应用的整个生命周期，重复创建 SqlSessionFactory 对象会造成数据库资源的过度消耗，因此，一般使用单例模式创建 SqlSessionFactory 对象，即每一个数据库对应一个 SqlSessionFactory 对象。

SqlSessionFactory 对象由 SqlSessionFactoryBuilder 对象创建，SqlSessionFactoryBuilder 对象通过它的 build()方法创建 SqlSessionFactory 对象，SqlSessionFactoryBuilder 对象的 build()方法有两种形式，具体如下所示。

第一种形式：

```
SqlSessionFactory build(InputStream inputStream [,String environment]
    [,properties props])
```

这种形式较为常用，参数 inputStream 封装了 xml 文件形式的配置信息；参数 environment 和参数 properties 为可选参数，其中，参数 environment 决定将要加载的环境，包括数据源和事务管理器；参数 props 决定将要加载的 properties 文件。

第二种形式：

```
SqlSessionFactory build(Configuration config)
```

其中，参数 config 需要预定义，它封装了绝大部分的配置信息，此处大家先初步了解即可。

创建 SqlSessionFactory 对象之后，接下来调用 SqlSessionFactory 对象的 openSession()方法创建 SqlSession 对象，SqlSessionFactory 对象的 openSession()方法有多种形式，具体如表 1.1 所示。

表 1.1　SqlSessionFactory 对象的 openSession()方法

方法名称	功能描述
SqlSession openSession()	开启一个事务，连接对象会从由活动环境配置的数据源实例中得到，事务隔离级别将会使用驱动或数据源的默认设置，预处理语句不会被复用，也不会批量处理更新
SqlSession openSession(Boolean autoCommit)	参数 autoCommit 可设置是否开启事务
SqlSession openSession(Connection connection)	参数 connection 可提供自定义连接
SqlSession openSession(TransactionIsolationLevel level)	参数 level 可设置隔离级别
SqlSession openSession(ExecutorType execType)	参数 ExecutorType 有三个可选值，其中，ExecutorType. SIMPLE 表示为每条语句的执行创建一条新的预处理语句；ExecutorType. REUSE 表示会复用预处理语句；ExecutorType. BATCH 表示会批量执行所有更新语句
SqlSession openSession(ExecutorType execType, boolean autoCommit)	参数 autoCommit 可设置是否开启事务，其他功能等同于不传入参数 autoCommit 时
SqlSession openSession(ExecutorType execType, Connection connection)	参数 connection 可提供自定义连接，其他功能等同于不传入参数 connection 时

2. SqlSession

SqlSession 对象是 MyBatis 中的核心类对象,它类似于 JDBC 中的 Connection 对象,其首要作用是执行持久化操作,具有强大功能,在开发过程中最为常用。

SqlSession 对象的生命周期贯穿数据库处理事务的过程,一定时间内没有使用的 SqlSession 对象要及时关闭,以免影响系统性能。SqlSession 对象是线程不安全的,也不能被共享,因此,开发者应重点关注多线程状态下的 SqlSession 对象,同时,操作时应注意隔离级别、数据库锁等高级特性。

SqlSession 对象中提供了执行 SQL 语句、提交或回滚事务、清理 Session 级的缓存以及使用映射器等功能的方法,具体如表 1.2 所示。

表 1.2 SqlSession 对象的方法

方法名称	功能描述
T selectOne(String statement)	执行单条记录的查询,返回一个映射查询结果的泛型对象
T selectOne(String statement, Object parameter)	执行单条记录的查询,可传入查询所需的参数对象,返回一个映射查询结果的泛型对象
List selectList(String statement)	执行多条记录的查询,返回一个映射查询结果的泛型对象的集合
List selectList(String statement, Object parameter)	执行多条记录的查询,可传入查询所需的参数对象,返回一个映射查询结果的泛型对象的集合
Map selectMap(String statement, String mapKey)	执行查询,返回一个映射查询结果的 Map
Map selectMap(String statement, Object parameter, String mapKey)	执行查询,可传入查询所需的参数对象,返回一个映射查询结果的 Map
int insert(String statement)	执行数据的插入,返回执行 SQL 语句影响的行数
int insert(String statement, Object parameter)	执行数据的插入,可传入插入所需的参数对象,返回执行 SQL 语句影响的行数
int update(String statement)	执行数据的更新,返回执行 SQL 语句影响的行数
int update(String statement, Object parameter)	执行数据的更新,可传入更新所需的参数对象,返回执行 SQL 语句影响的行数
int delete(String statement)	执行数据的删除,返回执行 SQL 语句影响的行数
int delete(String statement, Object parameter)	执行数据的删除,可传入删除所需的参数对象,返回执行 SQL 语句影响的行数
void commit()	提交事务
void commit(Boolean force)	强制提交事务
void rollback()	回滚事务
void rollback(Boolean force)	强制回滚事务
void clearCache()	清理 Session 级的缓存
void close()	关闭 SqlSession
T getMapper(Class type)	获取映射器

1.3 MyBatis 的下载和使用

由于 MyBatis 由第三方组织提供，因此在使用 MyBatis 之前首先要下载 jar 包。本书编写时 MyBatis 的最新版本为 3.4.6，因此书中基于该版本展开讲解。接下来，本节将讲解 MyBatis 相关 jar 包的下载方法，具体步骤如下。

（1）打开浏览器，访问 https://github.com/mybatis/mybatis-3/releases，浏览器跳转到下载页面，如图 1.4 所示。

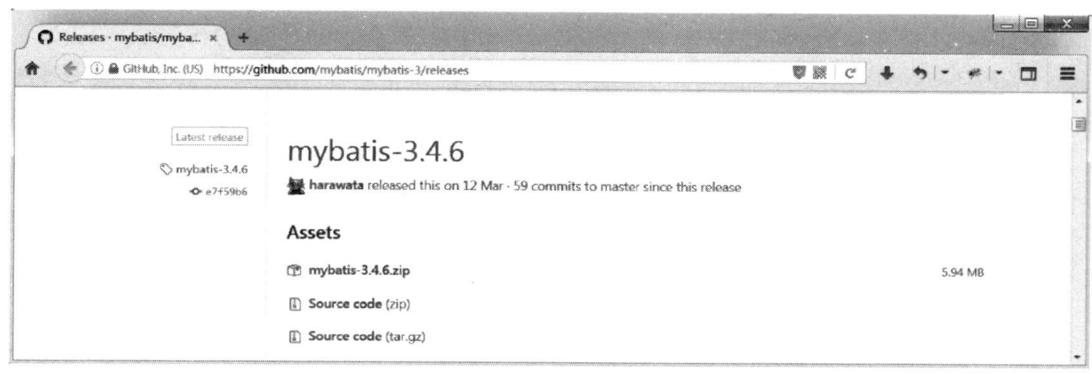

图 1.4　MyBatis 的下载页面

（2）单击页面中 mybatis-3.4.6.zip 超链接，将文件下载到指定目录。

（3）解压下载完成后的 mybatis-3.4.6.zip 文件，此时获得名称为 mybatis-3.4.6 的文件夹，打开该文件夹，可以看到 mybatis 的目录结构，具体如图 1.5 所示。

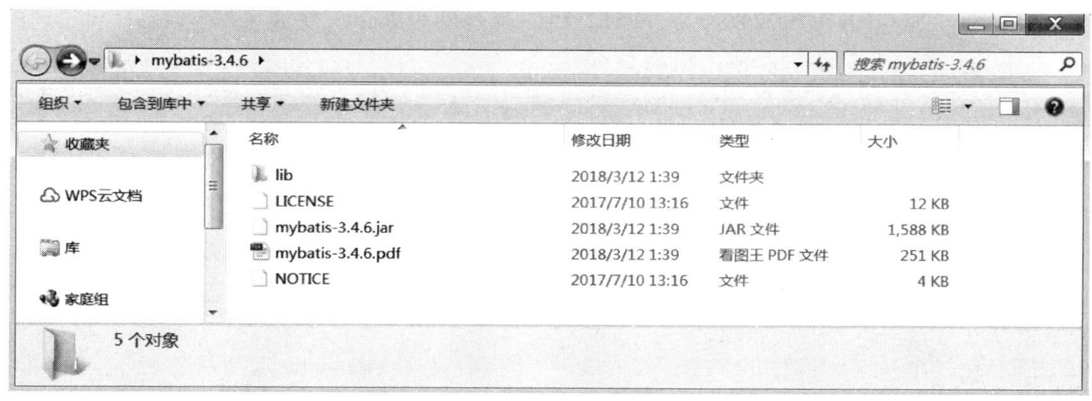

图 1.5　MyBatis 的目录结构

其中，lib 文件夹存放 MyBatis 运行依赖的 jar 包，mybatis-3.4.6.jar 是 MyBatis 的核心类库，mybatis-3.4.6.pdf 是 MyBatis 的参考文档，在使用时需要将 MyBatis 的相关 jar 包导入到工程中。

1.4 MyBatis的简单应用

1.4.1 搭建开发环境

1. 数据准备

(1) 在 MySQL 中创建数据库 chapter01 和数据表 student,SQL 语句如下所示。

```
DROP DATABASE IF EXISTS chapter01;
CREATE DATABASE chapter01;
USE chapter01;
CREATE TABLE student(
sid INT PRIMARY KEY AUTO_INCREMENT, #ID
sname VARCHAR(20),        #学生姓名
age VARCHAR(20),          #学生年龄
course VARCHAR(20)        #专业
);
```

(2) 向数据表 student 中插入数据,SQL 语句如下所示。

```
INSERT INTO student(sname,age,course) VALUES ('ZhangSan','20','Java');
INSERT INTO student(sname,age,course) VALUES ('liSi','21','Java');
INSERT INTO student(sname,age,course) VALUES ('WangWu','22','Java');
INSERT INTO student(sname,age,course) VALUES ('ZhaoLiu','22','Python');
INSERT INTO student(sname,age,course) VALUES ('SunQi','22','PHP');
INSERT INTO student(sname,age,course) VALUES ('ZhangSanSan','22','PHP');
```

(3) 通过 SQL 语句测试数据是否添加成功,执行结果如下所示。

```
mysql> SELECT * FROM student;
+-----+-------------+------+--------+
| sid | sname       | age  | course |
+-----+-------------+------+--------+
|   1 | ZhangSan    | 20   | Java   |
|   2 | liSi        | 21   | Java   |
|   3 | WangWu      | 22   | Java   |
|   4 | ZhaoLiu     | 22   | Python |
|   5 | SunQi       | 22   | PHP    |
|   6 | ZhangSanSan | 22   | PHP    |
+-----+-------------+------+--------+
6 rows in set (0.00 sec)
```

从以上执行结果可以看出,数据添加成功。

2. 创建工程

在 Eclipse 中新建 Web 工程 chapter01,将 MyBatis 的驱动 jar 包(mybatis-3.4.6.jar 和 lib 目录下的 jar 包)复制到 lib 目录下,完成 jar 包的导入。

1.4.2 创建 POJO 类

在工程 chapter01 的 src 目录下新建 com.qfedu.pojo 包，在 com.qfedu.pojo 包下新建 Java 类 Student，具体代码如例 1-1 所示。

【例 1-1】 Student.java

```
package com.qfedu.pojo;
public class Student {
    private int sid;
    private String sname;
    private String age;
    private String course;
    public Student() {
        super();
    }
    public Student(int sid, String sname, String age, String course) {
        super();
        this.sid = sid;
        this.sname = sname;
        this.age = age;
        this.course = course;
    }
    public int getSid() {
        return sid;
    }
    public void setSid(int sid) {
        this.sid = sid;
    }
    public String getSname() {
        return sname;
    }
    public void setSname(String sname) {
        this.sname = sname;
    }
    public String getAge() {
        return age;
    }
    public void setAge(String age) {
        this.age = age;
    }
    public String getCourse() {
        return course;
    }
    public void setCourse(String course) {
        this.course = course;
    }
    @Override
    public String toString() {
```

```
        return "Student [sid = " + sid + ", sname = " + sname + ", "
            + "age = " + age + ", " + "course = " + course + "]";
    }
}
```

Student 类提供了成员变量和 getter/setter 方法，MyBatis 将通过配置文件映射 Student 类和数据表 student 的关系。

1.4.3 创建配置文件

在 src 目录下新建配置文件 mybatis-config.xml，具体代码如例 1-2 所示。

【例 1-2】 mybatis-config.xml

```
<?xml version = "1.0" encoding = "UTF - 8" ?>
<!DOCTYPE configuration
PUBLIC " - //mybatis.org//DTD Config 3.0//EN"
"http://mybatis.org/dtd/mybatis - 3 - config.dtd">
<configuration>
    <! -- 配置环境 -->
    <environments default = "mysql">
        <environment id = "mysql">
            <transactionManager type = "JDBC"/>
            <dataSource type = "POOLED">
            <property name = "driver" value = "com.mysql.jdbc.Driver"/>
            <property name = "url"
                    value = "jdbc:mysql://localhost:3306/chapter01"/>
            <property name = "username" value = "root"/>
            <property name = "password" value = "root"/>
            </dataSource>
        </environment>
    </environments>
    <! -- 配置映射文件的位置 -->
    <mappers>
        <mapper resource = "com/qfedu/mapper/StudentMapper.xml"/>
    </mappers>
</configuration>
```

以上代码用于配置数据库的连接信息，<dataSource>元素的四个属性分别配置数据库的驱动、URL、用户名和密码。<mapper>元素用于配置映射文件的位置。

为了提升开发效率、减少编码错误，开发人员可以在 MyBatis 的使用手册 mybatis-3.4.6.pdf 的 2.1.2 节找到配置信息模板并复制到本地文件中，然后根据实际开发环境完善配置信息。

在 src 目录下新建 com.qfedu.mapper 包，在该包下新建映射文件 StudentMapper.xml，具体代码如例 1-3 所示。

【例 1-3】 StudentMapper.xml

```
<?xml version = "1.0" encoding = "UTF - 8" ?>
<!DOCTYPE mapper
```

```
PUBLIC " - //mybatis. org//DTD Mapper 3.0//EN"
"http://mybatis. org/dtd/mybatis - 3 - mapper. dtd">
<mapper namespace = "student">
    <select id = "findStudentBySid" parameterType = "Integer"
        resultType = "com. qfedu. pojo. Student">
        select * from Student where sid = #{sid}
    </select>
</mapper>
```

在以上代码中,< mapper >元素是映射文件的根元素,< mapper >元素的 namespace 属性指定该< mapper >元素的命名空间。< select >元素用于映射一个查询操作,其中,id 属性是该操作在该 Mapper 文件中的唯一标示,parameterType 属性用于指定传入参数的类型,resultType 属性用于指定返回结果的类型。在 SQL 语句中,# 具有占位符的功能,{sid}表示传入的参数。

此处需要注意的是,开发人员可以在 MyBatis 的使用手册 mybatis-3.4.6.pdf 的 2.1.5 节找到映射信息模板,然后直接复制模板信息并在此基础上编写 StudentMapper.xml 文件。

1.4.4 编写测试类

1. 测试通过 sid 字段查询学生信息

在 src 目录下新建包 com.qfedu.test,在该包下新建类 TestFindBySid,具体代码如例 1-4 所示。

【例 1-4】 TestFindBySid.java

```
package com. qfedu. test;
import java. io. * ;
import org. apache. ibatis. io. Resources;
import org. apache. ibatis. session. * ;
import com. qfedu. pojo. Student;
public class TestFindBySid {
    public static void main(String[ ] args) {
        //读取配置文件
        String resource = "mybatis - config. xml";
        try {
            InputStream in = Resources. getResourceAsStream(resource);
        //创建 SQLSessionFactory 对象
        SqlSessionFactory factory = new
            SqlSessionFactoryBuilder(). build(in);
        //创建 SqlSession 对象
        SqlSession sqlSession = factory. openSession();
        //调用 SqlSession 对象的 selectOne()方法执行查询
        Student student = sqlSession. selectOne("student. findStudentBySid", 1);

        System. out. println(student. toString());
        //关闭 SqlSession
```

```
            sqlSession.close();
        } catch (IOException e) {
            e.printStackTrace();
        }
    }
}
```

在以上代码中，首先获取配置文件的输入流，其次根据读取的配置信息创建 SqlSessionFactory 对象，然后通过 SqlSessionFactory 对象创建 SqlSession 对象，最后调用 SqlSession 对象的 selectOne()方法执行查询操作。SqlSession 对象的 selectOne()方法有两个参数，第一个参数匹配映射文件 StudentMapper. xml 中相应元素的属性值，其中，student 匹配< mapper >元素的 namespace 属性值，findStudentById 匹配< select >元素的 id 属性值，第二个参数表示 StudentMapper. xml 中< select >元素中 SQL 语句所需的参数。

执行 TestFindBySid 类，执行结果如图 1.6 所示。

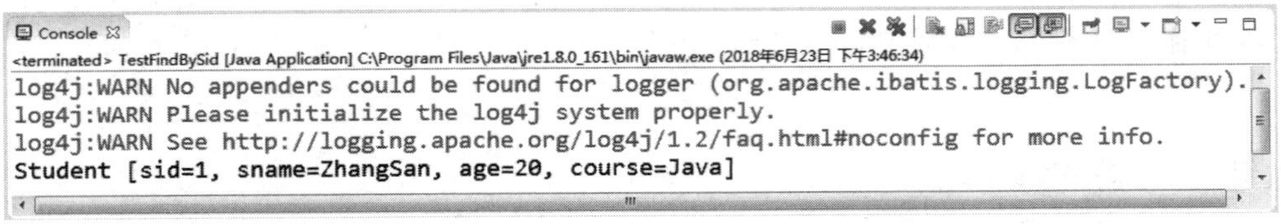

```
Console
<terminated> TestFindBySid [Java Application] C:\Program Files\Java\jre1.8.0_161\bin\javaw.exe (2018年6月23日 下午3:46:34)
log4j:WARN No appenders could be found for logger (org.apache.ibatis.logging.LogFactory).
log4j:WARN Please initialize the log4j system properly.
log4j:WARN See http://logging.apache.org/log4j/1.2/faq.html#noconfig for more info.
Student [sid=1, sname=ZhangSan, age=20, course=Java]
```

图 1.6　执行 TestFindBySid 类的结果

从图 1.6 中可以看出，程序输出了 sid 为 1 的学生信息。

2. 测试通过 sname 字段模糊查询学生信息

在 StudentMapper. xml 文件的< mapper >元素中加入映射信息，具体代码如下。

```
<select id = "findStudentBySname" parameterType = "String"
    resultType = "com.qfedu.pojo.Student">
    select * from student where sname like '% ${value} %'
</select>
```

在以上代码中，< select >元素用于映射一个查询操作，其中，id 属性是该操作在该 Mapper 文件中的唯一标示，parameterType 属性用于指定传入参数的类型，resultType 属性用于指定返回结果的类型。在 SQL 语句中，$ 具有拼接符的功能，{value}表示传入的参数。

在 com. qfedu. test 包下新建类 TestFindBySname，具体代码如例 1-5 所示。

【例 1-5】 TestFindBySname. java

```
package com.qfedu.test;
import java.io.*;
import java.util.List;
import org.apache.ibatis.io.Resources;
import org.apache.ibatis.session.*;
import com.qfedu.pojo.Student;
```

```
public class TestFindBySname {
    public static void main(String[ ] args) {
        //读取配置文件
        String resource = "mybatis-config.xml";
        try {
            InputStream in = Resources.getResourceAsStream(resource);
        //创建 SQLSessionFactory 对象
        SqlSessionFactory factory = new
            SqlSessionFactoryBuilder().build(in);
        //创建 SqlSession 对象
        SqlSession sqlSession = factory.openSession();
        //调用 SqlSession 的 selectList()方法执行查询
        List<Student> selectList = sqlSession.selectList
            ("student.findStudentBySname","ZhangSan");
        for (Student student : selectList) {
            System.out.println(student.toString());
        }
        //关闭 SqlSession
        sqlSession.close();
        } catch (IOException e) {
            e.printStackTrace();
        }
    }
}
```

在以上代码中，SqlSession 对象的 selectList()方法有两个参数，第一个参数匹配映射文件 StudentMapper.xml 中相应元素的属性值，第二个参数表示 StudentMapper.xml 中<select>元素中 SQL 语句所需的参数。

执行 TestFindBySname 类，执行结果如图 1.7 所示。

```
Console
<terminated> TestFindBySname [Java Application] C:\Program Files\Java\jre1.8.0_161\bin\javaw.exe (2018年8月24日 下午1:26:08)
log4j:WARN No appenders could be found for logger (org.apache.ibatis.logging.LogFactory).
log4j:WARN Please initialize the log4j system properly.
log4j:WARN See http://logging.apache.org/log4j/1.2/faq.html#noconfig for more info.
Student [id=1, sname=ZhangSan, age=20, course=Java]
Student [id=6, sname=ZhangSanSan, age=22, course=PHP]
```

图 1.7　执行 TestFindBySname 类的结果

从图 1.7 中可以看出，程序输出了字段 sname 包括 ZhangSan 的学生信息。

3. 测试添加学生信息

在 StudentMapper.xml 文件的<mapper>元素中加入映射信息，具体代码如下。

```
<insert id="addStudent" parameterType="com.qfedu.pojo.Student">
    insert into student(sname,age,course)values(#{sname},#{age},#{course})
</insert>
```

在以上代码中，<insert>元素用于映射一个插入操作，parameterType 属性用于指定传入参数的类型,#{sname}、#{age}、#{course}表示以占位符形式接收的参数。

在 com. qfedu. test 包下新建类 TestAdd,具体代码如例 1-6 所示。

【例 1-6】 TestAdd. java

```
package com.qfedu.test;
import java.io.*;
import org.apache.ibatis.io.Resources;
import org.apache.ibatis.session.*;
import com.qfedu.pojo.Student;
public class TestAdd {
    public static void main(String[] args) {
        //读取配置文件
        String resource = "mybatis-config.xml";
        try {
            InputStream in = Resources.getResourceAsStream(resource);
        //创建 SQLSessionFactory 对象
        SqlSessionFactory factory = new
            SqlSessionFactoryBuilder().build(in);
        //创建 SqlSession 对象
        SqlSession sqlSession = factory.openSession();
        Student student = new Student();
        student.setSname("ZhouBa");
        student.setAge("21");
        student.setCourse("Java");
        //调用 SqlSession 对象的 insert()方法执行插入
        int result = sqlSession.insert("student.addStudent", student);
        if (result>0) {
            System.out.println("成功插入" + result + "条数据");
        }else{
            System.out.println("插入操作失败");
        }
        //提交事务
        sqlSession.commit();
        //关闭 SqlSession
        sqlSession.close();
        } catch (IOException e) {
            e.printStackTrace();
        }
    }
}
```

在以上代码中,SqlSession 对象的 insert()方法有两个参数,第一个参数匹配映射文件 StudentMapper. xml 中相应元素的属性值,第二个参数传入了一个 POJO 类型的对象,该 POJO 对象中封装了 SQL 语句所需的参数信息。在执行 SQL 操作之后,程序将根据执行结果返回对应的提示信息。

执行 TestAdd 类,执行结果如图 1.8 所示。

从图 1.8 中可以看出,程序输出了成功插入数据的信息。

```
Console
<terminated> TestAdd [Java Application] C:\Program Files\Java\jre1.8.0_161\bin\javaw.exe (2018年6月25日 上午11:55:59)
log4j:WARN No appenders could be found for logger (org.apache.ibatis.logging.LogFactory).
log4j:WARN Please initialize the log4j system properly.
log4j:WARN See http://logging.apache.org/log4j/1.2/faq.html#noconfig for more info.
成功插入1条数据
```

图 1.8　执行 TestAdd 类的结果

4. 测试更新学生信息

在 StudentMapper. xml 文件的< mapper >元素中加入映射信息，具体代码如下。

```
<update id = "updateStudent" parameterType = "com.qfedu.pojo.Student" >
    update student set sname = #{sname},course = #{course} where sid = #{sid}
</update>
```

在以上代码中，< update >元素用于映射一个更新操作，parameterType 属性用于指定传入参数的类型，#{sid}表示以占位符形式接收的参数。

在 com. qfedu. test 包下新建类 TestUpdate，具体代码如例 1-7 所示。

【例 1-7】 TestUpdate. java

```
package com.qfedu.test;
import java.io.*;
import org.apache.ibatis.io.Resources;
import org.apache.ibatis.session.*;
import com.qfedu.pojo.Student;
public class TestUpdate {
    public static void main(String[] args) {
        //读取配置文件
        String resource = "mybatis-config.xml";
        try {
            InputStream in = Resources.getResourceAsStream(resource);
        //创建 SqlSessionFactory 对象
        SqlSessionFactory factory = new
            SqlSessionFactoryBuilder().build(in);
        //创建 SqlSession 对象
        SqlSession sqlSession = factory.openSession();
        Student student = new Student();
        student.setSid(7);
        student.setSname("WuJiu");
        student.setCourse("Python");
        //调用 SqlSession 对象的 update()方法执行更新
        int result = sqlSession.update("student.updateStudent", student);
        if (result>0) {
            System.out.println("成功更新" + result + "条数据");
        }else{
            System.out.println("更新操作失败");
        }
        //提交事务
```

```
            sqlSession.commit();
            //关闭 SqlSession
            sqlSession.close();
            } catch (IOException e) {
                e.printStackTrace();
            }
        }
}
```

在以上代码中,SqlSession 对象的 update()方法有两个参数,第一个参数匹配映射文件 StudentMapper. xml 中相应元素的属性值,第二个参数表示 SQL 语句所需的参数。

执行 TestUpdate 类,执行结果如图 1.9 所示。

```
Console
<terminated> TestUpdate [Java Application] C:\Program Files\Java\jre1.8.0_161\bin\javaw.exe (2018年6月25日 下午1:14:16)
log4j:WARN No appenders could be found for logger (org.apache.ibatis.logging.LogFactory).
log4j:WARN Please initialize the log4j system properly.
log4j:WARN See http://logging.apache.org/log4j/1.2/faq.html#noconfig for more info.
成功更新1条数据
```

图 1.9　执行 TestUpdate 类的结果

从图 1.9 中可以看出,程序输出了成功更新数据的信息。

5. 测试删除学生信息

在 StudentMapper. xml 文件的< mapper >元素中加入映射信息,具体代码如下。

```
<delete id = "deleteStudent" parameterType = "Integer">
    delete from student where sid = #{sid}
</delete>
```

在以上代码中,< delete >元素用于映射一个删除操作,parameterType 属性用于指定传入参数的类型,#{sid}表示以占位符形式接收的参数。

在 com. qfedu. test 包下新建类 TestDelete,具体代码如例 1-8 所示。

【例 1-8】　TestDelete. java

```
package com.qfedu.test;
import java.io.*;
import org.apache.ibatis.io.Resources;
import org.apache.ibatis.session.*;
public class TestDelete {
    public static void main(String[] args) {
        //读取配置文件
        String resource = "mybatis-config.xml";
        try {
            InputStream in = Resources.getResourceAsStream(resource);
        //创建 SqlSessionFactory 对象
        SqlSessionFactory factory = new
            SqlSessionFactoryBuilder().build(in);
        //创建 SqlSession 对象
```

```
            SqlSession sqlSession = factory.openSession();
            //调用 SqlSession 对象的 delete()方法执行删除
            int result = sqlSession.delete("student.deleteStudent", 7);
            if (result > 0) {
                System.out.println("成功删除" + result + "条数据");
            }else{
                System.out.println("删除操作失败");
            }
            //提交事务
            sqlSession.commit();
            //关闭 SqlSession
            sqlSession.close();
            } catch (IOException e) {
                e.printStackTrace();
            }
        }
}
```

在以上代码中,SqlSession 对象的 delete()方法有两个参数,第一个参数匹配映射文件 StudentMapper.xml 中相应元素的属性值,第二个参数表示<delete>元素中 SQL 语句所需的参数。

执行 TestDelete 类,执行结果如图 1.10 所示。

```
Console
<terminated> TestDelete [Java Application] C:\Program Files\Java\jre1.8.0_161\bin\javaw.exe (2018年6月25日 下午1:18:58)
log4j:WARN No appenders could be found for logger (org.apache.ibatis.logging.LogFactory).
log4j:WARN Please initialize the log4j system properly.
log4j:WARN See http://logging.apache.org/log4j/1.2/faq.html#noconfig for more info.
成功删除1条数据
```

图 1.10　执行 TestDelete 类的结果

从图 1.10 中可以看出,程序输出了成功删除数据的信息。

1.5　本章小结

本章主要介绍了 MyBatis 的基础知识,包括 MyBatis 的概念、功能结构、工作流程以及重要 API 等,最后通过一个案例详细讲解了使用 MyBatis 操作数据库的方法。通过本章知识的学习,大家应该能理解 MyBatis 的基础概念和原理,掌握 MyBatis 的开发流程并能够开发简单的 MyBatis 程序。

1.6　习　　题

1. 填空题

(1) ________的全称是 Object-Relation Mapping,即对象-关系映射。

(2) 近几年,在 Java 应用领域流行的 ORM 框架有________、________等。

(3) MyBatis 的前身是________组织的一个开源项目 iBatis。

(4) MyBatis 的功能架构由三层组成,包括________、________、________。

(5) ________对象的首要功能是创建 SqlSession 对象。

2. 选择题

(1) 关于 MyBatis 的功能,下列选项错误的是(　　)。

A. MyBatis 是一种"半自动化"的 ORM 框架

B. MyBatis 允许开发人员实现复杂查询,但不支持存储过程和视图

C. 当使用 MyBatis 开发时,可以根据开发需要灵活编写 SQL 语句

D. MyBatis 支持使用 XML 文件或注解完成配置和映射

(2) 在 MyBatis 的功能架构中,负责完成 SQL 语句解析和执行的是(　　)。

A. API 接口层　　B. 数据处理层

C. 基础支撑层　　D. 基础 API 层

(3) 关于 MyBatis 的工作流程,下列选项错误的是(　　)。

A. MyBatis 读取配置文件和映射文件,获取配置信息

B. MyBatis 根据配置信息创建 SqlSessionFactory 对象

C. SqlSession 是 MyBatis 提供给开发者使用的 API

D. 被处理的 SQL 信息被封装到一个 SqlSessionFactory 对象中

(4) 在 SqlSession 对象的方法中,用于获取映射器的是(　　)。

A. void commit()　　B. void clearCache()

C. T getMapper(Class type)　　D. void rollback()

(5) 在 SqlSession 对象的 openSession()方法中,不能作为参数 executorType 的可选值的是(　　)。

A. ExecutorType. SIMPLE　　B. ExecutorType. REUSE

C. ExecutorType. BATCH　　D. ExecutorType. MANY

3. 思考题

(1) 简述传统 JDBC 的劣势。

(2) 简述 MyBatis 的功能架构。

4. 编程题

通过 MyBatis 查询数据库 chapter01 中 student 表的所有学生信息,并将这些信息输出到控制台。

第2章 MyBatis 进阶

本章学习目标

- 理解 MyBatis 的配置和映射
- 掌握 MyBatis 配置文件的编写方法
- 掌握 MyBatis 映射文件的编写方法

通常情况下，MyBatis 通过其配置文件获取基础配置信息，通过映射文件匹配 POJO 对象和数据表之间的关系，这使得 MyBatis 封装了繁杂的技术细节并降低了程序的复杂性，对于开发人员来说，只要正确编写 MyBatis 的配置文件和映射文件，就能够通过 MyBatis 的 SqlSession 对象操作数据库，从而充分发挥 MyBatis 作为 ORM 框架的优势。接下来，本章将对 MyBatis 的配置文件和映射文件做深入讲解。

2.1 MyBatis 的配置文件

2.1.1 配置文件的结构

配置文件对 MyBatis 的整个运行体系产生影响，它包含了很多控制 MyBatis 功能的重要信息，是 MyBatis 实现功能的重要保证。在开发过程中，当需要更改 MyBatis 的配置信息时，只需更改配置文件中的相关元素及属性即可。

MyBatis 规定了其配置文件的层次结构，具体如下所示。

```
<?xml version = "1.0" encoding = "UTF - 8" ?>
<! DOCTYPE configuration
PUBLIC " - //mybatis.org//DTD Config 3.0//EN"
"http://mybatis.org/dtd/mybatis - 3 - config.dtd">
< configuration > <! -- 根元素 -- >
        < properties ><! --  属性  -- >
            < property name = "" value = ""/>
        </properties >
        < settings > <! --  设置 -- >
            < setting name = "" value = ""/>
        </settings >
        < typeAliases ><! --  类型别名 -- >
            < typeAlias type = ""/>
        </typeAliases >
        < typeHandlers ><! -- 类型处理器  -- >
```

```
                <typeHandler handler=""/>
            </typeHandlers>
            <objectFactory type=""/><!-- 对象工厂 -->
            <plugins><!-- 插件 -->
                <plugin interceptor=""></plugin>
            </plugins>
            <environments default=""><!-- 环境 -->
                <environment id=""><!-- 环境变量 -->
                    <transactionManager type=""/><!-- 事务管理器 -->
                    <dataSource type=""><!-- 数据源 -->
                        <property name="" value=""/>
                        <property name="" value=""/>
                        <property name="" value=""/>
                        <property name="" value=""/>
                    </dataSource>
                </environment>
            </environments>
            <databaseIdProvider type=""/><!-- 数据厂商标识 -->
            <mappers><!-- 映射文件 -->
                <mapper resource=""/>
            </mappers>
</configuration>
```

以上列出了 MyBatis 配置文件的元素，这些元素分别实现着支撑 MyBatis 运行的各项重要功能。此处需要注意的是，MyBatis 配置文件的元素在文件中的先后顺序是固定的，通常情况下，开发人员要按照官方提供的元素顺序编写配置文件，否则，MyBatis 会在解析配置文件时报错。

2.1.2 <properties>元素

<properties>是一个用于配置属性的元素，MyBatis 支持<properties>元素的两种配置方式：通过<property>子元素或通过 properties 文件，接下来本节将对这两种配置方式做详细讲解。

1. 通过<property>子元素

<properties>元素通过其子元素<property>完成属性传递，在 MyBatis 的配置文件中添加<properties>元素，具体代码如下。

```
<properties><!-- 属性 -->
        <property name="driver" value="com.mysql.jdbc.Driver"/>
        <property name="url"
            value="jdbc:mysql://localhost:3306/chapter02"/>
        <property name="username" value="root"/>
        <property name="password" value="root"/>
    </properties>
```

在完成上述配置后，< dataSource >元素的代码可直接引用< property >元素中的信息，具体代码如下。

```
<dataSource type = "POOLED"><! -- 数据源 -->
        <property name = "driver" value = " $ {driver}"/>
        <property name = "url" value = " $ {url}"/>
        <property name = "username" value = " $ {username}"/>
        <property name = "password" value = " $ {password}"/>
    </dataSource>
```

在以上代码中，${}表示引用< properties >的子元素< property >的内容，如此一来，< properties >通过子元素< property >实现参数传递。

2. properties 文件

在 src 目录下新建一个 db. properties 文件，具体代码如下。

```
jdbc.driver = com.mysql.jdbc.Driver
jdbc.url = jdbc:mysql://localhost:3306/chapter02
jdbc.username = root
jdbc.password = root
```

在 Mybatis 的配置文件中加入< properties >元素，具体代码如下。

```
<properties resource = "database.properties">
```

此时，< dataSource >元素的代码可直接引用< property >元素中的信息，具体如下。

```
<dataSource type = "POOLED"><! -- 数据源 -->
        <property name = "driver" value = " $ {jdbc.driver}"/>
        <property name = "url" value = " $ {jdbc.url}"/>
        <property name = "username" value = " $ {jdbc.username}"/>
        <property name = "password" value = " $ {jdbc.password}"/>
    </dataSource>
```

如此一来，< properties >元素通过 properties 文件实现参数配置。此处需要提醒大家的是，MyBatis 支持< property >子元素和 properties 文件两种配置形式同时出现。当上述两种配置形式同时出现时，MyBatis 会首先读取< properties >元素体内的内容，然后读取 properties 文件中的内容，如果有同名属性，后读取的内容会覆盖掉先读取的内容，即 properties 文件中的内容优先被程序采用。

2.1.3 < settings >元素

< settings >是 MyBatis 中较为复杂的配置元素，同时也包含重要的配置内容，这些内容控制着 MyBatis 运行时的状态和行为。因此，理解< settings >元素的常用配置内容有助于更好地使用 MyBatis 框架完成开发。< settings >元素包含的配置内容如表 2.1 所示。

表 2.1 <settings>元素的配置内容

设置参数	描　述	有效值	默认值
cacheEnabled	全局开启或关闭配置文件中所有映射器已配置的任何缓存	TRUE、FALSE	TRUE
lazyLoadingEnabled	延迟加载的全局开关。当开启时，所有关联对象都会延迟加载。特定关联关系中可通过设置 fetchType 属性来覆盖该项的开关状态	TRUE、FALSE	FALSE
aggressiveLazyLoading	当开启时，任何方法的调用都会加载该对象的所有属性；否则，每个属性会按需加载	TRUE、FALSE	FALSE
multipleResultSetsEnabled	是否允许单一语句返回多结果集	TRUE、FALSE	TRUE
useColumnLabel	使用列标签代替列名	TRUE、FALSE	TRUE
useGeneratedKeys	允许 JDBC 支持自动生成主键，如果设置为 TRUE，则强制使用自动生成主键	TRUE、FALSE	FALSE
autoMappingBehavior	指定 MyBatis 如何自动映射列到字段或属性。NONE 表示取消自动映射，PARTIAL 只会自动映射没有定义嵌套结果集映射的结果集，FULL 会自动映射任意复杂的结果集	NONE、PARTIAL、FULL	PARTIAL
autoMappingUnknownColumnBehavior	指定发现自动映射目标未知列或未知属性类型的行为。NONE 表示不做任何反应，WARNING 表示输出提醒日志，FAILING 表示映射失败、抛出 SqlSessionException	NONE、WARNING、FAILING	NONE
defaultExecutorType	配置默认的执行器。SIMPLE 就是普通的执行器；REUSE 执行器会重用预处理语句；BATCH 执行器将重用语句并执行批量更新	SIMPLE、REUSE、BATCH	SIMPLE
defaultStatementTimeout	设置超时时间，它决定驱动等待数据库响应的秒数	任意正整数	没有设置
defaultFetchSize	为驱动的结果集获取数量(fetchSize)设置一个提示值	任意正整数	没有设置
safeRowBoundsEnabled	允许在嵌套语句中使用分页(RowBounds)，如果允许使用则设置为 FALSE	TRUE、FALSE	FALSE
safeResultHandlerEnabled	允许在嵌套语句中使用分页(ResultHandler)，如果允许使用则设置为 FALSE	TRUE、FALSE	TRUE
mapUnderscoreToCamelCase	是否开启自动驼峰命名规则映射，即从经典数据库列名 A_COLUMN 到经典 Java 属性名 aColumn 的类似映射	TRUE、FALSE	FALSE

续表

设置参数	描　　述	有效值	默认值
localCacheScope	MyBatis 利用本地缓存机制防止循环引用和加速重复嵌套查询。如果设置值为 SESSION,那么会缓存一个会话中执行的所有查询。如果设置值为 STATEMENT,本地会话仅用在语句执行上,对相同 SqlSession 的不同调用将不会共享数据	SESSION、STATEMENT	SESSION
jdbcTypeForNull	当没有为参数提供特定的 JDBC 类型时,为空值指定 JDBC 类型	NULL、VARCHAR、OTHER	OTHER
lazyLoadTriggerMethods	指定哪个对象的方法触发一次延迟加载	方法名的 list 集合	equals, clone, hashCode, toString
defaultScriptingLanguage	指定动态 SQL 生成的默认语言	一个类型别名或完全限定类名	org. apache. ibatis. scripting. xmltags. XMLLanguageDriver
logPrefix	指定 MyBatis 增加到日志名称的前缀	任何字符串	没有设置
logImpl	指定 MyBatis 所用日志的具体实现,未指定时将自动查找	SLF4J、LOG4J LOG4J2、JDK_LOGGING、COMMONS_LOGGING、STDOUT_LOGGING、NO_LOGGING	没有设置
configurationFactory	指定一个提供 Configuration 对象的类。这个返回的 Configuration 对象用来加载被反序列化对象的懒加载属性值。这个类必须包含一个签名方法 static Configuration getConfiguration()	类型别名或全类名	没有设置

表 2.1 中列出了常用的配置项,此处给出一个配置样例,具体如下。

```
<settings>
    <setting name="cacheEnabled" value="true"/>
    <setting name="lazyLoadingEnabled" value="true"/>
    <setting name="multipleResultSetsEnabled" value="true"/>
    <setting name="useColumnLabel" value="true"/>
    <setting name="useGeneratedKeys" value="false"/>
    <setting name="autoMappingBehavior" value="PARTIAL"/>
    <setting name="autoMappingUnknownColumnBehavior" value="WARNING"/>
```

```
        <setting name = "defaultExecutorType" value = "SIMPLE"/>
        <setting name = "defaultStatementTimeout" value = "25"/>
        <setting name = "defaultFetchSize" value = "100"/>
        <setting name = "safeRowBoundsEnabled" value = "false"/>
        <setting name = "mapUnderscoreToCamelCase" value = "false"/>
        <setting name = "localCacheScope" value = "SESSION"/>
        <setting name = "jdbcTypeForNull" value = "OTHER"/>
        <setting name = "lazyLoadTriggerMethods"
                value = "equals,clone,hashCode,toString"/>
</settings>
```

以上是< settings >配置元素的具体使用方法，大家结合表 2.1 理解并掌握< settings >元素中的常用参数值及其含义。

2.1.4 < typeAliases >元素

由于类的完全限定名较长，为了简化开发、降低编码的烦琐度，MyBatis 支持使用别名。别名就是为类设置一个简短的名字，别名存在的意义是为了减少类完全限定名造成的冗余，方便开发人员编程。

别名的设置一般通过配置文件中的< typeAliases >元素进行，具体示例代码如下。

```
<typeAliases>
        <typeAlias  alias = "student" type = "com.qfedu.pojo.Studnet"/>
    </typeAliases>
```

在以上示例代码中，< typeAliases >元素包含一个子元素< typeAlias >，< typeAlias >元素的 type 属性用于指定被设定别名类的完全限定名，alias 属性用于指定类的别名。完成配置后，可以在很多场景下以别名 student 代替 com.qfedu.pojo.Student，由此一来，开发人员的工作量大大降低，也提升了程序的可读性和维护性。

除此之外，MyBatis 还支持通过扫描包的形式设置别名，具体示例代码如下。

```
<typeAliases>
        <package name = "com.qfedu.pojo"/>
    </typeAliases>
```

在以上示例代码中，< typeAliases >元素包含一个子元素< package >，< package >元素的 name 属性用于指定将要被自动扫描的类包。完成配置后，com.qfedu.pojo 类包下的所有 POJO 类都被以首字母小写的非限定类名作为别名，在本例中，com.qfedu.pojo.Student 类的别名是 student。

当以扫描包的形式设置别名时，如果开发人员不愿使用 MyBatis 默认的别名，那么可通过注解的形式来自定义别名，例如，如果要将 com.qfedu.pojo.Student 类的别名设置为 stu，可在 POJO 类代码中加入注解，具体如下。

```
package com.qfedu.pojo;
import org.apache.ibatis.type.Alias;
```

```
@Alias(value = "stu")
public class Student {
… …
}
```

在以上代码中,@Alias 注解的 value 属性用于指定 POJO 类的别名。

为了方便开发,MyBatis 为一些常用的 Java 类型提供了别名,具体如表 2.2 所示。

表 2.2 MyBatis 为常用 Java 类提供的别名

别　　名	映射的类型	别　　名	映射的类型
_byte	Byte	float	Float
_long	Long	boolean	Boolean
_short	Short	date	Date
_int	Int	decimal	BigDecimal
_integer	Int	bigdecimal	BigDecimal
_double	Double	object	Object
_float	Float	map	Map
_boolean	Boolean	hashmap	HashMap
string	String	list	List
byte	Byte	arraylist	ArrayList
long	Long	collection	Collection
short	Short	iterator	Iterator
int	Integer	float	Float
integer	Integer	boolean	Boolean
double	Double	date	Date

表 2.2 列举了 MyBatis 提供的常用 Java 类的别名,这些别名由 MyBatis 默认设置,无须定义即可在 MyBatis 中直接使用。

2.1.5 <typeHandlers>元素

在程序运行过程中,当 MyBatis 为 SQL 语句设置参数或从结果集中取值时,都要通过 typeHandler 完成类型转换。typeHandler,即类型处理器,它的核心功能是根据需要将数据由 Java 类型转化成 JDBC 类型,或者由 JDBC 类型转化为 Java 类型。

MyBatis 内部定义了一系列的 typeHandler,其中,常用的 typeHandler 如表 2.3 所示。

表 2.3 常用的 typeHandler

类型处理器	Java 类型	JDBC 类型
BooleanTypeHandler	java. lang. Boolean,boolean	数据库兼容的 BOOLEAN
ByteTypeHandler	java. lang. Byte,byte	数据库兼容的 NUMERIC 或 BYTE
ShortTypeHandler	java. lang. Short,short	数据库兼容的 NUMERIC 或 SHORT INTEGER
IntegerTypeHandler	java. lang. Integer,int	数据库兼容的 NUMERIC 或 INTEGER
LongTypeHandler	java. lang. Long,long	数据库兼容的 NUMERIC 或 LONG INTEGER
FloatTypeHandler	java. lang. Float,float	数据库兼容的 NUMERIC 或 FLOAT

续表

类型处理器	Java 类型	JDBC 类型
DoubleTypeHandler	java.lang.Double,double	数据库兼容的 NUMERIC 或 DOUBLE
BigDecimalTypeHandler	java.math.BigDecimal	数据库兼容的 NUMERIC 或 DECIMAL
StringTypeHandler	java.lang.String	CHAR,VARCHAR
ClobTypeHandler	java.lang.String	CLOB,LONGVARCHAR
NStringTypeHandler	java.lang.String	NVARCHAR,NCHAR
NClobTypeHandler	java.lang.String	NCLOB
ByteArrayTypeHandler	byte[]	数据库兼容的字节流类型
BlobTypeHandler	byte[]	BLOB,LONGVARBINARY
DateTypeHandler	java.util.Date	TIMESTAMP
DateOnlyTypeHandler	java.util.Date	DATE
TimeOnlyTypeHandler	java.util.Date	TIME

表 2.3 列举出了 MyBatis 内部定义的 typeHandler,这些 typeHandler 无须显式声明,MyBatis 会自动探测数据类型并完成转换。

通常情况下,MyBatis 内部定义的 typeHandler 可以满足大多数场景的需要,但是,如果出现这些 typeHandler 无法满足需求的特殊情景,开发人员必须通过自定义 typeHandler 来解决。自定义 typeHandler 分为两个环节:首先要编写 typeHandler 类,其次要完成配置。自定义的 typeHandler 类要实现 TypeHandler 接口或继承 BaseTypeHandler 类,编写完 typeHandler 类之后,要将该 typeHandler 类配置到 MyBatis 的配置文件中。

MyBatis 通过< typeHandlers >元素配置 typeHandler,具体示例代码如下。

```
<typeHandlers>
        <typeHandler javaType = "String" jdbcType = "VARCHAR"
handler = "com.qfedu.handler.StudenttypeHandler"/>
    </typeHandlers>
```

在以上示例代码中,< typeHandler >元素指定一个类型处理器,其中,handler 属性指定 typeHandler 类的完全限定类名,javaType 属性指定一个 Java 类型,jdbcType 属性指定一个 JDBC 类型。

当一个类包下有多个 typeHandler 类时,可以通过自动扫包的方式注册 typeHandler,具体示例代码如下。

```
<typeHandlers>
    <package name = "com.qfedu.handler"/>
    </typeHandlers>
```

在以上示例代码中,< package >元素指定要自动扫描的类包,位于该类包下的 typeHandler 类将被 MyBatis 识别,此处需要提醒大家的是,这时要以注解的方式指定 Java 类型和 JDBC 类型。

2.1.6 <ObjectFactory>元素

MyBatis 通过 ObjectFactory(对象工厂)创建结果集对象,在默认情况下,MyBatis 通过其定义的 DefaultObjectFactory 类完成相关的工作。但是,在实际开发中,当需要干预结果集对象的创建过程时,就需要自定义 ObjectFactory。

MyBatis 支持自定义 ObjectFactory,自定义 ObjectFactory 分为两个环节：首先要编写 ObjectFactory 类,其次要完成配置。自定义的 ObjectFactory 类通常要实现 ObjectFactory 接口或继承 DefaultObjectFactory 类,编写完 ObjectFactory 类之后,要将该 ObjectFactory 类配置到 MyBatis 的配置文件中。

MyBatis 通过<ObjectFactory>元素配置<ObjectFactory>,具体示例代码如下。

```
<objectFactory type = "com.qfedu.factory.StudentObjectFactory"/>
```

在以上示例代码中,<objectFactory>元素用于指定 ObjectFactory,其中,type 属性指定 ObjectFactory 类的完全限定类名。

2.1.7 <environments>元素

在 MyBatis 的体系中,运行环境的主要作用是配置数据库信息。

MyBatis 支持多种环境,使用不同的环境可以操作不同的数据库,并且 MyBatis 可以将相同的 SQL 映射应用于多种数据库。通过修改运行环境,MyBatis 能够实现匹配数据库的常见需求,例如,在开发、测试和生产环境间切换数据库,让多个数据库使用相同的 SQL 映射等。

MyBatis 的运行环境信息包括事务管理器和数据源。在 MyBatis 的配置文件中,MyBatis 通过<environment>元素定义一个运行环境,进而通过<transactionManager>元素配置事务管理器,通过<dataSource>元素配置数据源,具体示例代码如下。

```
<environments default = "development">
<environment id = "development">
        <transactionManager type = "JDBC">
        </transactionManager>
        <dataSource type = "POOLED">
            <property name = "driver" value = " ${driver}"/>
            <property name = "url" value = " ${url}"/>
            <property name = "username" value = " ${username}"/>
            <property name = "password" value = " ${password}"/>
        </dataSource>
    </environment>
</environments>
```

在以上示例代码中,<environments>元素是配置运行环境的根元素,其 default 属性用于指定默认环境的 id 值,一个<environments>元素下可以有多个<environment>子元素;<environment>元素用于定义一个运行环境,其 id 属性用于设置所定义环境的 id 值;<transactionManager>元素用于配置事务管理器,其 type 属性用于指定事务管理器的类型；< dataSource >元素用于配置数据源，其 type 属性用于指定数据源的类型。

事务管理器有两种类型：JDBC 和 MANAGED，其中，JDBC 事务管理器直接使用了 JDBC 的提交和回滚设置，它依赖于从数据源获取的连接来管理事务作用域；MANAGED 事务管理器从来不提交或回滚一个连接，它是让容器来管理事务的整个生命周期。由于 MyBatis 配置文件中< transactionManager >元素的 type 属性指定事务管理器的类型，因此，JDBC 和 MANAGED 可作为该 type 属性的可选值。

MyBatis 提供了三种不同类型的数据源：UNPOOLED、POOLED 和 JNDI，它们通过标准的 JDBC 数据源接口来获取 JDBC 连接对象的资源。由于 MyBatis 配置文件中< transactionManager >元素的 type 属性指定事务管理器的类型，因此，UNPOOLED、POOLED 和 JNDI 可作为该 type 属性的可选值。接下来分别对这三种不同类型的数据源做详细讲解。

1. UNPOOLED

非连接池类型，该类型数据源只是在每次被请求时才会打开和关闭连接，它适用于对性能要求不高的简单应用程序。

UNPOOLED 类型的数据源需要配置五种属性，如表 2.4 所示。

表 2.4　UNPOOLED 数据源需配置的属性

属　　性	说　　明
driver	JDBC 驱动的 Java 类的完全限定名，不是 JDBC 驱动中可能包含的数据源类
url	数据库的 JDBC URL
username	登录数据库的用户名
password	登录数据库的密码
defaultTransactionIsolationLevel	默认的连接事务隔离级别

2. POOLED

连接池类型，该类型数据源利用“池”的概念将 JDBC 连接对象组织起来，避免了创建新的连接实例时所必需的初始化和认证时间。POOLED 是一种使并发 Web 应用快速响应请求的流行处理方式，由于它维持有一定活跃量的连接对象，因此，它适用于对性能有一定要求的应用程序。

除了上述 UNPOOLED 类型中提及的属性，POOLED 数据源还可以配置更多的属性，如表 2.5 所示。

表 2.5　POOLED 数据源需配置的属性

属　　性	说　　明
poolMaximumActiveConnections	在任意时间可以存在的活动连接数量，默认值是 10
poolMaximumIdleConnections	在任意时间可能存在的空闲连接数
poolMaximumCheckoutTime	在被强制返回之前，连接池中的连接被检出(checked out)时间，默认值为 20000ms(即 20s)
poolTimeToWait	如果获取连接花费了相当长的时间，连接池会打印状态日志并重新尝试获取一个连接，默认值为 20000ms(即 20s)

续表

属　　性	说　　明
poolPingQuery	发送到数据库的侦测查询，用来检验连接是否正常工作并准备接受请求，默认是“NO PING QUERY SET”，这会导致多数数据库驱动失败时带有一个恰当的错误消息
poolPingEnabled	是否启用侦测查询。若开启，需要设置 poolPingQuery 属性为一个可执行的 SQL 语句，默认值为 false
poolPingConnectionsNotUsedFor	配置 poolPingQuery 的频率。可以被设置为和数据库连接超时时间一样，来避免不必要的侦测，默认值为 0(即所有连接每一时刻都被侦测，只有当 poolPingEnabled 为 true 时适用)

3. JNDI

该类型数据源能在如 EJB 或应用服务器之类的容器中使用，容器可以集中或在外部配置数据源，然后放置一个 JNDI 上下文的引用。

JNDI 类型数据源配置只需要两个属性，如表 2.6 所示。

表 2.6　JNDI 类型数据源需配置的属性

属　　性	说　　明
initial_context	这个属性用来在 InitialContext 中寻找上下文，为可选属性，如果忽略，直接从 InitialContext 中寻找
data_source	引用数据源实例位置的上下文的路径。提供了 initial_context 配置时会在其返回的上下文中进行查找，没有提供时则直接在 InitialContext 中查找

2.1.8　<mappers>元素

在 MyBatis 的配置文件中，<mappers>元素用于引入映射文件。映射文件包含了 POJO 对象和数据表之间的映射信息，<mappers>元素引导 MyBatis 找到映射文件并解析其中的映射信息。

通过<mappers>元素引入映射文件有四种方法，具体如下。

1）使用相对于类路径的资源引用

```
<mappers>
<mapper resource = "com/qfedu/mapper/StudentMapper.xml"/>
</mappers>
```

2）使用完全限定资源定位符(本地文件路径)

```
<mappers>
<mapper url = "file:///D:/com/qfedu/mapper/StudentMapper.xml"/>
</mappers>
```

3）使用映射器接口实现类的完全限定类名

```
<mappers>
```

```
<mapper class = "com.qfedu.mapper.StudentMapper"/>
</mappers>
```

4）将包内的映射器接口实现全部注册为映射器

```
<mappers>
<package name = "com.qfedu.mapper"/>
</mappers>
```

2.2 MyBatis 的映射文件

2.2.1 映射文件的结构

映射文件是 MyBatis 中的又一重要组件，它包含了各类 SQL 语句、参数、结果集、映射规则等信息。使用映射文件，开发人员可以灵活编写 SQL 语句以满足不同场景的需要，同时，也可将 SQL 语句从代码中分离，进而能够通过 POJO 对象完成对数据库的读写。

MyBatis 规定了其映射文件的层次结构，具体如下所示。

```
<?xml version = "1.0" encoding = "UTF - 8" ?>
<!DOCTYPE mapper
PUBLIC " - //mybatis.org//DTD Mapper 3.0//EN"
"http://mybatis.org/dtd/mybatis - 3 - mapper.dtd">
<mapper namespace = "student">
        <cache></cache>
        <cache - ref namespace = ""/>
        <parameterMap type = "" id = "">
            <parameter property = ""  javaType = "" resultMap = ""
                typeHandler = ""/>
        </parameterMap>
        <sql id = "" >
        </sql>
        <resultMap type = "" id = "">
        </resultMap>
        <select id = "" parameterType = "" resultType = "">
        </select>
        <insert id = "" parameterType = "" >
        </insert>
        <update id = "" parameterType = "" >
        </update>
        <delete id = "" parameterType = "">
        </delete>
    </mapper>
```

以上列出了 MyBatis 映射文件的元素，开发人员应熟练掌握这些元素的功能和使用方法。

2.2.2 <select>元素

<select>是 MyBatis 中最常用的元素之一，主要用于映射查询语句，它包含了 SQL 语句、参数类型和返回值类型等信息。

通过<select>元素映射查询语句，具体示例代码如下。

```
<select id = "findStudentBySid" parameterType = "Integer"
    resultType = "com.qfedu.pojo.Student">
        select * from Student where sid = #{sid}
    </select>
```

在以上示例代码中，<select>元素映射一个查询语句，id 属性用于指定该映射关系在当前命名空间的唯一标识符，parameterType 用于指定传入参数的完全限定类名或别名，resultType 用于指定返回结果的完全限定类名或别名，SQL 语句中#{sid}表示通过占位符的形式接收参数 sid。

为了更加灵活地映射查询语句，<select>元素中提供了一系列属性，具体如表 2.7 所示。

表 2.7 <select>元素的属性

属性	说明
id	用于指定 SQL 语句在命名空间中的唯一标识符，可以被用来引用这条语句，如果 id 属性在当前命名空间中不唯一，那么 MyBatis 会抛出异常
parameterType	用于指定 SQL 语句所需参数的类的完全限定名或别名。该属性是可选的，因为 MyBatis 可以通过 TypeHandler 推断出具体传入语句的参数，默认值为 unset
resultType	用于指定执行这条 SQL 语句返回的类的完全限定名或别名。如果是集合，则指定集合可以包含的类型，而不能是集合本身。resultType 不能和 resultMap 同时使用
resultMap	映射集的引用，执行强大的映射功能，可以配置映射规则、级联等。resultMap 不能和 resultType 同时使用
flushCache	用于指定 SQL 语句被调用之后，是否清空 MyBatis 的本地缓存和二级缓存，默认值为 false
useCache	用于指定 MyBatis 二级缓存的开关，默认值为 true
timeout	用于设置在抛出异常之前，驱动程序等待数据库返回请求结果的秒数
fetchSize	获取记录的总条数设定
statementType	用于设置 MyBatis 分别使用 Statement、PreparedStatement 或 CallableStatement 中的哪一个，可选值为 STATEMENT、PREPARED 或 CALLABLE 的一个，默认值为 PREPARED
resultSetType	用于设置 resultSet 的类型，可选值是 FORWARD_ONLY、SCROLL_SENSITIVE 或 SCROLL_INSENSITIVE 中的一个
resultOrdered	这个设置仅适用于嵌套结果 select 语句。如果为 true，则假设包含嵌套结果集或是分组，当返回一个主结果行时，就不能对前面结果集进行引用，这确保在获取嵌套结果集时不至于导致内存不够用
resultSets	这个设置仅适用多结果集，它将列出语句执行后每个结果集的名称，名称是逗号分隔的

表 2.7 列举出了< select >元素的属性，开发人员可根据具体需要酌情使用。

2.2.3 < insert >元素、< update >元素和< delete >元素

< insert >元素用于映射插入语句，< update >元素用于映射更新语句，< delete >元素用于映射删除语句。与< select >元素的结构组成类似，这三个元素也包含了 SQL 语句、参数类型等信息。

通过< insert >元素映射插入语句，具体示例代码如下。

```
<insert id = "addStudent" parameterType = "com.qfedu.pojo.Student" >
        insert into student(sname,age,course)values(#{sname},#{age},#{course})
    </insert>
```

通过< update >元素映射更新语句，具体示例代码如下。

```
<update id = "updateStudent" parameterType = "com.qfedu.pojo.Student" >
        update student set sname  = #{sname},course  =  #{course} where sid  =  #{sid}
    </update>
```

通过< delete >元素映射删除语句，具体示例代码如下。

```
<delete id = "deleteStudent" parameterType = "Integer">
        delete from  student where sid = #{sid}
    </delete>
```

在以上示例代码中，id 属性用于指定该映射关系在当前命名空间的唯一标识符，parameterType 用于指定传入参数的完全限定类名或别名，SQL 语句中的#{sname}、#{age}、#{course}和#{sid}表示通过占位符的形式接收相应的参数。由此可见，< insert >元素、< update >元素、< delete >元素的属性和< select >元素的属性基本类似，此处需要说明的是，表 2.7 的列举内容也同样适用于< insert >元素、< update >元素和< delete >元素。

除此之外，< insert >元素和< update >元素还包含有三个特有的属性，这三个属性具有一些特殊的功能，具体如表 2.8 所示。

表 2.8 < insert >元素和< update >元素的特有属性

属　性	说　明
useGeneratedKeys	是否开启 JDBC 的 getGeneratedKeys 方法获取由数据库内部生成的主键，默认值为 false，该方法仅适用于 insert 和 update 语句
keyProperty	用于唯一标记一个属性，MyBatis 通过 getGeneratedKeys 的返回值或 insert 语句的 selectKey 子元素设置它的键值，如果是复合主键，要把每一个名称用逗号分开，该方法仅适用于 insert 和 update 语句
keyColumn	此属性用于设置第几列是主键，当主键列不是表中的第一列时需要设置，在需要主键联合时，值可以用逗号分开，该方法仅适用于 insert 和 update 语句

表 2.8 列举出了< insert >元素、< update >元素的特有属性，开发人员可根据具体需要酌情使用。

2.2.4 <sql>元素

<sql>元素用于定义可重用的 SQL 代码片段。在 MyBatis 的应用中，如果根据业务要求编写多条 SQL 语句，而这些 SQL 语句包含了相同的代码片段，那么就可以把这些相同的片段抽取并通过<sql>元素定义，进而直接重用这些 SQL 代码片段。

通过<sql>元素定义代码片段，具体示例代码如下。

```
<sql id = "studentCols">
        sid,sname,age,course
    </sql>
```

其中，id 属性用于指定该代码片段在命名空间的唯一标识，id 属性的声明便于<select>和<insert>等元素引用该代码片段。

完成代码片段的定义后，可以在<select>和<insert>等元素中使用这个代码片段，具体示例代码如下。

```
<select id = "findStudent" parameterType = "Integer"
        resultType = "com.qfedu.pojo.Student">
        select  <include refid = "studentCols"/> from student where sid = #{sid}
    </select>
```

在以上示例代码中，<include>元素引入<sql>元素定义的 SQL 代码片段，其 refid 属性匹配<sql>元素的 id 属性。

2.2.5 <ResultMap>元素

<ResultMap>元素用于映射结果集，它避免了 JDBC ResultSets 的冗余和烦琐，同时，它封装并实现了一些 JDBC 没有提供的功能。

<ResultMap>元素包含了一些子元素，这些元素的层次结构如下所示。

```
<resultMap id = "" type = "">
<constructor>
        <idArg />
<arg />
    </constructor>
        <id/>
  <result property = "" column = ""/>
  <association property = "" javaType = "">
        <id property = "" column = ""/>
        <result property = "" column = ""/>
  </association>
  <collection property = "" ofType = "">
</collection>
  <discriminator javaType = "" column = "">
        <case value = "" resultType = ""/>
  </discriminator>
</resultMap>
```

在< ResultMap >元素的子元素中,< constructor >元素用于配置构造方法,当 POJO 对象中不存在没有参数的构造方法时,可以通过< constructor >元素注入配置信息;< id >元素用于指定主键,多个主键可称为联合主键;< result >元素用于配置 POJO 对象与数据表字段名的映射关系;< association >元素用于配置一对一关联,< collection >元素用于配置一对多关联,< discriminator >元素用于配置结果集处理方法,这三个元素将在以后的章节详细讲解,大家先对它们做初步了解。

2.3 本章小结

本章首先讲解了 MyBatis 中配置文件的作用、结构以及重要元素,其中,重点讲解了各个元素的功能和用法;然后讲解了 MyBatis 的映射文件,包括配置文件的结构、配置文件中重要元素的功能和用法等。通过本章知识的学习,大家应该能理解配置文件和映射文件在 MyBatis 体系中的角色和地位,掌握配置文件和映射文件的编写和使用,能够灵活利用配置文件和映射文件编写一些 MyBatis 程序。

2.4 习题

1. 填空题

(1) 使用< properties >元素配置属性可通过________和________两种方式。

(2) 在 MyBatis 的配置文件中,< typeAlias >元素用于为类设置________。

(3) 在 MyBatis 的配置文件中,< transactionManager >元素用于指定________。

(4) 在 MyBatis 的映射文件中,用于映射查询语句的元素是________。

(5) 在 MyBatis 的映射文件中,用于定义可重用 SQL 代码片段的元素是________。

2. 选择题

(1) 关于 MyBatis 的配置文件,下列选项错误的是(　　)。

A. MyBatis 通过其配置文件获取基础配置信息

B. MyBatis 的配置相对简单,在编写配置文件时无须关心元素顺序

C. MyBatis 配置文件的根元素是< configuration >元素

D. 在编写 MyBatis 配置文件时,不是所有的元素都必须要配置

(2) 在< settings >元素的属性中,负责指定允许 JDBC 支持自动生成主键的是(　　)。

A. useColumnLabel　　B. useGeneratedKeys

C. localCacheScope　　D. jdbcTypeForNull

(3) 关于 MyBatis 的映射文件,下列选项错误的是(　　)。

A. 一个 MyBatis 配置文件中可引入多个映射文件

B. 在编写 MyBatis 的映射文件时,开发人员无须关心元素顺序

C. < mappers >元素是 MyBatis 映射文件的根元素

D. 在编写 MyBatis 映射文件时,不是所有的元素都必须要配置

(4) 在下列选项中,与 MyBatis 提供的数据源类型无关的是(　　)。

A. UNPOOLED　　B. POOLED

C. JNDI　　　　　　　　　　　　　　D. MANAGED

(5) 在下列选项中，不属于< insert >元素和< update >元素特有属性的是(　　)。

A. useGeneratedKeys　　　　　　　　B. keyProperty

C. keyColumn　　　　　　　　　　　D. flushCache

3. 思考题

(1) 简述 MyBatis 配置文件的功能与结构。

(2) 简述 MyBatis 映射文件的功能与结构。

4. 编程题

通过 MyBatis 查询数据库 chapter02 中 student 表的所有学生信息，并将这些信息输出到控制台，要求使用< ResultMap >元素。

第3章 MyBatis的关联映射

本章学习目标

- 理解表与表之间的关系
- 掌握一对一关系的映射方法
- 掌握一对多关系的映射方法
- 掌握多对多关系的映射方法

在实际开发中，根据业务需要，数据表之间往往会存在某种关联关系，例如，一对一、一对多等。当程序操作数据库时，如果被操作的表与其他表相关联，那么处理这些表中数据时必须要考虑它们之间的关联关系。为此，作为流行的 ORM 框架，MyBatis 提供了映射表之间关联关系的功能，如此一来，使用 MyBatis 能够以简洁的方式操作多张数据表。接下来，本章将对 MyBatis 的关联映射做详细讲解。

3.1 表与表之间的关系

在学习 MyBatis 的关联映射前，首先要了解表与表之间的关系。表与表之间的关系主要包括一对一、一对多和多对多等，接下来本节将对这几种关系做详细讲解。

在一对一关系中，一方数据表中的一条记录最多可以和另一方数据表中的一条记录相关。例如，现实生活中学生与校园卡就属于一对一的关系，一个学生只能拥有一张校园卡，一张校园卡只能属于一个学生，如图 3.1 所示。

在一对多关系中，主键数据表中的一条记录可以和另外一张数据表的多条记录相关，例如，现实生活中班级与学生的关系就属于一对多的关系，一个班级可以有很多学生，一个学生只能属于一个班级，如图 3.2 所示。

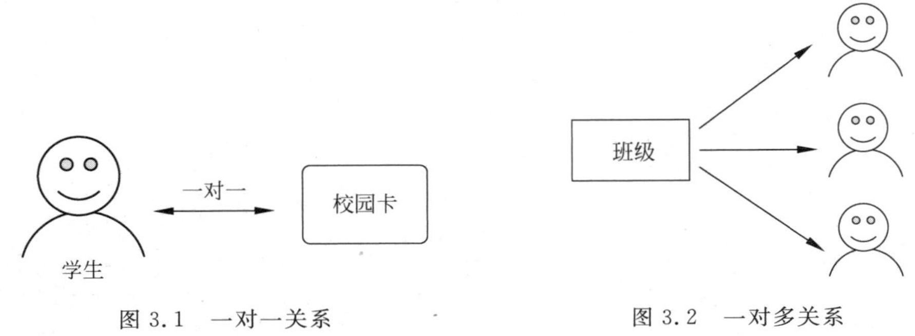

图 3.1　一对一关系　　图 3.2　一对多关系

在多对多关系中，两个数据表里的每条记录都可以和另一张数据表里任意数量的记录相关，例如，现实生活中学生与教师就属于多对多的关系，一名学生可以由多名教师授课，一名教师可以为多名学生授课，具体如图 3.3 所示。

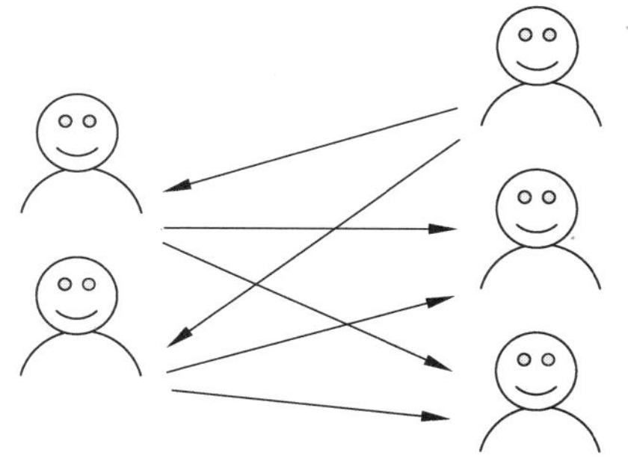

图 3.3　多对多关系

如果直接通过 SQL 语句维护数据表，在维护一对一的表关系时，通常采用在任意一方引入对方的主键作为外键的方式；在维护一对多的表关系时，通常采用在“多”方加入“一”方主键作为外键的方式；在维护多对多的表关系时，通常采用中间表的方式。

然而在企业项目开发中，程序通常是以操作 POJO 对象的形式来操作数据库的，为了满足程序需要，MyBatis 提供了关系映射的功能，这使得开发人员能够以操作 POJO 对象的形式处理表关系。

MyBatis 中支持一对一、一对多、多对多等多种映射方式，在实际应用中，MyBatis 通过映射文件中的< resultMap >元素实现关联映射，接下来本章将展开具体讲解。

3.2　一　对　一

在完成数据表设计之后，如果使用 MyBatis 处理一对一的表关系，需要在 MyBatis 的映射文件中添加< association >元素。

< association >元素是映射文件< resultMap >元素的子元素，其配置代码如下所示。

```
<resultMap type = "" id = "">
        <id property = "" column = ""/>
        <result property = "" column = ""/>
        <association property = "" column = ""  javaType = "" select = "" >
        </association>
    </resultMap>
```

< association >元素提供了一系列属性用于维护数据表关系，具体如表 3.1 所示。

表 3.1　< association >元素的属性

属　　性	说　　明
property	用于指定所映射的类的属性
column	用于指定数据表中对应的字段，与 property 属性指定的类的属性相对应
javaType	用于指定所映射的类的属性的类型
jdbcType	用于指定数据表中对应的字段的类型
fetchType	用于指定在关联查询时是否延迟加载
select	用于指定引入嵌套查询的子 SQL 语句
autoMapping	用于指定是否自动映射
typeHandler	用于指定一个类型处理器

表 3.1 列举了<association>元素的属性，开发人员可根据需要酌情使用。

前文中已经讲过，在处理表关系时，学生与校园卡是一对一的关联关系，接下来，本节将以此为例演示 MyBatis 实现一对一的关联映射。

1. 数据准备

(1) 在 MySQL 中创建数据库 chapter03，在数据库 chapter03 中创建数据表 stu_card、数据表 stu，其中，数据表 stu_card 用于映射校园卡信息，数据表 stu 用于映射学生信息，SQL 语句如下所示。

```
DROP DATABASE IF EXISTS chapter03;
CREATE DATABASE chapter03;
USE chapter03;
CREATE TABLE stu_card(
cid INT PRIMARY KEY AUTO_INCREMENT, # ID
balance DOUBLE              #卡内余额
);
CREATE TABLE stu(
sid INT PRIMARY KEY AUTO_INCREMENT, # ID
sname VARCHAR(20),         #学生姓名
age VARCHAR(20),           #学生年龄
course VARCHAR(20),        #科目
cardid INT UNIQUE,         #校园卡 ID
FOREIGN KEY(cardid) REFERENCES stu_card(cid)   #外键
);
```

(2) 向数据表 stu_card 中插入数据，SQL 语句如下所示。

```
INSERT INTO stu_card (balance) VALUES (1000.5);
INSERT INTO stu_card (balance) VALUES (5000.5);
```

(3) 向数据表 stu 中插入数据，SQL 语句如下所示。

```
INSERT INTO stu(sname,age,course,cardid)VALUES('ZhangSan','20','Java',1);
INSERT INTO stu(sname,age,course,cardid)VALUES('liSi','21','Java',2);
```

(4) 通过 SQL 语句测试数据是否插入成功，执行结果如下所示。

```
mysql> select * from stu_card;
+-----+---------+
| cid | balance |
+-----+---------+
|   1 |  1000.5 |
|   2 |  5000.5 |
+-----+---------+
2 rows in set (0.00 sec)

mysql> select * from stu;
```

```
+-----+----------+------+--------+--------+
| sid | sname    | age  | course | cardid |
+-----+----------+------+--------+--------+
|   1 | ZhangSan | 20   | Java   |      1 |
|   2 | liSi     | 21   | Java   |      2 |
+-----+----------+------+--------+--------+
2 rows in set (0.00 sec)
```

从以上执行结果可以看出，数据插入成功。

2. 创建工程

在 Eclipse 中新建 Web 工程 chapter03，将 MyBatis 的驱动 jar 包（mybatis-3.4.6.jar 和 lib 目录下的 jar 包）复制到 lib 目录下，完成 jar 包的导入。

3. 创建 POJO 类

在工程 chapter03 的 src 目录下创建 com.qfedu.pojo 包，在 com.qfedu.pojo 包下分别创建两个 POJO 类：Stu 和 StuCard，如例 3-1 和例 3-2 所示。

【例 3-1】 StuCard.java

```
package com.qfedu.pojo;
public class StuCard {
    private int cid;
    private double balance;
    public StuCard() {
        super();
    }
    public StuCard(int cid, double balance) {
        super();
        this.cid = cid;
        this.balance = balance;
    }
    public int getCid() {
        return cid;
    }
    public void setCid(int cid) {
        this.cid = cid;
    }
    public double getBalance() {
        return balance;
    }
    public void setBalance(double balance) {
        this.balance = balance;
    }
    @Override
    public String toString() {
        return "StuCard [cid=" + cid + ", balance=" + balance + "]";
    }
}
```

【例 3-2】 **Stu.java**

```
package com.qfedu.pojo;
public class Stu {
    private int sid;
    private String sname;
    private String age;
    private String course;
    private StuCard sc;
    public int getSid() {
        return sid;
    }
    public void setSid(int sid) {
        this.sid = sid;
    }
    public String getSname() {
        return sname;
    }
    public void setSname(String sname) {
        this.sname = sname;
    }
    public String getAge() {
        return age;
    }
    public void setAge(String age) {
        this.age = age;
    }
    public String getCourse() {
        return course;
    }
    public void setCourse(String course) {
        this.course = course;
    }
    public StuCard getSc() {
        return sc;
    }
    public void setSc(StuCard sc) {
        this.sc = sc;
    }
    @Override
    public String toString() {
        return "Stu [sid=" + sid + ", sname=" + sname + ", age=" + age + ",
              course=" + course + ", sc=" + sc + "]";
    }
}
```

以上代码是两个 POJO 类，其中 Stu 类封装了学生信息，StuCard 类封装了校园卡信息，除了 getter 和 setter 方法之外，为方便输出信息，这两个类还提供了 toString 方法。

4. 创建映射文件

在工程 chapter03 的 src 目录下分别创建 com. qfedu. mapper 包，在 com. qfedu. mappr 包下创建两个映射文件：StuCardMapper. xml 和 StuMapper. xml，如例 3-3 和例 3-4 所示。

【例 3-3】 StuCardMapper. xml

```
<?xml version = "1.0" encoding = "UTF - 8" ?>
<!DOCTYPE mapper
PUBLIC " - //mybatis.org//DTD Mapper 3.0//EN"
"http://mybatis.org/dtd/mybatis - 3 - mapper.dtd">
< mapper namespace = "stucard">
    < select id = "findStuCardBycid" parameterType = "Integer"
          resultType = "com.qfedu.pojo.StuCard">
        select * from stu_card where cid = #{cid}
    </select >
</mapper >
```

【例 3-4】 StuMapper. xml

```
<?xml version = "1.0" encoding = "UTF - 8" ?>
<!DOCTYPE mapper
PUBLIC " - //mybatis.org//DTD Mapper 3.0//EN"
"http://mybatis.org/dtd/mybatis - 3 - mapper.dtd">
< mapper namespace = "stu">
    < select id = "findStudentBySid" parameterType = "Integer"
          resultMap = "stuResultsMap">
        select s.*, c.balance from stu  s, stu_card c where s.cardid = c.cid
               and s.sid = #{sid}
    </select >
    < resultMap type = "stu" id = "stuResultsMap">
        < id column = "sid" property = "sid"/>
        < result column = "sname" property = "sname"/>
        < result column = "age" property = "age"/>
        < result column = "course" property = "course"/>
        < association property = "sc" javaType = "StuCard">
            < id column = "cid" property = "cid"/>
            < result column = "balance" property = "balance"/>
        </association >
    </resultMap >
</mapper >
```

在以上代码中，< select >元素用于映射查询，< resultMap >元素用于结果集处理，在< resultMap >元素的子元素中，< id >元素用于映射主键，< result >元素用于映射其他字段，< association >元素用于映射 stu 表和 stu_card 表的关联关系。

5. 创建配置文件

在工程 chapter03 的 src 目录下创建配置文件 mybatis-config. xml，如例 3-5 所示。

【例 3-5】 **mybatis-config. xml**

```
<?xml version = "1.0" encoding = "UTF - 8" ?>
<! DOCTYPE configuration
PUBLIC " - //mybatis.org//DTD Config 3.0//EN"
"http://mybatis.org/dtd/mybatis - 3 - config.dtd">
< configuration >
    <! -- 配置别名   -->
    < typeAliases >
        < package name = "com.qfedu.pojo"/>
    </typeAliases >
    <! -- 配置环境   -->
    < environments default = "mysql">
        < environment id = "mysql">
            < transactionManager type = "JDBC"/>
            < dataSource type = "POOLED">
            < property name = "driver" value = "com.mysql.jdbc.Driver"/>
            < property name = "url"
                value = "jdbc:mysql://localhost:3306/chapter03"/>
            < property name = "username" value = "root"/>
            < property name = "password" value = "root"/>
            </dataSource >
        </environment >
    </environments >
    <! -- 引入映射文件 -->
    < mappers >
        < mapper resource = "com/qfedu/mapper/StuMapper.xml"/>
        < mapper resource = "com/qfedu/mapper/StuCardMapper.xml"/>
    </mappers >
</configuration >
```

在以上配置文件中,< typeAliases >用于指定别名,本次采用的是扫描包的方法;< mappers >元素用于引入映射文件。

6. 编写测试类

在工程 chapter03 的 src 目录下创建 com. qfedu. test 包,在 com. qfedu. test 包下创建测试类 TestOneToOne,如例 3-6 所示。

【例 3-6】 **TestOneToOne. java**

```
package com.qfedu.test;
import java.io.InputStream;
import org.apache.ibatis.io.Resources;
import org.apache.ibatis.session.* ;
import com.qfedu.pojo.Stu;
public class TestOneToOne {
    public static void main(String[ ] args) throws Exception {
        InputStream in =
            Resources.getResourceAsStream("mybatis - config.xml");
        SqlSessionFactory sessionFactory = new
```

```
                SqlSessionFactoryBuilder().build(in);
        SqlSession session = sessionFactory.openSession();
        //使用 MyBatis 查询结果 sid 为 1 的学生信息(包括学生的校园卡信息)
        Stu stu = session.selectOne("stu.findStudentBySid",1);
        System.out.println(stu);
        session.close();
    }
}
```

在以上代码中,先通过 SqlSession 对象获取查询结果,此结果包含有处理表关系后的信息,然后将结果输出。

执行测试类 TestOneToOne,执行结果如图 3.4 所示。

```
Console ☒
<terminated> TestOneToOne [Java Application] C:\Program Files\Java\jre1.8.0_161\bin\javaw.exe (2018年7月3日 下午6:26:27)
log4j:WARN No appenders could be found for logger (org.apache.ibatis.logging.LogFactory).
log4j:WARN Please initialize the log4j system properly.
log4j:WARN See http://logging.apache.org/log4j/1.2/faq.html#noconfig for more info.
Stu [sid=1, sname=ZhangSan, age=20, course=Java, sc=StuCard [cid=null, balance=1000.5]]
```

图 3.4 执行 TestOneToOne 类的结果

从图 3.4 可以看出,程序显示出包括校园卡在内的学生信息,由此可见,MyBatis 成功处理一对一关系映射。

3.3 一 对 多

相比于一对一,一对多的表关系在开发中更为常见,处理过程也略微复杂。通常情况下,如果使用 MyBatis 处理一对多的关系,需要在 MyBatis 的映射文件中添加< collection >元素。

< collection >元素是映射文件< resultMap >元素的子元素,其配置代码如下所示。

```
<resultMap type = "" id = "">
        <id property = "" column = ""/>
        <result property = "" column = ""/>
        <collection property = "" column = "" javaType = "">
            <id column = "" property = ""/>
            <result column = "" property = "" />
        </collection>
</resultMap>
```

在以上配置代码中,< id >元素用于映射主键,< column >元素用于映射普通字段,< collection >元素用于映射一对多的关系。在< collection >的属性中,property 属性用于指定 POJO 对象的成员属性,column 属性用于指定对应的字段。

前文中已经讲过,在处理表关系时,班级与学生是一对多的关联关系,接下来,本书将以此为例演示 MyBatis 实现一对多的关联映射。

1. 数据准备

(1) 在数据库 chapter03 中创建数据表 stu_class、数据表 stu_info，其中，数据表 stu_class 用于映射班级信息，数据表 stu_info 用于映射学生信息，SQL 语句如下所示。

```
USE chapter03;
CREATE TABLE stu_class(
cid INT PRIMARY KEY AUTO_INCREMENT, #ID
cname VARCHAR(20),          #班级名称
sum INT                     #总人数
);
CREATE TABLE stu_info(
sid INT PRIMARY KEY AUTO_INCREMENT, #ID
sname VARCHAR(20),          #学生姓名
age VARCHAR(20),            #学生年龄
course VARCHAR(20),         #科目
classid INT,                #班级 ID
FOREIGN KEY(classid) REFERENCES stu_class(cid)   #外键
);
```

(2) 向数据表 stu_class 中插入数据，SQL 语句如下所示。

```
INSERT INTO stu_class(cname,sum) VALUES ('Java1801',50);
INSERT INTO stu_class(cname,sum) VALUES ('Java1802',46);
```

(3) 向数据表 stu_info 中插入数据，SQL 语句如下所示。

```
INSERT INTO stu_info(sname,age,course,classid) VALUES
('ZhangSan','20','Java',1);
INSERT INTO stu_info(sname,age,course,classid) VALUES
('liSi','21','Java',2);
INSERT INTO stu_info(sname,age,course,classid) VALUES
('WangWu','20','Java',1);
INSERT INTO stu_info(sname,age,course,classid) VALUES
('ZhaoLiu','19','Java',1);
```

(4) 通过 SQL 语句测试数据是否插入成功，执行结果如下所示。

```
mysql> select * from stu_class;
+-----+----------+------+
| cid | cname    | sum  |
+-----+----------+------+
|   1 | Java1801 |  50  |
|   2 | Java1802 |  46  |
+-----+----------+------+
2 rows in set (0.00 sec)

mysql> select * from stu_info;
```

```
+-----+----------+------+--------+---------+
| sid | sname    | age  | course | classid |
+-----+----------+------+--------+---------+
|   1 | ZhangSan | 20   | Java   |       1 |
|   2 | liSi     | 21   | Java   |       2 |
|   3 | WangWu   | 20   | Java   |       1 |
|   4 | ZhaoLiu  | 19   | Java   |       1 |
+-----+----------+------+--------+---------+
4 rows in set (0.00 sec)
```

从以上执行结果可以看出，数据插入成功。

2. 创建 POJO 类

在 src 目录下的 com.qfedu.pojo 包下分别创建两个 POJO 类：StuInfo 和 StuClass，如例 3-7 和例 3-8 所示。

【例 3-7】 StuInfo.java

```java
package com.qfedu.pojo;
public class StuInfo {
    private int sid;
    private String sname;
    private String age;
    private String course;
    public int getSid() {
        return sid;
    }
    public void setSid(int sid) {
        this.sid = sid;
    }
    public String getSname() {
        return sname;
    }
    public void setSname(String sname) {
        this.sname = sname;
    }
    public String getAge() {
        return age;
    }
    public void setAge(String age) {
        this.age = age;
    }
    public String getCourse() {
        return course;
    }
    public void setCourse(String course) {
        this.course = course;
    }
    @Override
    public String toString() {
```

```
        return "Stu [sid=" + sid + ", sname=" + sname + ", age=" + age + ","
            + " course=" + course + "]";
    }
}
```

【例 3-8】 **StuClass.java**

```
package com.qfedu.pojo;
import java.util.List;
public class StuClass {
    private Integer cid;
    private String cname;
    private Integer sum;
    private List<StuInfo> stuInfoList;
    public Integer getCid() {
        return cid;
    }
    public void setCid(Integer cid) {
        this.cid = cid;
    }
    public String getCname() {
        return cname;
    }
    public void setCname(String cname) {
        this.cname = cname;
    }
    public Integer getSum() {
        return sum;
    }
    public void setSum(Integer sum) {
        this.sum = sum;
    }
    public List<StuInfo> getStuInfoList() {
        return stuInfoList;
    }
    public void setStuInfoList(List<StuInfo> stuInfoList) {
        this.stuInfoList = stuInfoList;
    }
}
```

以上代码是两个 POJO 类，其中 StuInfo 类封装了学生信息，StuClass 类封装了班级信息。

3. 创建映射文件

在 com.qfedu.mappr 包下创建映射文件 StuClassMapper.xml，如例 3-9 所示。

【例 3-9】 **StuClassMapper.xml**

```
<?xml version="1.0" encoding="UTF-8" ?>
<!DOCTYPE mapper
```

```
PUBLIC "-//mybatis.org//DTD Mapper 3.0//EN"
"http://mybatis.org/dtd/mybatis-3-mapper.dtd">
<mapper namespace="stuClass">
    <select id="findStuClassByCid" parameterType="Integer"
        resultMap="stuClassResultsMap">
        select c.*,s.sid,s.sname,s.age,s.course from stu_class c , stu_info
            s where s.classid = c.cid and c.cid = #{cid}
    </select>
    <resultMap type="stuClass" id="stuClassResultsMap">
        <id column="cid" property="cid"/>
        <result column="cname" property="cname"/>
        <result column="sum" property="sum"/>
        <collection property="stuInfoList" ofType="stuInfo">
            <id column="sid" property="sid"/>
            <result column="sname" property="sname"/>
            <result column="age" property="age"/>
            <result column="course" property="course"/>
        </collection>
    </resultMap>
</mapper>
```

在以上代码中,<select>元素用于映射查询,<resultMap>元素用于结果集处理,在<resultMap>元素的子元素中,<id>元素用于映射主键,<result>元素用于映射其他字段,<collection>元素用于映射 stu_class 表和 stu_info 表的关联关系。

4. 在配置文件中引入最新的映射文件

打开 src 目录下的 mybatis-config. xml 文件,修改该文件的<mappers>元素,将映射文件 StuClassMapper. xml 引入到配置文件,修改后的代码如下所示。

```
<mappers>
        <mapper resource="com/qfedu/mapper/StuMapper.xml"/>
        <mapper resource="com/qfedu/mapper/StuCardMapper.xml"/>
        <mapper resource="com/qfedu/mapper/StuClassMapper.xml"/>
</mappers>
```

5. 编写测试类

在工程 chapter03 的 src 目录下创建 com. qfedu. test 包,在 com. qfedu. test 包下创建测试类 TestOneToMany,如例 3-10 所示。

【例 3-10】 TestOneToMany. java

```
package com.qfedu.test;
import java.io.InputStream;
import java.util.List;
import org.apache.ibatis.io.Resources;
import org.apache.ibatis.session.*;
import com.qfedu.pojo.*;
public class TestOneToMany {
```

```
    public static void main(String[ ] args) throws Exception {
        InputStream in =
              Resources.getResourceAsStream("mybatis-config.xml");
        SqlSessionFactory sessionFactory = new
              SqlSessionFactoryBuilder().build(in);
        SqlSession session = sessionFactory.openSession();
        //使用 MyBatis 查询 cid 为 1 的班级信息(包括该班级的学生信息)
        StuClass stuClass =
              session.selectOne("stuClass.findStuClassByCid",1);
        System.out.println("班级 ID: " + stuClass.getCid() + "\n 班级名称: " +
            stuClass.getCname() + "\n 班级人数: " + stuClass.getSum() + "\n 学生信息: ");
        List<StuInfo> stuInfoList = stuClass.getStuInfoList();
        for (StuInfo stuInfo : stuInfoList) {
            System.out.println(stuInfo);
        }
        session.close();
    }
}
```

在以上代码中，先通过 SqlSession 对象获取查询结果，此结果包含有处理表关系后的信息，然后将结果输出。

执行测试类 TestOneToMany，执行结果如图 3.5 所示。

```
Console
<terminated> TestOneToMany [Java Application] C:\Program Files\Java\jre1.8.0_161\bin\javaw.exe (2018年7月4日 下午4:13:35)
班级ID: 1
班级名称: Java1801
班级人数: 50
学生信息:
Stu [sid=1, sname=ZhangSan, age=20, course=Java]
Stu [sid=3, sname=WangWu, age=20, course=Java]
Stu [sid=4, sname=ZhaoLiu, age=19, course=Java]
```

图 3.5　执行 TestOneToMany 类的结果

从图 3.5 中可以看出，程序显示出包括学生信息在内的班级信息，由此可见，MyBatis 成功实现一对多关系映射。

3.4　多　对　多

通常情况下，多对多表关系要转化为一对多的形式进行处理，这种转化是通过一张中间表来实现的。在使用 MyBatis 处理多对多关系时，需要先将多对多关系转化为一对多关系，然后使用<collection>元素完成映射。

3.3 节中已经讲过，在处理表关系时，学生与教师是多对多的关联关系，接下来，本节将以此为例演示 MyBatis 实现多对多的关联映射。

1. 数据准备

(1) 在数据库 chapter03 中创建数据表 class_info、数据表 teach_info，其中，数据表 class_info 用于映射班级信息，数据表 teach_info 用于映射教师信息，SQL 语句如下所示。

```
USE chapter03;
CREATE TABLE class_info(
cid INT PRIMARY KEY AUTO_INCREMENT, # ID
cname VARCHAR(20),        #班级名称
sum INT                   #总人数
);
CREATE TABLE teach_info(
tid INT PRIMARY KEY AUTO_INCREMENT, # ID
tname VARCHAR(20),        #教师姓名
age VARCHAR(20),          #教师年龄
course VARCHAR(20)        #课程
);
```

（2）向数据表 class_info 中插入数据，SQL 语句如下所示。

```
INSERT INTO class_info(cname,sum) VALUES ('Java1801',50);
INSERT INTO class_info(cname,sum) VALUES ('Java1802',46);
```

（3）向数据表 teach_info 中插入数据，SQL 语句如下所示。

```
INSERT INTO teach_info(tname,age,course) VALUES('ZhangSan','31','JavaSE');
INSERT INTO teach_info(tname,age,course) VALUES('LiSi','33','JavaWeb');
```

（4）建立一张中间表 class_teach 并向表中插入数据，SQL 语句如下所示。

```
CREATE TABLE class_teach(
id INT PRIMARY KEY AUTO_INCREMENT, # ID
class_id  INT, #班级 ID
teach_id  INT, #教师 ID
FOREIGN KEY(class_id) REFERENCES class_info(cid),   #外键
FOREIGN KEY(teach_id) REFERENCES teach_info(tid)    #外键
);
INSERT INTO class_teach(class_id,teach_id)VALUES(1,1);
INSERT INTO class_teach(class_id,teach_id)VALUES(2,1);
INSERT INTO class_teach(class_id,teach_id)VALUES(1,2);
```

（5）通过 SQL 语句测试数据是否插入成功，执行结果如下所示。

```
mysql> select * from class_info;
+-----+-----------+------+
| cid | cname     | sum  |
+-----+-----------+------+
|   1 | Java1801  |  50  |
|   2 | Java1802  |  46  |
+-----+-----------+------+
2 rows in set (0.00 sec)

mysql> select * from teach_info;
```

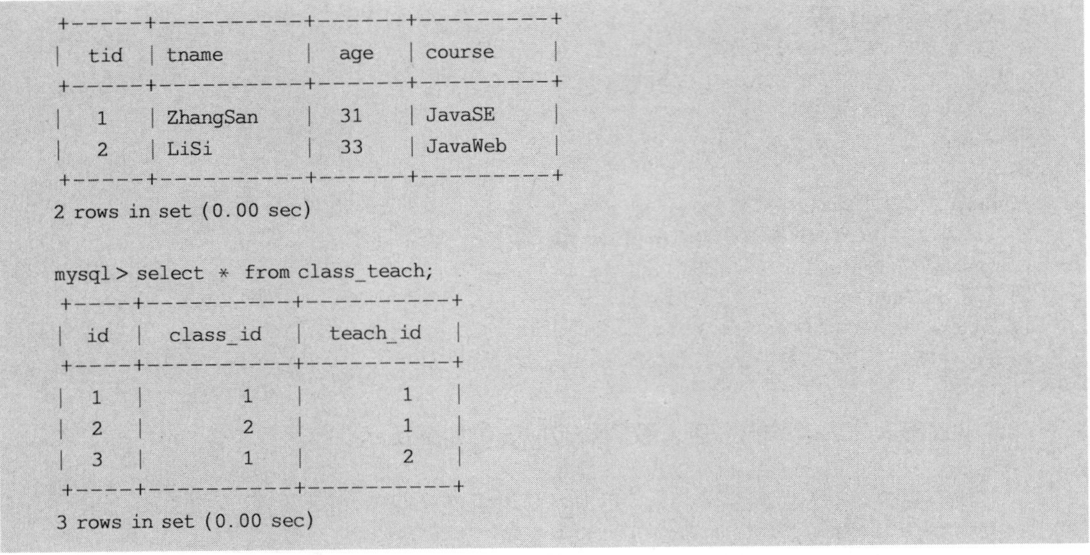

```
+-----+-----------+------+---------+
| tid | tname     | age  | course  |
+-----+-----------+------+---------+
|  1  | ZhangSan  | 31   | JavaSE  |
|  2  | LiSi      | 33   | JavaWeb |
+-----+-----------+------+---------+
2 rows in set (0.00 sec)

mysql> select * from class_teach;
+----+----------+----------+
| id | class_id | teach_id |
+----+----------+----------+
| 1  |        1 |        1 |
| 2  |        2 |        1 |
| 3  |        1 |        2 |
+----+----------+----------+
3 rows in set (0.00 sec)
```

从以上执行结果可以看出，数据插入成功。

2. 创建 POJO 类

在 src 目录的 com.qfedu.pojo 包下分别创建两个 POJO 类：TeachInfo 和 ClassInfo，如例 3-11 和例 3-12 所示。

【例 3-11】 TeachInfo.java

```
package com.qfedu.pojo;
import java.util.List;
public class TeachInfo {
    private Integer tid;
    private String tname;
    private String age;
    private String course;
    private List<ClassInfo> classInfoList;
    public Integer getTid() {
        return tid;
    }
    public void setTid(Integer tid) {
        this.tid = tid;
    }
    public String getTname() {
        return tname;
    }
    public void setTname(String tname) {
        this.tname = tname;
    }
    public String getAge() {
        return age;
    }
    public void setAge(String age) {
```

```java
        this.age = age;
    }
    public String getCourse() {
        return course;
    }
    public void setCourse(String course) {
        this.course = course;
    }
    public List<ClassInfo> getClassInfoList() {
        return classInfoList;
    }
    public void setClassInfoList(List<ClassInfo> classInfoList) {
        this.classInfoList = classInfoList;
    }
    @Override
    public String toString() {
        return "TeachInfo [tid=" + tid + ", tname=" + tname + ", age=" + age
         + ", course=" + course + ", classInfoList=" + classInfoList + "]";
    }
}
```

【例 3-12】 ClassInfo.java

```java
package com.qfedu.pojo;
import java.util.List;
public class ClassInfo {
    private Integer cid;
    private String cname;
    private Integer sum;
    private List<TeachInfo> teachInfoList;
    public Integer getCid() {
        return cid;
    }
    public void setCid(Integer cid) {
        this.cid = cid;
    }
    public String getCname() {
        return cname;
    }
    public void setCname(String cname) {
        this.cname = cname;
    }
    public Integer getSum() {
        return sum;
    }
    public void setSum(Integer sum) {
        this.sum = sum;
    }
    public List<TeachInfo> getTeachInfoList() {
```

```
        return teachInfoList;
    }
    public void setTeachInfoList(List<TeachInfo> teachInfoList) {
        this.teachInfoList = teachInfoList;
    }
    @Override
    public String toString() {
        return "ClassInfo [cid=" + cid + ", cname=" + cname + ", sum=" + sum
            + ", teachInfoList=" + teachInfoList + "]";
    }
}
```

以上代码是两个 POJO 类，其中 ClassInfo 类封装了班级信息，TeachInfo 类封装了教师信息，除了 getter 和 setter 方法之外，为方便输出信息，这两个类还提供了 toString()方法。

3. 创建映射文件

在 com.qfedu.mappr 包下创建映射文件 ClassInfoMapper.xml，如例 3-13 所示。

【例 3-13】 ClassInfoMapper.xml

```
<?xml version="1.0" encoding="UTF-8" ?>
<!DOCTYPE mapper
PUBLIC "-//mybatis.org//DTD Mapper 3.0//EN"
"http://mybatis.org/dtd/mybatis-3-mapper.dtd">
<mapper namespace="classInfo">
    <select id="findClassInfoByCid" parameterType="Integer"
         resultMap="classInfoResultsMap">
        select c.*,t.* from class_info  c ,teach_info t, class_teach ct where
            c.cid=ct.class_id
        and ct.teach_id = t.tid and c.cid = #{cid}
    </select>
    <resultMap type="classInfo" id="classInfoResultsMap">
        <id column="cid" property="cid"/>
        <result column="cname" property="cname"/>
        <result column="sum" property="sum"/>
        <collection property="teachInfoList" ofType="teachInfo">
            <id column="tid" property="tid"/>
            <result column="tname" property="tname"/>
            <result column="age" property="age" />
            <result column="course"  property="course"/>
        </collection>
    </resultMap>
</mapper>
```

在以上代码中，<select>元素用于映射查询，<resultMap>元素用于结果集处理，在<resultMap>元素的子元素中，<id>元素用于映射主键，<result>元素用于映射其他字段，<collection>元素用于映射 class_info 表和 teach_info 表的关联关系。

4. 在配置文件中引入最新的映射文件

打开 src 目录下的 mybatis-config.xml 文件，修改该文件的<mappers>元素，将映射文件 ClassInfoMapper.xml 引入到配置文件，修改后的代码如下所示。

```
<mappers>
        <mapper resource = "com/qfedu/mapper/StuMapper.xml"/>
        <mapper resource = "com/qfedu/mapper/StuCardMapper.xml"/>
        <mapper resource = "com/qfedu/mapper/StuClassMapper.xml"/>
        <mapper resource = "com/qfedu/mapper/ClassInfoMapper.xml"/>
    </mappers>
```

5. 编写测试类

在 com.qfedu.test 包下创建测试类 TestManyToMany，如例 3-14 所示。

【例 3-14】 TestManyToMany.java

```
package com.qfedu.test;
import java.io.InputStream;
import java.util.List;
import org.apache.ibatis.io.Resources;
import org.apache.ibatis.session.*;
import com.qfedu.pojo.*;
public class TestManyToMany {
    public static void main(String[] args) throws Exception {
        InputStream in =
            Resources.getResourceAsStream("mybatis-config.xml");
        SqlSessionFactory sessionFactory = new
            SqlSessionFactoryBuilder().build(in);
        SqlSession session = sessionFactory.openSession();
        //使用 MyBatis 查询 cid 为 1 的班级信息(包括该班级的教师信息)
        ClassInfo  classInfo =
            session.selectOne("classInfo.findClassInfoByCid",1);
        System.out.println("班级 ID: " + classInfo.getCid() + "\n 班级名称: " +
      classInfo.getCname() + "\n 班级人数: " + classInfo.getSum() + "\n 教师信息: ");
        List<TeachInfo> teachInfoList = classInfo.getTeachInfoList();
        for (TeachInfo teachInfo : teachInfoList) {
            System.out.println(teachInfo.toString());
        }
        session.close();
    }
}
```

在以上代码中，先通过 SqlSession 对象获取查询结果，此结果包含有处理表关系后的信息，然后将结果输出。

执行测试类 TestManyToMany，执行结果如图 3.6 所示。

从图 3.6 中可以看出，程序显示出包括学生信息在内的班级信息，由此可见，MyBatis 成功实现关系映射。

```
Console ✕
<terminated> TestManyToMany [Java Application] C:\Program Files\Java\jre1.8.0_161\bin\javaw.exe (2018年7月10日 上午11:24:25)
班级ID：1
班级名称：Java1801
班级人数：50
教师信息：
TeachInfo [tid=1, tname=ZhangSan, age=31, course=JavaSE, classInfoList=null]
TeachInfo [tid=2, tname=LiSi, age=33, course=JavaWeb, classInfoList=null]
```

图 3.6　执行 TestManyToMany 类的结果

3.5　主键映射

数据表的主键用于标识该表中每条记录的唯一性，使用 MyBatis 操作数据库应考虑到多表关联下的主键映射问题。

在实际开发中，当往数据表中插入数据时，如果对于插入数据的主键没有特殊要求，那么可以采用不返回主键值的方式配置插入语句，这样能够避免额外的 SQL 开销。但是，实际开发中常遇到一些多表关联下的操作，例如，在一次操作中插入具有一对多关系的表数据，当插入多方的数据时，需要获取刚刚插入的一方的主键值，此时就要采用插入后获取主键值的方式配置。

获取主键值可以通过配置映射文件中的< insert >元素完成，< insert >元素的 useGeneratedKeys 属性用于获取数据库内部产生的主键值，keyProperty 属性用于指定主键，

此处需要注意的是，如果操作的数据库不支持主键自增功能时，要使用< insert >元素的< selectKey >子元素获取主键值。

接下来，本节以一对多条件下的数据表 stu_class 为例演示主键映射。

1. 修改映射文件

修改映射文件 StuClassMapper.xml，添加< insert >元素，具体代码如下所示。

```
<insert id = "addStuClass01" parameterType = "stuClass"
useGeneratedKeys = "true"  keyProperty = "cid">
        insert into stu_class(cname, sum) values ( #{cname}, #{sum})
    </insert>
```

在以上代码中，useGeneratedKeys 属性用于获取数据库内部产生的主键值，keyProperty 属性用于指定主键。

2. 编写测试类

在 com.qfedu.test 包下创建测试类 TestAddStuClass，如例 3-15 所示。

【例 3-15】　TestAddStuClass.java

```
package com.qfedu.test;
import java.io.InputStream;
import org.apache.ibatis.io.Resources;
```

```
import org.apache.ibatis.session.*;
import com.qfedu.pojo.StuClass;
public class TestAddStuClass {
    public static void main(String[] args) throws Exception {
        InputStream in =
            Resources.getResourceAsStream("mybatis-config.xml");
        SqlSessionFactory sessionFactory = new
            SqlSessionFactoryBuilder().build(in);
        SqlSession session = sessionFactory.openSession();
        StuClass stuClass = new StuClass();
        stuClass.setCname("Java1803");
        stuClass.setSum(50);
        //使用 MyBatis 插入一条班级信息
        int result = session.insert("stuClass.addStuClass01",stuClass );
        if (result>0) {
            System.out.println("成功插入" + result + "条数据");
            System.out.println("插入数据的主键 cid 为:" + stuClass.getCid());
        }else{
            System.out.println("插入操作失败");
        }
        //提交事务
        session.commit();
        //关闭 SqlSession
        sessicn.close();
    }
}
```

在以上代码中，程序向数据库中插入一条班级信息，由于映射文件中配置了< insert >元素的 useGeneratedKeys 属性，程序可直接通过 stuClass 对象的 getCid()方法获取新添加记录的主键 cid 值。

执行测试类 TestAddStuClass，执行结果如图 3.7 所示。

```
Console
<terminated> TestAddStuClass [Java Application] C:\Program Files\Java\jre1.8.0_161\bin\javaw.exe (2018年7月10日 下午8:41:26)
log4j:WARN No appenders could be found for logger (org.apache.ibatis.logging.LogFactory).
log4j:WARN Please initialize the log4j system properly.
log4j:WARN See http://logging.apache.org/log4j/1.2/faq.html#noconfig for more info.
成功插入1条数据
插入数据的主键cid为:3
```

图 3.7　执行 TestAddStuClass 类的结果

从图 3.7 中可以看出，控制台输出成功插入数据的提示信息以及新添加记录的主键 cid 值，由此可见，MyBatis 成功实现多表关联场景下的主键映射。

除此之外，还可使用< insert >元素的< selectKey >子元素获取主键值，这种方式不但适用于数据库支持主键自增的场景，而且适用于数据库不支持主键自增的场景，接下来通过实例具体演示< selectKey >子元素的使用。

1. 修改映射文件

修改映射文件 StuClassMapper.xml，添加< insert >元素，具体代码如下所示。

```
<insert id="addStuClass02" parameterType="stuClass">
        insert into stu_class(cname,sum) values (#{cname},#{sum})
        <selectKey keyColumn="cid" keyProperty="cid" resultType="Integer"
            order="AFTER">
          select LAST_INSERT_ID()
        </selectKey>
    </insert>
```

在以上代码中，<selectKey>元素的 keyProperty 属性用于指定数据表主键，keyColumn 属性用于指定 POJO 对象的成员属性，resultType 属性用于指定返回值类型，order 属性的设定和使用的数据库有关，如果使用 MySQL 数据库，则 order 属性值是 AFTER。

2. 修改测试类

修改测试类 TestAddStuClass，将其第 17 行代码修改为如下所示。

```
int result = session.insert("stuClass.addStuClass02",stuClass);
```

执行测试类 TestAddStuClass，执行结果如图 3.8 所示。

```
Console
<terminated> TestAddStuClass [Java Application] C:\Program Files\Java\jre1.8.0_161\bin\javaw.exe (2018年7月11日 下午1:25:05)
log4j:WARN No appenders could be found for logger (org.apache.ibatis.logging.LogFactory).
log4j:WARN Please initialize the log4j system properly.
log4j:WARN See http://logging.apache.org/log4j/1.2/faq.html#noconfig for more info.
成功插入1条数据
插入数据的主键cid为:4
```

图 3.8 重新执行 TestAddStuClass 类的结果

从图 3.8 中可以看出，控制台输出成功插入数据的提示信息以及新添加记录的主键 cid 值，由此可见，通过<selectKey>也可实现返回主键值。

3.6 本章小结

本章首先介绍了表与表之间的关系，其中包括一对一、一对多、多对多等；然后分别讲解了如何使用 MyBatis 实现不同表关系的关联映射，最后详细讲解了如何使用 MyBatis 实现主键映射。通过本章知识的学习，大家应该理解表与表之间的关联关系，掌握处理表关系的思路和方法，能够使用 MyBatis 框架完成关联状态下的多表操作。

3.7 习题

1. 填空题

(1) 表与表之间的关系包括________、________、________等。

(2) 在一对一表关系中，每张数据表的关键字在对应的关系表中至多存在________条记录。

(3) <association>元素的________属性用于指定数据表中对应字段的类型。

(4) <collection>元素的________子元素用于映射代表多方的数据表的主键。

(5) 在处理多对多表关系时,通常要将其转化成________的形式进行处理。

2. 选择题

(1) 关于数据表的关联关系,下列选项错误的是(　　)。

A. MyBatis 支持一对一的表关系映射

B. 班级和学生的关系属于一对多的关系

C. 在一对多表关系中,主键数据表中的一个记录可与其关系表中的多个记录相关

D. MyBatis 无法支持多对多的表关系映射

(2) 在<association>元素的属性中,负责指定是否自动映射的是(　　)。

A. fetchType　　B. select

C. autoMapping　　D. typeHandler

(3) 在 MyBatis 的映射文件中,用于映射一对一关系的元素是(　　)。

A. <association>　　B. <collection>

C. <typeAlias>　　D. <discriminator>

(4) 在 MyBatis 的映射文件中,用于映射一对多关系的元素是(　　)。

A. <association>　　B. <collection>

C. <typeAlias>　　D. <discriminator>

(5) 在<selectKey>元素的属性中,用于指定数据表主键字段的是(　　)。

A. keyColumn　　B. keyProperty

C. resultType　　D. order

3. 思考题

(1) 简述一对多表关系的处理过程。

(2) 简述多对多表关系的处理过程。

4. 编程题

创建一张订单表,创建一张商品表,通过 MyBatis 查询 id 为 1 的订单表中的所有商品信息,注意多表关联。

第 4 章　动态 SQL 和注解

本章学习目标

- 理解动态 SQL 的功能
- 掌握动态 SQL 中常用元素的使用
- 理解 MyBatis 中的常用注解
- 掌握 MyBatis 中常用注解的使用

在传统的 JDBC 编程中，拼装并执行 SQL 语句是一个相当冗杂且烦琐的过程，如果处理业务关系复杂的场景，这种劣势会更加明显。作为获得广泛应用的 ORM 框架，MyBatis 提供了更为灵活高效的方式来处理 SQL 语句，这主要体现在它的动态 SQL 机制。通过动态 SQL，开发人员可根据需要动态组装 SQL 语句，除此之外，MyBatis 还支持通过注解配置 SQL 语句，接下来，本章将对 MyBatis 的动态 SQL 和注解做详细讲解。

4.1　动态 SQL

4.1.1　动态 SQL 简介

动态 SQL 是 MyBatis 提供的拼接 SQL 语句的强大机制。在 MyBatis 的映射文件中，开发人员可通过动态 SQL 元素灵活组装 SQL 语句，这在很大程度上避免了单一 SQL 语句的反复堆砌，提高了 SQL 语句的复用性。

MyBatis 提供了一系列的动态 SQL 元素，其中常用的元素如表 4.1 所示。

表 4.1　常用的动态 SQL 元素

设 置 参 数	描　　述
if	相当于判断语句，常用于单条件分支判断
choose (when, otherwise)	相当于判断语句，常用于多条件分支判断
trim (where, set)	辅助元素，用于处理一些拼装问题
foreach	相当于循环语句，在 in 语句等列举条件时常用

表 4.1 列举出了常用的动态 SQL 元素，每个元素分别实现不同的功能，接下来，本书将对这些元素做详细讲解。

4.1.2　<if>元素

<if>元素主要用于条件判断，它类似于 Java 代码中的 if 语句，通常与 test 属性联合

使用。

在实际开发中，< if >元素的任务是根据需求动态控制 where 子句的内容。例如，现有一张学生表，程序可通过学生姓名、年龄、科目等字段查询学生信息，如果将科目作为非必选的查询条件，满足既可以通过姓名和科目查询学生信息，又可以直接通过姓名查询学生信息，此时需要使用到< if >元素。

接下来，本节通过以上实例演示< if >元素的使用。

1. 数据准备

（1）在数据库 chapter04 中创建数据表 student，该表用于映射学生信息，SQL 语句如下所示。

```
DROP DATABASE IF EXISTS chapter04;
CREATE DATABASE chapter04;
USE chapter04;
CREATE TABLE student(
sid INT PRIMARY KEY AUTO_INCREMENT, # ID
sname VARCHAR(20),    #学生姓名
age VARCHAR(20),      #学生年龄
course VARCHAR(20)    #科目
);
```

（2）向数据表 student 中插入数据，SQL 语句如下所示。

```
INSERT INTO student(sname,age,course) VALUES ('ZhangSan','20','Java');
INSERT INTO student(sname,age,course) VALUES ('LiSi','21','Java');
INSERT INTO student(sname,age,course) VALUES ('LiSi','20','Python');
INSERT INTO student(sname,age,course) VALUES ('WangWu','19','Java');
```

（3）通过 SQL 语句测试数据是否插入成功，执行结果如下所示。

```
mysql > select * from student;
+-----+----------+------+--------+
| sid | sname    | age  | course |
+-----+----------+------+--------+
|   1 | ZhangSan | 20   | Java   |
|   2 | LiSi     | 21   | Java   |
|   3 | LiSi     | 20   | Python |
|   4 | WangWu   | 19   | Java   |
+-----+----------+------+--------+
4 rows in set (0.00 sec)
```

从以上执行结果可以看出，数据插入成功。此处需要提醒大家的是，数据表 student 中插入的两条 sname 字段均为 LiSi 的学生信息，但它们的其他字段值不同。

2. 创建工程

在 Eclipse 中新建 Web 工程 chapter04，将 MyBatis 的驱动 jar 包（mybatis-3.4.6.jar 和 lib 目录下的 jar 包）复制到 lib 目录下，完成 jar 包的导入。

3. 创建 POJO 类

在工程 chapter04 的 src 目录下创建 com. qfedu. pojo 包，在 com. qfedu. pojo 包下创建 POJO 类 Student，该类的代码与本书第 1 章中的例 1-1 所示代码相同，此处不再重复列出。

4. 创建配置文件

在 src 目录下新建配置文件 mybatis-config. xml，该文件的代码与第 3 章例 3-5 所示代码大致相同，由于要连接的数据库发生变化，此处需要将数据库名称修改为 chapter04，要修改的代码如下所示。

```
<property name = "url" value = "jdbc:mysql://localhost:3306/chapter04"/>
```

除此之外，由于映射文件发生变化，需要修改< mappers >元素的内容，要修改的代码如下。

```
<mappers>
        <mapper resource = "com/qfedu/mapper/StudentMapper.xml"/>
    </mappers>
```

5. 创建映射文件

在工程 chapter04 的 src 目录下创建 com. qfedu. mapper 包，在 com. qfedu. mapper 包下创建 StudentMapper. xml 文件，由于要满足科目作为非必选的查询条件，映射文件中使用< if >元素动态拼装 SQL 语句，具体代码如例 4-1 所示。

【例 4-1】 StudentMapper. xml

```
<?xml version = "1.0" encoding = "UTF - 8" ?>
<!DOCTYPE mapper
PUBLIC " - //mybatis.org//DTD Mapper 3.0//EN"
"http://mybatis.org/dtd/mybatis - 3 - mapper.dtd">
<mapper namespace = "student">
    <select id = "findStudentBySnameAndCourse"
         parameterType = "student"
         resultType = "student">
        select * from student where sname = #{sname}
        <!-- 根据条件动态拼装 SQL 语句 -->
        <if test = " null!= course and ''!= course">
        and course = #{course}
        </if>
    </select>
</mapper>
```

在以上代码中，< if >元素用于条件判断，当 course 的值不为 null 或不为空字符串时，where 子句中会追加将 course 作为判断条件的内容。

6. 编写测试类

在 src 目录下新建包 com. qfedu. test，在该包下新建类 TestIf，具体代码如例 4-2 所示。

【例 4-2】 **TestIf.java**

```
1 package com.qfedu.test;
2 import java.io.*;
3 import java.util.List;
4 import org.apache.ibatis.io.Resources;
5 import org.apache.ibatis.session.*;
6 import com.qfedu.pojo.Student;
7 public class TestIf {
8     public static void main(String[] args) {
9         String resource = "mybatis-config.xml";
10        try {
11            InputStream in = Resources.getResourceAsStream(resource);
12        SqlSessionFactory factory = new
13            SqlSessionFactoryBuilder().build(in);
14        SqlSession sqlSession = factory.openSession();
15        Student student = new Student();
16        student.setSname("LiSi");
17        List<Student> selectList =
18        sqlSession.selectList
19            ("student.findStudentBySnameAndCourse", student);
20        for (Student stu : selectList) {
21            System.out.println(stu.toString());
22        }
23        sqlSession.close();
24        } catch (IOException e) {
25            e.printStackTrace();
26        }
27    }
28 }
```

在以上代码中，先创建 Student 对象，接下来为 Student 对象的相关成员属性赋值，然后调用 sqlSession 的 selectList()方法查询学生信息，此时 Student 对象为 selectList()方法的参数。

执行 TestIf 类，执行结果如图 4.1 所示。

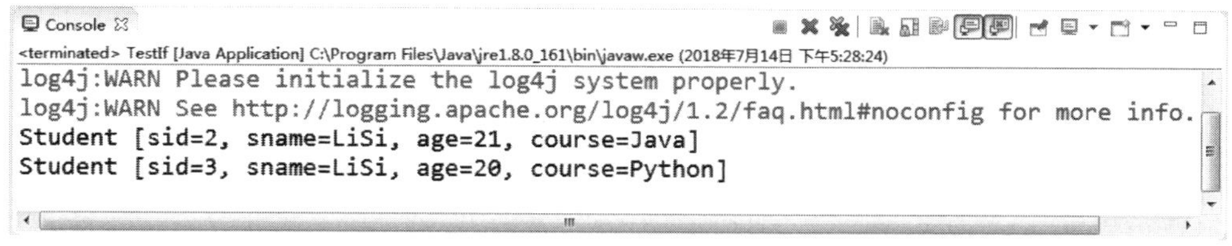

图 4.1 执行 TestIf 类的结果

从图 4.1 中可以看出，程序输出了所有 sname 为 LiSi 的学生信息。这就说明，当 Student 对象作为 sqlSession 的 selectList()方法的参数时，如果 Student 对象的 course 属性为空，那么映射文件< if >元素所包含的语句将不会被执行。

修改例 4-2 所示的 TestIf 类，在第 16 行后添加一行代码，具体如下。

```
student.setCourse("Java");
```

以上代码为 Student 对象的 course 属性赋值，此时 Student 对象的 course 属性值为 Java。

重新执行 TestIf 类，执行结果如图 4.2 所示。

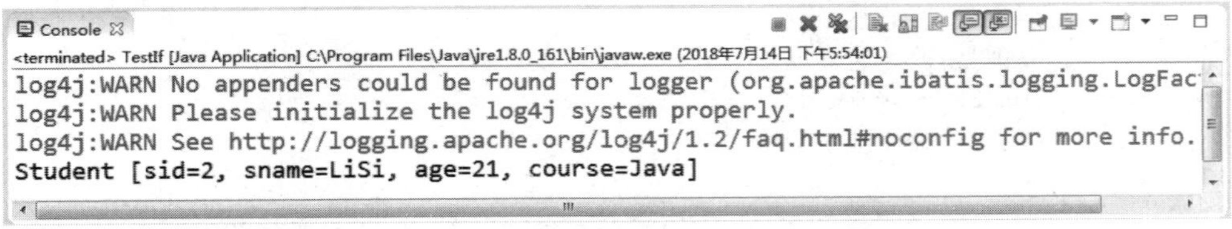

```
Console ⊠
<terminated> TestIf [Java Application] C:\Program Files\Java\jre1.8.0_161\bin\javaw.exe (2018年7月14日 下午5:54:01)
log4j:WARN No appenders could be found for logger (org.apache.ibatis.logging.LogFac
log4j:WARN Please initialize the log4j system properly.
log4j:WARN See http://logging.apache.org/log4j/1.2/faq.html#noconfig for more info.
Student [sid=2, sname=LiSi, age=21, course=Java]
```

图 4.2　重新执行 TestIf 类的结果

从图 4.2 中可以看出，程序只输出一条 sname 为 LiSi 的学生信息，并且该信息的 course 为 Java。这就说明，由于传入 selectList()方法的 Student 对象的 course 属性被赋值，映射文件中< if >元素包含的内容被执行，因此，程序只输出一条满足查询条件的信息。

4.1.3　< choose >、< when >和< otherwise >元素

与< if >元素的功能类似，< choose >元素同样用于条件判断，但不同的是，< choose >元素适用于多个判断条件的场景，它类似于 Java 代码中的 switch 语句。

< choose >元素包含< when >和< otherwise >两个子元素，其中，一个< choose >元素中至少要包含 1 个< when >子元素、0 个或 1 个< otherwise >子元素。当程序中的业务关系相对复杂时，MyBatis 可通过< choose >元素动态控制 SQL 语句中的内容。例如，现在要从数据表 student 中查询学生信息，如果 sid 不为空，则根据 sid 查询学生信息；如果 sid 为空、sname 不为空，则根据 sname 模糊查询学生信息；如果 sid 为空、sname 为空，则查询所有 course 值为 Java 的学生信息。

接下来，本节通过以上实例演示< choose >元素的使用。

1. 修改映射文件

在 StudentMapper.xml 文件的< mapper >元素中加入映射信息，具体代码如下。

```
<select id = "findStudentByChoose"  parameterType = "student"
resultType = "student">
        select * from student where 1 = 1
        <choose>
            <! -- 如果 sid 不为 null 或空字符串  -->
            <when test = "null!= sid and''!= sid">
                and sid = #{sid}
            </when>
            <! -- 如果 sname 不为 null 或空字符串  -->
            <when test = "null!= sname and ''!= sname">
                and sname like '% ${sname} %'
            </when>
            <! -- 如果以上两个条件都不满足,则执行下列内容  -->
```

```
            <otherwise>
                and course = 'Java'
            </otherwise>
        </choose>
    </select>
```

在以上代码中,< choose >元素用于根据条件组装 SQL 语句,第一个<. when >元素的 test 属性为 True 时,则追加该< when >元素包含的 SQL 语句,否则,进入第二个<. when >元素的条件判断;当前两个<. when >元素的 test 属性为 False 时,则直接追加< otherwise >元素所包含的 SQL 语句。

2. 编写测试类

在 com. qfedu. test 包下新建类 TestChoose,具体代码如例 4-3 所示。

【例 4-3】 TestChoose. java

```
package com.qfedu.test;
import java.io.*;
import java.util.List;
import org.apache.ibatis.io.Resources;
import org.apache.ibatis.session.*;
import com.qfedu.pojo.Student;
public class TestChoose {
    public static void main(String[] args) {
        String resource = "mybatis-config.xml";
        try {
            InputStream in = Resources.getResourceAsStream(resource);
        SqlSessionFactory factory = new
            SqlSessionFactoryBuilder().build(in);
        SqlSession sqlSession = factory.openSession();
        Student student = new Student();
        student.setId(2);
        List<Student> selectList =
        sqlSession.selectList("student.findStudentByChoose",student);
        for (Student stu : selectList) {
            System.out.println(stu.toString());
        }
        sqlSession.close();
        } catch (IOException e) {
            e.printStackTrace();
        }
    }
}
```

在以上代码中,先创建 Student 对象,接下来为 Student 对象的相关成员属性赋值,然后调用 sqlSession 的 selectList()方法查询学生信息,此时 Student 对象为 selectList()方法的参数。

执行 TestChoose 类,执行结果如图 4.3 所示。

从图 4.3 中可以看出,程序输出了 sid 为 2 的学生信息。这就说明,当 Student 对象作

```
Console
<terminated> TestChoose [Java Application] C:\Program Files\Java\jre1.8.0_161\bin\javaw.exe (2018年7月16日 下午4:07:49)
log4j:WARN No appenders could be found for logger (org.apache.ibatis.logging.LogFactory).
log4j:WARN Please initialize the log4j system properly.
log4j:WARN See http://logging.apache.org/log4j/1.2/faq.html#noconfig for more info.
Student [sid=2, sname=LiSi, age=21, course=Java]
```

图 4.3　执行 TestChoose 类的结果

为 sqlSession 的 selectList()方法的参数时，如果 Student 对象的 sid 属性不为空，那么 < choose >元素中< when >元素所包含的 SQL 语句将不会被执行。

修改例 4-3 所示的 TestChoose 类，将第 16 行代码替换，替换后的代码如下所示。

```
student.setSname("LiSi");
```

以上代码为 Student 对象的 sname 属性赋值，此时，Student 对象的 sname 属性值为 LiSi。

重新执行 TestChoose 类，执行结果如图 4.4 所示。

```
Console
<terminated> TestChoose [Java Application] C:\Program Files\Java\jre1.8.0_161\bin\javaw.exe (2018年7月16日 下午4:20:57)
log4j:WARN Please initialize the log4j system properly.
log4j:WARN See http://logging.apache.org/log4j/1.2/faq.html#noconfig for more info.
Student [sid=2, sname=LiSi, age=21, course=Java]
Student [sid=3, sname=LiSi, age=20, course=Python]
```

图 4.4　重新执行 TestChoose 类的结果

从图 4.4 中可以看出，程序输出所有 sname 为 LiSi 的学生信息。这就说明，由于传入 selectList()方法的 Student 对象的 sname 属性被赋值，< choose >元素中第二个< when >元素包含的内容被执行。

再次修改例 4-3 所示的 TestChoose 类，删除第 16 行代码，不再为 sname 属性赋值。

执行 TestChoose 类，执行结果如图 4.5 所示。

```
Console
<terminated> TestChoose [Java Application] C:\Program Files\Java\jre1.8.0_161\bin\javaw.exe (2018年7月16日 下午4:32:17)
log4j:WARN No appenders could be found for logger (org.apache.ibatis.logging.LogFactory).
log4j:WARN Please initialize the log4j system properly.
log4j:WARN See http://logging.apache.org/log4j/1.2/faq.html#noconfig for more info.
Student [sid=1, sname=ZhangSan, age=20, course=Java]
Student [sid=2, sname=LiSi, age=21, course=Java]
Student [sid=4, sname=WangWu, age=19, course=Java]
```

图 4.5　再次执行 TestChoose 类的结果

从图 4.5 中可以看出，程序输出所有 course 为 Java 的学生信息。这就说明，< choose >元素中< otherwise >元素所包含的内容被执行。

4.1.4　< where >元素

< where >元素用于拼接 SQL 语句中的关键字，通常与< if >元素联合使用。如果< where >元素包含的子元素有返回值时，程序就会在相应的 SQL 语句后追加一个 where

子句，如果 where 后跟有以 and 或 or 开头的内容时，则将 and 或 or 删除。

接下来通过 4.1.3 节所讲的实例来演示< where >元素的使用。

该实例映射信息中有一条 SQL 语句：select * from student where 1=1，由于该 SQL 语句要后跟以 and 开头的内容，因此，where 后接 1=1 来保证 SQL 语句正确拼接，否则，就可能会出现 select * from student where and course = 'Java'之类的错误写法。

除了 where 1=1 的方式，< where >元素也可以实现相同的功能。修改 4.1.3 节所讲实例的映射信息，修改后的代码如下所示。

```
<select id="findStudentByChoose"  parameterType="student"
    resultType="student">
        select * from student
        <where>
            <if test="null!= sid and''!= sid">
                and sid = #{sid}
            </if>
            <if test="null!= sname and ''!= sname">
                and sname like '%${sname}%'
            </if>
            <if test="null == course">
                and course = 'Java'
            </if>
        </where>
</select>
```

在以上代码中，< where >元素用于简化 where 语句中 where 条件判断，并可以处理 and 条件。

执行 TestChoose 类，执行结果如图 4.6 所示。

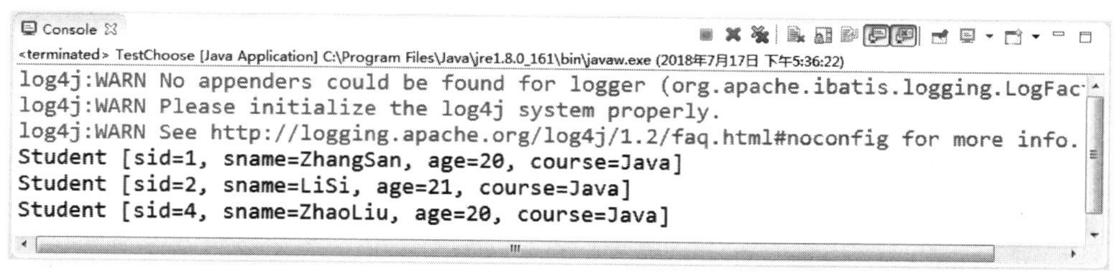

图 4.6　修改后执行 TestChoose 类的结果

从图 4.6 可以看出，程序输出所有 course 为 Java 的学生信息。这就说明，< where >元素实现了对 where 子句的动态拼装。

4.1.5　< set >元素

< set >元素和< where >元素的功能基本相似，不同的是，< set >元素用于 update 语句中，可以动态包含需要更新的列。如果< set >元素包含的子元素有返回值时，程序就会在相应的 SQL 语句后追加一个 set 子句，如果 set 后跟有以逗号结尾的内容时，则将逗号删除。

在实际开发中，根据业务需求，有时需要更新数据库中某条数据记录的字段，这时就需

要用到< set >元素。例如,学生表 student 中包含 sname、age、course 等字段,MyBatis 程序可通过< set >元素动态设置 SET 关键字并消除无关的逗号。

接下来,本节通过一个实例演示< set >元素的使用。

1. 修改映射文件

在 StudentMapper. xml 文件的< mapper >元素中加入映射信息,具体代码如下。

```
<update id = "updateStudent" parameterType = "student">
    update student
    <set>
        <if test = "null!= sname and ''!= sname">
            sname  =  #{sname},
        </if>
        <if test = "null!= age and ''!= age">
            age   =  #{age},
        </if>
    </set>
    where sid  =  #{sid}
</update>
```

在以上代码中,当 sname 不为 null 或空字符串时,则拼接第一个< if >元素包含的 SQL 语句;当 age 不为 null 或空字符串时,则拼接第二个< if >元素包含的 SQL 语句;此处需要大家注意的是,当拼接第二个< if >元素包含的 SQL 语句时,其结尾处的逗号将被自动删除。

2. 编写测试类

在包 com. qfedu. test 下新建类 TestSet,具体代码如例 4-4 所示。

【例 4-4】 TestSet. java

```
package com.qfedu.test;
import java.io.*;
import org.apache.ibatis.io.Resources;
import org.apache.ibatis.session.*;
import com.qfedu.pojo.Student;
public class TestSet {
    public static void main(String[] args) {
        String resource  =  "mybatis-config.xml";
        try {
            InputStream in  =  Resources.getResourceAsStream(resource);
        SqlSessionFactory factory  =  new
            SqlSessionFactoryBuilder().build(in);
        SqlSession sqlSession  =  factory.openSession();
        Student student  =  new Student();
        //为 Student 对象的成员属性赋值
        student.setId(4);
        student.setSname("ZhaoLiu");
        student.setAge("20");
        //调用 sqlSession 的 update()方法
```

```
        int result = sqlSession.update("student.updateStudent",student);
        if (result>0) {
            System.out.println("成功更新" + result + "条数据");
        }else{
            System.out.println("更新操作失败");
        }
        sqlSession.commit();
        sqlSession.close();
        } catch (IOException e) {
            e.printStackTrace();
        }
    }
}
```

在以上代码中，先创建 Student 对象，接下来为 Student 对象的相关成员属性赋值，然后调用 sqlSession 的 update()方法更新学生信息，此时 Student 对象为 update()方法的参数。

执行 TestSet 类，执行结果如图 4.7 所示。

```
Console
<terminated> TestSet [Java Application] C:\Program Files\Java\jre1.8.0_161\bin\javaw.exe (2018年7月17日 下午4:04:23)
log4j:WARN No appenders could be found for logger (org.apache.ibatis.logging.LogFactory).
log4j:WARN Please initialize the log4j system properly.
log4j:WARN See http://logging.apache.org/log4j/1.2/faq.html#noconfig for more info.
成功更新1条数据
```

图 4.7 执行 TestSet 类的结果

从图 4.7 中可以看出，程序输出了成功更新一条数据的信息。此时，查询数据表 student 中的信息，查询结果如下所示。

通过 SQL 语句测试数据是否更新成功，执行结果如下所示。

```
mysql> select * from student;
+-----+----------+-------+--------+
| sid | sname    | age   | course |
+-----+----------+-------+--------+
|   1 | ZhangSan | 20    | Java   |
|   2 | LiSi     | 21    | Java   |
|   3 | LiSi     | 20    | Python |
|   4 | ZhaoLiu  | 20    | Java   |
+-----+----------+-------+--------+
4 rows in set (0.00 sec)
```

从以上执行结果可以看出，sid 为 4 的学生信息已被更新。由此可见，<set>元素实现了对 update 语句的动态拼装。

4.1.6 <trim>元素

<trim>元素用于删除多余关键字，它可以直接实现<where>元素和<set>元素的

功能。

<trim>元素包含有 4 个属性，具体如表 4.2 所示。

表 4.2 <trim>元素的属性

设置参数	描　　述
prefix	指定给 SQL 语句增加的前缀
prefixOverrides	指定 SQL 语句中要去掉的前缀字符串
suffix	指定给 SQL 语句增加的后缀
suffixOverrides	指定 SQL 语句中要去掉的后缀字符串

表 4.2 列举出了<trim>元素的属性，在实现<where>元素对应的功能时，通常使用 prefix 和 prefixOverrides 属性，在实现<set>元素对应的功能时，通常使用 prefix 和 suffixOverrides 属性。

接下来，本节将演示以<trim>元素实现<where>元素功能。

修改 4.1.4 节所讲实例的映射信息，修改后的代码如下所示。

```
<select id = "findStudentByChoose"  parameterType = "student"
    resultType = "student">
       select * from student
       <trim prefix = "where" prefixOverrides = "and">
           <if test = "null!= sid and''!= sid">
               and sid = #{sid}
           </if>
           <if test = "null!= sname and ''!= sname">
               and sname like '% ${sname} %'
           </if>
           <if test = "null == course">
               and course = 'Java'
           </if>
       </trim>
   </select>
```

在以上代码中，<trim>元素的 prefix 属性指定要添加<trim>元素所包含内容的前缀为 where，prefixOverrides 属性指定去除<trim>元素所包含内容的前缀 and。

执行 TestChoose 类，执行结果如图 4.8 所示。

```
Console
<terminated> TestChoose [Java Application] C:\Program Files\Java\jre1.8.0_161\bin\javaw.exe (2018年7月17日 下午5:36:22)
log4j:WARN No appenders could be found for logger (org.apache.ibatis.logging.LogFac
log4j:WARN Please initialize the log4j system properly.
log4j:WARN See http://logging.apache.org/log4j/1.2/faq.html#noconfig for more info.
Student [sid=1, sname=ZhangSan, age=20, course=Java]
Student [sid=2, sname=LiSi, age=21, course=Java]
Student [sid=4, sname=ZhaoLiu, age=20, course=Java]
```

图 4.8　执行 TestChoose 类的结果

从图 4.8 可以看出，程序输出所有 course 为 Java 的学生信息。这就说明，<trim>元素实现了与<where>元素相同的功能。

除此之外，<trim>元素还可以实现与<set>元素相同的功能。

修改 4.1.5 节所讲实例的映射信息，修改后的代码如下所示。

```
<update id = "updateStudent" parameterType = "student">
        update student
        <trim prefix = "set" suffixOverrides = ",">
            <if test = "null!= sname and ''!= sname">
                sname  =  #{sname},
            </if>
            <if test = "null!= age and ''!= age">
                age   =  #{age},
            </if>
        </trim>
        where sid  =  #{sid}
</update>
```

在以上代码中，<trim>元素的 prefix 属性指定要添加<trim>元素所包含内容的前缀为 set，suffixOverrides 属性指定去除<trim>元素所包含内容的后缀为逗号。

修改例 4-4 所示的 TestSet 类，将第 17 行、第 18 行代码修改为如下所示。

```
student.setSname("SunQi");
  student.setAge("21");
```

执行 TestSet 类，执行结果如图 4.9 所示。

```
Console ⊠
<terminated> TestSet [Java Application] C:\Program Files\Java\jre1.8.0_161\bin\javaw.exe (2018年7月17日 下午5:44:53)
log4j:WARN No appenders could be found for logger (org.apache.ibatis.logging.LogFactory).
log4j:WARN Please initialize the log4j system properly.
log4j:WARN See http://logging.apache.org/log4j/1.2/faq.html#noconfig for more info.
成功更新1条数据
```

图 4.9 修改后执行 TestSet 类的结果

通过 SQL 语句测试数据是否更新成功，执行结果如下所示。

```
mysql> select * from student;
+-----+----------+------+--------+
| sid | sname    | age  | course |
+-----+----------+------+--------+
|   1 | ZhangSan | 20   | Java   |
|   2 | LiSi     | 21   | Java   |
|   3 | LiSi     | 20   | Python |
|   4 | SunQi    | 21   | Java   |
+-----+----------+------+--------+
4 rows in set (0.00 sec)
```

从以上执行结果可以看出,sid 为 4 的学生信息已被修改。由此可见,< trim >元素实现了与< set >元素相同的功能。

4.1.7 < foreach >元素

< foreach >元素主要用于遍历,能够支持数组、List 或 Set 接口的集合。< foreach >元素包含有五个属性,具体如表 4.3 所示。

表 4.3 < foreach >元素的属性

设置参数	描述
collection	指定传递进来的参数名称,可以代表 List、Set 等集合
item	指定循环中当前的元素
index	指定当前元素在集合的位置下标
open、close	指定包装集合元素的符号
separator	指定各个元素的间隔符

在实际开发中,< foreach >元素通常用于 SQL 语句中的 in 关键字。例如,要从数据表 student 中查询出 sid 为 1、2、3 的学生信息,如果采用单条查询的方式,势必会造成资源的浪费,此时就可以通过< foreach >元素完成批量查询。

接下来,本节将演示< foreach >元素的使用。

1. 修改映射文件

在 StudentMapper.xml 文件的< mapper >元素中加入映射信息,具体代码如下。

```
<select id = "findStudentByForeach"    resultType = "student">
        select   * from student where sid in
        <foreach item = "sid"   index = "index" collection = "list" open = "("
              separator = "," close = ")" >
            #{sid}
        </foreach>
    </select>
```

在以上代码中,< foreach >元素用于遍历集合,其中,item 属性指定循环中的当前元素为 sid,index 属性指定当前元素在集合中的下标,collection 属性指定传入的参数为 list,open 和 close 属性指定以"()"将集合元素包装起来,separator 属性指定以","将集合元素分割。

2. 编写测试类

在包 com.qfedu.test 下新建类 TestForeach,具体代码如例 4-5 所示。

【例 4-5】 TestForeach.java

```
package com.qfedu.test;
import java.io.*;
import java.util.*;
import org.apache.ibatis.io.Resources;
import org.apache.ibatis.session.*;
```

```java
import com.qfedu.pojo.Student;
public class TestForeach {
    public static void main(String[] args) {
        //读取配置文件
        String resource = "mybatis-config.xml";
        try {
            InputStream in = Resources.getResourceAsStream(resource);
        //创建 SQLSessionFactory 对象
        SqlSessionFactory factory = new
            SqlSessionFactoryBuilder().build(in);
        //创建 SqlSession 对象
        SqlSession sqlSession = factory.openSession();
        ArrayList<Integer> list = new ArrayList<Integer>();
        list.add(1);
        list.add(2);
        list.add(3);
        //调用 SqlSession 对象的 selectList()方法执行查询
        List<Student> stuList =
            sqlSession.selectList("student.findStudentByForeach", list);
        for (Student stu : stuList) {
            System.out.println(stu.toString());
        }
        //关闭 SqlSession
        sqlSession.close();
        } catch (IOException e) {
            e.printStackTrace();
        }
    }
}
```

执行 TestForeach 类,执行结果如图 4.10 所示。

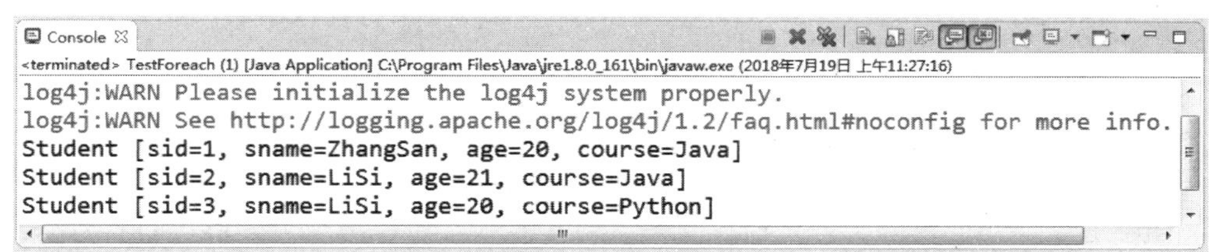

图 4.10　执行 TestForeach 类的结果

从图 4.10 中可以看出,程序输出了 sid 为 1、2、3 的学生信息。

4.1.8　<bind>元素

在实际开发中,由于不同数据库支持的 SQL 语法略有不同,如果需要更换数据库,那么程序中的相应 SQL 语句就需要重写,这就会带来维护成本的上升。此时,可以通过<bind>元素来解决此类问题。

<bind>元素将 OGNL 表达式的值绑定到一个变量,通过<bind>元素,对变量的值的

引用变得简单，而且，MyBatis 程序的可移植性得到提升。

接下来，本节将演示<bind>元素的使用。

1. 修改映射文件

在 StudentMapper.xml 文件的<mapper>元素中加入映射信息，具体代码如下。

```
<select id = "findStudentByBind"   parameterType = "student"
resultType = "student">
        select * from student   where
        <bind name = "sname_pattern" value = "'%' + sname + '%'"/>
        <if test = "null!= sname_pattern and ''!= sname_pattern">
                sname like   #{sname_pattern}
        </if>
</select>
```

在以上代码中，<bind>元素用于将 OGNL 表达式的值绑定到一个变量，其中，name 属性指定变量名，value 属性指定变量的值。

2. 编写测试类

在包 com.qfedu.test 下新建类 TestBind，具体代码如例 4-6 所示。

【例 4-6】 TestBind.java

```java
package com.qfedu.test;
import java.io.*;
import java.util.List;
import org.apache.ibatis.io.Resources;
import org.apache.ibatis.session.*;
import com.qfedu.pojo.Student;
public class TestBind {
    public static void main(String[] args) {
        String resource  =  "mybatis-config.xml";
        try {
            InputStream in  =  Resources.getResourceAsStream(resource);
        SqlSessionFactory factory  =  new
            SqlSessionFactoryBuilder().build(in);
        SqlSession sqlSession  =  factory.openSession();
        Student student  =  new Student();
        student.setSname("LiSi");
        List<Student> selectList  =
            sqlSession.selectList("student.findStudentByBind",student);
        for (Student stu : selectList) {
            System.out.println(stu.toString());
        }
        sqlSession.close();
        } catch (IOException e) {
            e.printStackTrace();
        }
    }
}
```

执行 TestBind 类,执行结果如图 4.11 所示。

```
Console ⌧
<terminated> TestBind [Java Application] C:\Program Files\Java\jre1.8.0_161\bin\javaw.exe (2018年7月23日 下午8:23:02)
log4j:WARN Please initialize the log4j system properly.
log4j:WARN See http://logging.apache.org/log4j/1.2/faq.html#noconfig for more info.
Student [sid=2, sname=LiSi, age=21, course=Java]
Student [sid=3, sname=LiSi, age=20, course=Python]
```

图 4.11　执行 TestBind 类的结果

从图 4.11 中可以看出,程序输出了 sid 为 2、3 的学生信息。

4.2 注　　解

4.2.1 简介

在 MyBatis 中,除了 XML 的映射方式,MyBatis 还支持通过注解实现 POJO 对象和数据表之间的关系映射。使用注解时,一般将 SQL 语句直接写在接口上。与 XML 的映射方式相比,注解相对简单并且不会造成大量的开销。

MyBatis 提供了若干注解,其中常用的注解如表 4.4 所示。

表 4.4　常用的注解

注　　解	描　　述
@Select	查询操作
@Insert	插入操作
@Update	更新操作
@Delete	删除操作
@Param	传入参数

以上列举出了 MyBatis 提供的常用注解,接下来,本书将对这些注解进行详细讲解。

4.2.2 @Select 注解

@Select 注解用于映射查询语句,其作用等同于 xml 文件中的< select >元素。接下来,本节将通过一个实例演示@Select 注解的使用。

1. 创建 Mapper 接口

在 com.qfedu.mapper 包下创建接口 StudentMapper,具体代码如例 4-7 所示。

【例 4-7】 StudentMapper.java

```
package com.qfedu.mapper;
import org.apache.ibatis.annotations.Select;
import com.qfedu.pojo.Student;
public interface StudentMapper {
    @Select("select * from student where sid = #{sid}")
    Student selectStudent(int sid);
}
```

在以上代码中，@Select 注解的参数是一个查询语句，当程序调用@Select 注解下一行的 selectStudent()方法时，@Select 注解中映射的查询语句将被执行。

在创建完 StudentMapper 接口后，在配置文件 mybatis-config.xml 的< mapper >元素加入相应的配置信息，具体代码如下。

```
<mapper class = "com.qfedu.mapper.StudentMapper"/>
```

2. 创建测试类

在 com.qfedu.test 包下创建测试类 TestFindBySid，具体代码如例 4-8 所示。

【例 4-8】 TestFindBySid.java

```
package com.qfedu.test;
import java.io.*;
import org.apache.ibatis.io.Resources;
import org.apache.ibatis.session.*;
import com.qfedu.mapper.StudentMapper;
import com.qfedu.pojo.Student;
public class TestFindBySid {
    public static void main(String[] args) {
        String resource = "mybatis-config.xml";
        try {
            InputStream in = Resources.getResourceAsStream(resource);
        SqlSessionFactory factory = new
            SqlSessionFactoryBuilder().build(in);
        SqlSession sqlSession = factory.openSession();
        StudentMapper mapper = sqlSession.getMapper(StudentMapper.class);
        Student student = mapper.selectStudent(1);
        System.out.println(student.toString());
        sqlSession.close();
        } catch (IOException e) {
            e.printStackTrace();
        }
    }
}
```

在以上代码中，通过调用 SqlSession 对象的 getMapper ()方法获取 StudentMapper 对象，进而执行 StudentMapper 对象的 selectStudent()方法。

执行 TestFindBySid 类，执行结果如图 4.12 所示。

```
Console
<terminated> TestFind [Java Application] C:\Program Files\Java\jre1.8.0_161\bin\javaw.exe (2018年7月24日 下午4:49:42)
log4j:WARN No appenders could be found for logger (org.apache.ibatis.logging.LogFactory).
log4j:WARN Please initialize the log4j system properly.
log4j:WARN See http://logging.apache.org/log4j/1.2/faq.html#noconfig for more info.
Student [sid=1, sname=ZhangSan, age=20, course=Java]
```

图 4.12 执行结果

从图 4.12 中可以看出，程序输出了 sid 为 1 的学生信息。

4.2.3 @Insert 注解

@Insert 注解用于映射插入语句,其作用等同于 xml 文件中的< insert >元素。接下来,本节将通过一个实例演示@Insert 注解的使用。

1. 修改 Mapper 接口

在 StudentMapper 接口中添加向数据库插入数据的方法,具体代码如下所示。

```
@Insert("insert into student(sname,age,course) "
        + " values(#{sname}, #{age}, #{course})")
int insertStudent(Student student);
```

在以上代码中,@Insert 注解的参数是一个插入语句,当程序调用@Insert 注解下一行的 insertStudent ()方法时,@ Insert 注解中映射的插入语句将被执行。

2. 创建测试类

在 com.qfedu.test 包下创建测试类 TestInsert,具体代码如例 4-9 所示。

【例 4-9】 TestInsert.java

```
package com.qfedu.test;
import java.io.*;
import org.apache.ibatis.io.Resources;
import org.apache.ibatis.session.*;
import com.qfedu.mapper.StudentMapper;
import com.qfedu.pojo.Student;
public class TestInsert {
    public static void main(String[] args) {
        String resource = "mybatis-config.xml";
        try {
            InputStream in = Resources.getResourceAsStream(resource);
        SqlSessionFactory factory = new
            SqlSessionFactoryBuilder().build(in);
        SqlSession sqlSession = factory.openSession();
        Student student = new Student();
        student.setSname("ZhouBa");
        student.setAge("21");
        student.setCourse("Java");
        StudentMapper mapper = sqlSession.getMapper(StudentMapper.class);
        int result = mapper.insertStudent(student);
        if (result>0) {
            System.out.println("成功插入" + result + "条数据");
        }else{
            System.out.println("插入操作失败");
        }
        sqlSession.commit();
        sqlSession.close();
        } catch (IOException e) {
            e.printStackTrace();
        }
```

```
    }
}
```


执行 TestFindBySid 类，执行结果如图 4.13 所示。

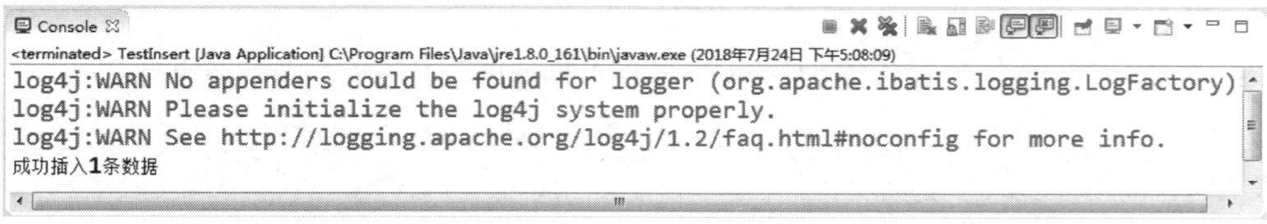

```
Console
<terminated> TestInsert [Java Application] C:\Program Files\Java\jre1.8.0_161\bin\javaw.exe (2018年7月24日 下午5:08:09)
log4j:WARN No appenders could be found for logger (org.apache.ibatis.logging.LogFactory)
log4j:WARN Please initialize the log4j system properly.
log4j:WARN See http://logging.apache.org/log4j/1.2/faq.html#noconfig for more info.
成功插入1条数据
```

图 4.13　执行 TestFindBySid 类的结果

从图 4.13 中可以看出，程序输出了成功添加 1 条数据的信息。

4.2.4　@Update 注解

@Update 注解用于映射更新语句，其作用等同于 xml 文件中的< update >元素。接下来，本节将通过一个实例演示@Update 注解的使用。

1. 修改 Mapper 接口

在 StudentMapper 接口中添加更新数据库中数据的方法，具体代码如下所示。

```
@Update("update student "
        + "set sname = #{sname},age = #{age} where sid = #{sid}")
int updateStudent(Student student);
```

在以上代码中，@Update 注解的参数是一个更新语句，当程序调用@Update 注解下一行的 updateStudent()方法时，@Update 注解中映射的更新语句将被执行。

2. 创建测试类

在 com.qfedu.test 包下创建测试类 TestUpdate，具体代码如例 4-10 所示。

【例 4-10】　TestUpdate.java

```
package com.qfedu.test;
import java.io.*;
import org.apache.ibatis.io.Resources;
import org.apache.ibatis.session.*;
import com.qfedu.mapper.StudentMapper;
import com.qfedu.pojo.Student;
public class TestUpdate {
    public static void main(String[] args) {
        String resource = "mybatis-config.xml";
        try {
            InputStream in = Resources.getResourceAsStream(resource);
        SqlSessionFactory factory = new
            SqlSessionFactoryBuilder().build(in);
        SqlSession sqlSession = factory.openSession();
        Student student = new Student();
        student.setId(5);
```

```
            student.setSname("WuJiu");
            student.setCourse("Python");
            StudentMapper mapper = sqlSession.getMapper(StudentMapper.class);
            int result   = mapper.updateStudent(student);
            if (result > 0) {
                System.out.println("成功更新" + result + "条数据");
            }else{
                System.out.println("更新操作失败");
            }
            sqlSession.commit();
            sqlSession.close();
            } catch (IOException e) {
                e.printStackTrace();
            }
        }
}
```

执行 TestUpdate 类，执行结果如图 4.14 所示。

```
Console ☒
<terminated> TestUpdate (2) [Java Application] C:\Program Files\Java\jre1.8.0_161\bin\javaw.exe (2018年7月24日 下午5:49:54)
log4j:WARN No appenders could be found for logger (org.apache.ibatis.logging.LogFactory)
log4j:WARN Please initialize the log4j system properly.
log4j:WARN See http://logging.apache.org/log4j/1.2/faq.html#noconfig for more info.
成功更新1条数据
```

图 4.14　执行 TestUpdate 类的结果

从图 4.14 中可以看出，程序输出了成功更新 1 条数据的信息。

4.2.5　@Delete 注解

@Delete 注解用于映射删除语句，其作用等同于 xml 文件中的< delete >元素。接下来，本节将通过一个实例演示@Delete 注解的使用。

1. 修改 Mapper 接口

在 StudentMapper 接口中添加删除数据库中数据的方法，具体代码如下所示。

```
@Delete("delete from student where sid = #{sid}")
int deleteStudent(intsid);
```

在以上代码中，@Delete 注解的参数是一个删除语句，当程序调用@Delete 注解下一行的 deleteStudent ()方法时，@Delete 注解中映射的删除语句将被执行。

2. 创建测试类

在 com.qfedu.test 包下创建测试类 TestDelete，具体代码如例 4-11 所示。

【例 4-11】 TestDelete.java

```
package com.qfedu.test;
import java.io.*;
import org.apache.ibatis.io.Resources;
```

```java
import org.apache.ibatis.session.*;
import com.qfedu.mapper.StudentMapper;
public class TestDelete {
    public static void main(String[] args) {
        String resource = "mybatis-config.xml";
        try {
            InputStream in = Resources.getResourceAsStream(resource);
        SqlSessionFactory factory = new
            SqlSessionFactoryBuilder().build(in);
        SqlSession sqlSession = factory.openSession();
        StudentMapper mapper = sqlSession.getMapper(StudentMapper.class);
        int result = mapper.deleteStudent(5);
        if (result>0) {
            System.out.println("成功删除" + result + "条数据");
        }else{
            System.out.println("删除操作失败");
        }
        sqlSession.commit();
        sqlSession.close();
        } catch (IOException e) {
            e.printStackTrace();
        }
    }
}
```

执行 TestDelete 类,执行结果如图 4.15 所示。

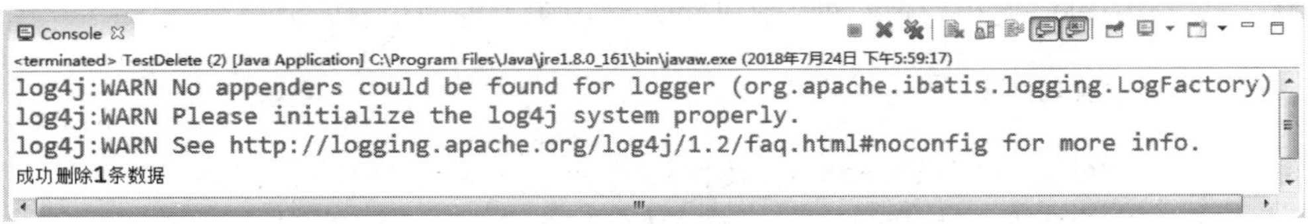

```
Console
<terminated> TestDelete (2) [Java Application] C:\Program Files\Java\jre1.8.0_161\bin\javaw.exe (2018年7月24日 下午5:59:17)
log4j:WARN No appenders could be found for logger (org.apache.ibatis.logging.LogFactory)
log4j:WARN Please initialize the log4j system properly.
log4j:WARN See http://logging.apache.org/log4j/1.2/faq.html#noconfig for more info.
成功删除1条数据
```

图 4.15 执行 TestDelete 类的结果

从图 4.15 中可以看出,程序输出了成功删除 1 条数据的信息。

4.2.6 @Param 注解

@Param 注解的功能是指定参数,通常用于 SQL 语句中参数比较多的情况。接下来,本节将通过一个实例演示@Param 注解的使用。

1. 修改 Mapper 接口

在 StudentMapper 接口中添加多条件查询的方法,具体代码如下所示。

```java
@Select("select * from student where sname = #{param01} "
        + "and course = #{param02}")
Student selectBySnameAndCourse(@Param("param01")String sname ,
@Param("param02")String course);
```

在以上代码中,@Param 注解指定了两个参数,它们分别是 param01 和 param02,这两个参数可以在 SQL 语句中被直接引用。

2. 创建测试类

在 com. qfedu. test 包下创建测试类 TestSelect,具体代码如例 4-12 所示。

【例 4-12】 TestSelect. java

```
package com.qfedu.test;
import java.io.*;
import org.apache.ibatis.io.Resources;
import org.apache.ibatis.session.*;
import com.qfedu.mapper.StudentMapper;
import com.qfedu.pojo.Student;
public class TestSelect {
    public static void main(String[] args) {
        String resource = "mybatis-config.xml";
        try {
            InputStream in = Resources.getResourceAsStream(resource);
        SqlSessionFactory factory = new
            SqlSessionFactoryBuilder().build(in);
            SqlSession sqlSession = factory.openSession();
        StudentMapper mapper = sqlSession.getMapper(StudentMapper.class);
        Student student = mapper.selectBySnameAndCourse("LiSi","Java");
        System.out.println(student.toString());
        sqlSession.close();
        } catch (IOException e) {
            e.printStackTrace();
        }
    }
}
```

执行 TestSelect 类,执行结果如图 4.16 所示。

```
Console
<terminated> TestSelect [Java Application] C:\Program Files\Java\jre1.8.0_161\bin\javaw.exe (2018年7月24日 下午6:51:23)
log4j:WARN No appenders could be found for logger (org.apache.ibatis.logging.LogFactory)
log4j:WARN Please initialize the log4j system properly.
log4j:WARN See http://logging.apache.org/log4j/1.2/faq.html#noconfig for more info.
Student [sid=2, sname=LiSi, age=21, course=Java]
```

图 4.16 执行 TestSelect 类的结果

从图 4.16 中可以看出,程序输出了 sid 为 2 的学生信息。

4.3 本 章 小 结

本章首先介绍了 MyBatis 中动态 SQL 的相关知识,详细演示了常见动态 SQL 元素的使用方法,然后介绍了 MyBatis 中注解的概念及具体使用方法。通过本章知识的学习,大家应该能理解动态 SQL 和注解的概念，掌握动态 SQL 和注解的使用方法，能够根据不同的需求场景灵活实现表与 POJO 对象的映射关系。

4.4 习　　题

1. 填空题

(1) ________元素主要用于条件判断，它类似于 Java 代码中的 if 语句，通常与 test 属性联合使用。

(2) ________元素适用于多个判断条件的场景，它类似于 Java 代码中的 switch 语句。

(3) ________元素用于遍历集合，通常用于 SQL 语句中的 in 关键字。

(4) ________注解用于映射查询语句，其作用等同于 xml 文件中的< select >元素。

(5) ________注解用于映射插入语句，其作用等同于 xml 文件中的< insert >元素。

2. 选择题

(1) 关于常用的动态 SQL 元素，下列选项错误的是(　　)。

A. < if >元素主要用于条件判断，它类似于 Java 代码中的 if 语句，通常与 test 属性联合使用

B. < choose >元素包含< when >和< otherwise >两个子元素，其中，一个< choose >元素中至少要包含 1 个< when >子元素、1 个< otherwise >子元素

C. 在实际开发中，< if >元素的任务是根据需求动态控制 where 子句的内容

D. 当程序中的业务关系相对复杂时，MyBatis 可通过< choose >元素动态控制 SQL 语句中的内容

(2) 在实际开发中，根据业务需求，有时需要更新数据库中某条数据记录的字段，这时就需要用到(　　)元素。

A. < bind >　　B. < select >

C. < set >　　D. < update >

(3) (　　)素主要用于遍历，能够支持数组、List 或 Set 接口的集合。

A. < foreach >　　B. < trim >

C. < typeAlias >　　D. < where >

(4) 关于注解的说法，下列选项错误的是(　　)。

A. @Select 注解用于映射查询语句，其作用等同于 xml 文件中的< select >元素

B. @update 注解用于映射插入语句，其作用等同于 xml 文件中的< update >元素

C. @delete 注解用于映射删除语句，其作用等同于 xml 文件中的< delete >元素

D. @Param 注解的功能是指定参数，通常用于 SQL 语句中参数比较多的情况

(5) (　　)元素用于拼接 SQL 语句中的关键字，通常与< if >元素联合使用。如果它包含的标签中有返回值，它就插入一个'where'，如果标签返回的内容是以 and 或 or 开头的，则它会剔除掉 and 或 or。

A. < set >　　B. < choose >

C. < where >　　D. < otherwise >

3. 思考题

（1）简述什么是动态 SQL。

（2）MyBatis 提供的常用注解有哪些？

4. 编程题

编写一段登录功能的代码，要求用户在输入学号和密码时，如果不为空，才能继续验证登录信息，否则登录失败。

第 5 章 MyBatis 缓存处理

本章学习目标

- 理解 MyBatis 的一级缓存机制
- 掌握 MyBatis 处理一级缓存的方法
- 理解 MyBatis 的二级缓存机制
- 掌握 MyBatis 处理二级缓存的方法
- 掌握 MyBatis 集成 EhCache 缓存的方法

为了降低高并发访问给数据库带来的压力，大型企业项目中都会使用缓存。使用缓存可以降低磁盘 IO、减少程序与数据库的交互，帮助程序迅速获取所需数据，提升系统响应速度，对优化系统整体性能具有重要意义。作为 ORM 框架，MyBatis 支持缓存并提供了强大的缓存特性，接下来，本章将对 MyBatis 的缓存处理机制做详细讲解。

5.1 MyBatis 的缓存机制

通常情况下，缓存将程序使用频率较高的数据存储在内存中，它具有可被快速读取和使用的特点。

MyBatis 支持缓存，它在内存中开辟一块区域，该区域用于保存程序对数据库的操作信息和数据库返回的数据，如果下一次程序再执行相同的操作，那么直接从缓存中读取数据而不是从数据库读取，如此一来，大大提升数据库性能并减轻系统压力。

MyBatis 支持的缓存分为一级缓存和二级缓存，如图 5.1 所示。

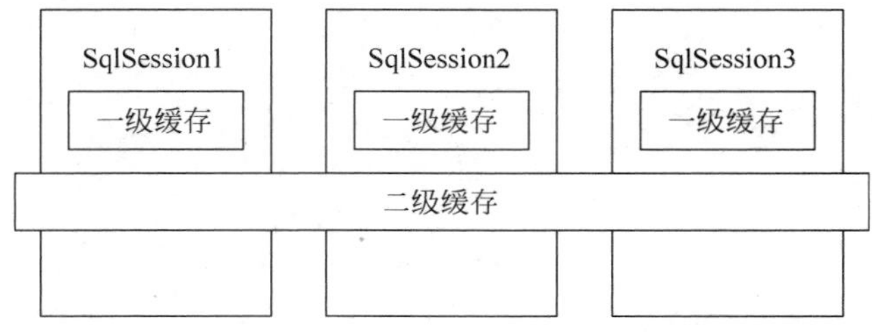

图 5.1 MyBatis 支持的缓存

其中，一级缓存是 SqlSession 级别的缓存。当创建 SqlSession 对象操作数据库时，MyBatis 在 SqlSession 对象中引入一个 HashMap 对象作为存储数据的区域，HashMap 对象的 key 是由 SQL 语句、条件、statement 等信息组成一个唯一值。HashMap 对象的 value 是查询出的结果对象。不同 SqlSession 对象之间的 HashMap 对象互不影响。

二级缓存是 Mapper 级别的缓存，多个 SqlSession 去操作同一个 Mapper 的 SQL 语句，二级缓存是跨 SqlSession 的，多个 SqlSession 可以共用二级缓存。与一级缓存的 HashMap 对象相同，二级缓存的 HashMap 对象的 key 由 SQL 语句、条件、statement 等信息组成，HashMap 对象的 value 是查询出的结果对象。

在实际应用中，一级缓存是默认开启的，二级缓存需要手动开启。

5.2 一级缓存

5.2.1 一级缓存的原理

在 MyBatis 中，如果一级缓存开启，当同一个 SqlSession 对象多次执行完全相同的 SQL 语句时，在第一次执行完成后，MyBatis 会将查询结果写入到一级缓存，此后，如果程序没有执行插入、更新、删除操作，当第二次执行相同的查询语句时，MyBatis 会直接读取一级缓存中的数据。

在实际开发中，如果多次执行同一条 SQL 语句，例如，从数据表 student 中查询出 sid 为 1 的学生信息，具体过程如图 5.2 所示。

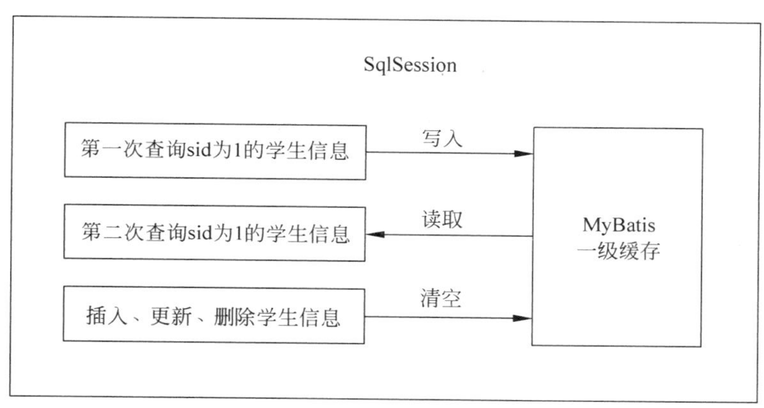

图 5.2 MyBatis 一级缓存

从图 5.2 中可以看出，当程序第一次查询 sid 为 1 的学生信息时，程序会将结果写入 MyBatis 一级缓存，当程序第二次查询 sid 为 1 的学生信息时，MyBatis 将直接从一级缓存中读取。当程序对数据库执行了 DML 操作，MyBatis 会清空一级缓存中的内容以防止程序误读。

5.2.2 一级缓存的应用

上个小节讲解了 MyBatis 一级缓存的原理，接下来，本节通过一个实例演示 MyBatis 一级缓存的应用。

1. 数据准备

创建数据库 chapter05，在数据库 chapter05 中创建数据表 student，向数据表 student 中插入学生信息，SQL 语句参照第 4 章 4.1.2 小节中内容，此处不再具体列出。

2. 创建工程

在 Eclipse 中新建 Web 工程 chapter05，将 MyBatis 的驱动 jar 包(mybatis-3. 4. 6. jar、mysql-connector-java-5. 1. 10-bin. jar 和 lib 目录下的 jar 包)复制到 lib 目录下，完成 jar 包的导入。

3. 创建 POJO 类

在工程 chapter05 的 src 目录下创建 com. qfedu. pojo 包，在 com. qfedu. pojo 包下创建 POJO 类 Student，具体代码参照第 1 章例 1-1 所示内容，此处不再重复列出。

4. 创建配置文件

在 src 目录下新建配置文件 mybatis-config. xml，该文件的代码与第 3 章例 3-5 所示代码大致相同，由于要连接的数据库发生变化，此处需要将数据库名称修改为 chapter05，另外 < mapper ></ mapper >内的内容也需要修改，要修改的代码如下所示。

```
<property name = "url" value = "jdbc:mysql://localhost:3306/chapter05"/>
<mappers>
        <mapper resource = "com/qfedu/mapper/StudentMapper.xml" />
    </mappers>
```

5. 创建映射文件

在工程 chapter05 的 src 目录下创建 com. qfedu. mapper 包，在 com. qfedu. mapper 包下创建 StudentMapper. xml 文件，具体代码如例 5-1 所示。

【例 5-1】 StudentMapper. xml

```
<?xml version = "1.0" encoding = "UTF - 8" ?>
<! DOCTYPE mapper
PUBLIC " - //mybatis.org//DTD Mapper 3.0//EN"
"http://mybatis.org/dtd/mybatis - 3 - mapper.dtd">
<mapper namespace = "student">
    <! -- 根据 sid 查询学生信息 -->
    <select id = "findStudentBySid" parameterType = "student"
        resultType = "student">
        select * from student where sid = #{sid}
    </select>
    <! -- 根据 sid 更新学生信息 -->
    <update id = "updateStudent" parameterType = "student">
        update student set sname = #{sname},age = #{age} where sid = #{sid}
    </update>
</mapper>
```

6. 创建 log4j. properties 文件

由于需要通过 log4j 日志组件查看 MyBatis 一级缓存的工作状态，因此，在 src 目录下创建 log4j. properties 文件，具体代码如例 5-2 所示。

【例 5-2】 log4j. properties

```
# 全局日志配置
log4j.rootLogger = ERROR, stdout
```

```
3  # MyBatis 日志配置
4  log4j.logger.student = DEBUG
5  # 控制台输出配置
6  log4j.appender.stdout = org.apache.log4j.ConsoleAppender
7  log4j.appender.stdout.layout = org.apache.log4j.PatternLayout
8  log4j.appender.stdout.layout.ConversionPattern = %5p [%t] - %m%n
```

在以上代码中，第 4 行用于声明 MyBatis 日志的配置信息，其中 student 为映射文件中<mapper>元素的 namespace 属性值，DEBUG 为 MyBatis 输出日志信息的级别。此处需要提醒大家的是，MyBatis 使用手册中提供有 log4j.properties 文件的示例模板，为节省时间，实际开发时只需复制到本地并根据需要修改即可。

7. 编写测试类

在 src 目录下新建包 com.qfedu.test，在该包下新建类 TestCache01，具体代码如例 5-3 所示。

【例 5-3】 TestCache01.java

```
package com.qfedu.test;
import java.io.*;
import org.apache.ibatis.io.Resources;
import org.apache.ibatis.session.*;
import com.qfedu.pojo.Student;
public class TestCache01 {
    public static void main(String[] args) {
        String resource = "mybatis-config.xml";
        try {
            InputStream in = Resources.getResourceAsStream(resource);
        SqlSessionFactory factory = new
            SqlSessionFactoryBuilder().build(in);
        SqlSession sqlSession = factory.openSession();
        Student student = new Student();
        student.setSid(1);
        Student stu01 = sqlSession.selectOne
            ("student.findStudentBySid", student);
        System.out.println(stu01.toString());
        Student stu02 = sqlSession.selectOne
            ("student.findStudentBySid", student);
        System.out.println(stu02.toString());
        sqlSession.close();
        } catch (IOException e) {
            e.printStackTrace();
        }
    }
}
```

在以上代码中，程序先后两次执行相同的查询语句，由于 MyBatis 的一级缓存机制，当程序第二次执行该查询语句时，可以直接从缓存中获取数据。

执行 TestCache01 类，执行结果如图 5.3 所示。

```
Console ⌧
<terminated> TestCache01 [Java Application] C:\Program Files\Java\jre1.8.0_161\bin\javaw.exe (2018年7月26日 下午6:16:10)
DEBUG [main] - ==>  Preparing: select * from student where sid = ?
DEBUG [main] - ==> Parameters: 1(Integer)
DEBUG [main] - <==      Total: 1
Student [sid=1, sname=ZhangSan, age=20, course=Java]
Student [sid=1, sname=ZhangSan, age=20, course=Java]
```

图 5.3　执行 TestCache01 类的结果

从执行结果可以看出，控制台输出了所有的 DEBUG 信息。通过分析 DEBUG 信息可以发现，当程序执行第一次查询时，程序发送了 SQL 语句，当程序再次执行相同的查询时，程序没有发出 SQL 语句，这就说明，程序直接从一级缓存中获取了数据。

在 com.qfedu.test 包下新建类 TestCache02，具体代码如例 5-4 所示。

【例 5-4】　TestCache02.java

```
package com.qfedu.test;
import java.io.IOException;
import java.io.InputStream;
import org.apache.ibatis.io.Resources;
import org.apache.ibatis.session.*;
import com.qfedu.pojo.Student;
public class TestCache02 {
    public static void main(String[] args) {
        String resource = "mybatis-config.xml";
        try {
            InputStream in = Resources.getResourceAsStream(resource);
        SqlSessionFactory factory = new
            SqlSessionFactoryBuilder().build(in);
        SqlSession sqlSession = factory.openSession();
        Student student01 = new Student();
        student01.setSid(1);
        Student stu01 =  sqlSession.selectOne
            ("student.findStudentBySid",student01);
        System.out.println(stu01.toString());
        Student student02 = new Student();
        student02.setSid(3);
        student02.setSname("ZhaoLiu");
        student02.setAge("19");
        sqlSession.update("student.updateStudent",student02);
        sqlSession.commit();
        Student stu02 = sqlSession.selectOne
             ("student.findStudentBySid",student01);
        System.out.println(stu02.toString());
        sqlSession.close();
        } catch (IOException e) {
            e.printStackTrace();
```

```
        }
    }
}
```

在以上代码中，程序先查询 sid 为 1 的学生信息，接下来更新 sid 为 3 的学生信息，然后再次查询 sid 为 1 的学生信息。由于程序执行了 DML 操作，当再次查询 sid 为 1 的学生信息时，只能从数据库中获取。

执行 TestCache02 类，执行结果如图 5.4 所示。

```
Console
<terminated> TestCache02 [Java Application] C:\Program Files\Java\jre1.8.0_161\bin\javaw.exe (2018年7月26日 下午6:41:51)
DEBUG [main] - ==>  Preparing: select * from student where sid = ?
DEBUG [main] - ==> Parameters: 1(Integer)
DEBUG [main] - <==      Total: 1
Student [sid=1, sname=ZhangSan, age=20, course=Java]
DEBUG [main] - ==>  Preparing: update student set sname = ?,age = ? where sid = ?
DEBUG [main] - ==> Parameters: ZhaoLiu(String), 19(String), 3(Integer)
DEBUG [main] - <==    Updates: 1
DEBUG [main] - ==>  Preparing: select * from student where sid = ?
DEBUG [main] - ==> Parameters: 1(Integer)
DEBUG [main] - <==      Total: 1
Student [sid=1, sname=ZhangSan, age=20, course=Java]
```

图 5.4　执行 TestCache02 类的结果

从执行结果可以看出，控制台输出了所有的 DEBUG 信息。通过分析 DEBUG 信息可以发现，当程序第二次查询 sid 为 1 的学生信息时，程序向数据库发送了 1 条 SQL 语句，由此可见，MyBatis 的一级缓存被清空，程序通过向数据库发送 SQL 语句来获取数据。

5.3　二级缓存

5.3.1　二级缓存的原理

二级缓存的作用域是 Mapper，与一级缓存相比，二级缓存的作用域范围更大，它可以跨多个 SqlSession 对象，并且二级缓存可以自定义缓存源。

当开启二级缓存时，MyBatis 以 namespace 区分 Mapper，如果多个 SqlSession 对象使用同一个 Mapper 的相同查询语句去操作数据库，在第一个 SqlSession 对象执行完成后，MyBatis 会将查询结果写入二级缓存，此后，如果程序没有执行插入、更新、删除操作，当第二个 SqlSession 对象执行相同的查询语句时，MyBatis 会直接读取二级缓存中的数据。

在实际开发中，经常会遇到多个 SqlSession 在同一个 Mapper 中执行操作，例如，SqlSession1、SqlSession2 分别执行相同的查询，SqlSession3 执行插入、更新或删除，具体过程如图 5.5 所示。

从图 5.5 中可以看出，当 SqlSession1 执行 StudentMapper 查询时，程序会将结果写入 MyBatis 二级缓存，当 SqlSession2 执行相同的查询语句时，MyBatis 将直接从二级缓存中读取数据，当 SqlSession3 对数据库执行插入、更新、删除操作时，MyBatis 会清空二级缓存中的内容以防止程序误读。

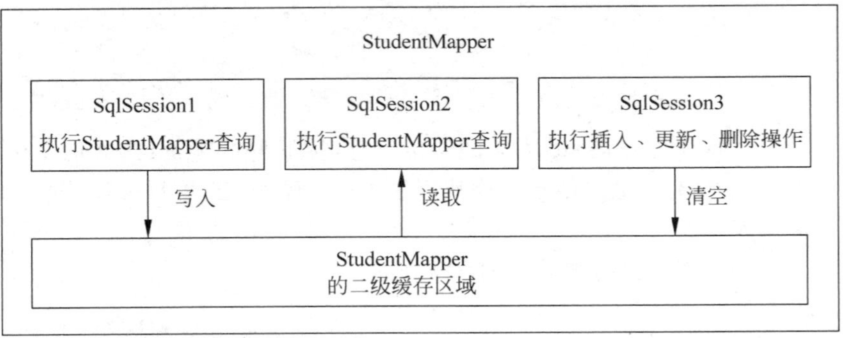

图 5.5　MyBatis 二级缓存

5.3.2　二级缓存的配置

与一级缓存不同，MyBatis 的二级缓存需要手动开启，开启二级缓存通常要完成以下两个步骤。

1. 开启二级缓存的全局配置

使用二级缓存首先要打开全局配置，这可以通过 MyBatis 配置文件中的< settings >元素来完成，具体代码如下。

```
<settings>
        <setting name = "cacheEnabled" value = "true"/>
    </settings>
```

以上代码将 cacheEnabled 属性的值设置为 true，此时二级缓存的全局配置处于开启状态，由于 MyBatis 默认为 cacheEnabled 属性的值为 true，因此，这段代码也可直接省去。

2. 开启当前 Mapper 的 namespace 的二级缓存

开启全局配置以后，接下来需要开启当前 Mapper 的 namespace 的二级缓存，这可以通过 MyBatis 映射文件中的< cache >元素来完成，具体代码如下。

```
<cache></cache>
```

以上代码开启了当前 Mapper 的 namespace 的二级缓存，此时二级缓存的特性处于默认状态，可实现的功能如下。

- 映射语句文件中的所有 select 语句将会被缓存
- 映射语句文件中的所有 insert、update 和 delete 语句会刷新缓存
- 缓存会使用 LRU(最近最少使用)算法收回
- 没有刷新间隔，缓存不会以任何时间顺序来刷新
- 缓存会存储列表集合或对象的 1024 个引用
- 缓存是可读/可写的缓存，这意味着对象检索不是共享的，缓存可以安全地被调用者修改，而不干扰其他调用者或线程所做的潜在修改

以上是二级缓存在默认状态下的特性，如果需要调整上述特性，可通过设置< cache >元

素的属性来实现，< cache >元素提供了若干属性，具体如表 5.1 所示。

表 5.1 < cache >元素的属性

属性名称	描　述
eviction	收回策略。该属性有 LRU、FIFO、SOFT、WEAK 共 4 个可选值，其中，LRU 表示最近最少策略，用于移除最长时间不被使用的对象；FIFO 表示先进先出策略，用于按对象进入缓存的顺序移除它们；SOFT 表示软引用策略，用于移除基于垃圾回收器状态和软引用规则的对象；WEAK 表示弱引用策略，用于更积极地移除基于垃圾收集器状态和弱引用规则的对象
flushInterval	刷新间隔。该属性可以被设置为任意的正整数，而且它们代表一个合理的毫秒形式的时间段。默认情况是不设置，也就是没有刷新间隔，只在调用语句时刷新
size	引用数目。可以被设置为任意正整数，要记住缓存的对象数目和运行环境的可用内存资源数目，默认值为 1024
readOnly	只读。属性可以被设置为 true 或 false。当缓存设置为只读时，缓存对象不能被修改，但此时缓存性能较高。当缓存设置为可读写时，性能稍低，但是安全性高

表 5.1 列举出了< cache >元素的属性，开发人员可以根据需要调整二级缓存的特性。

5.3.3 二级缓存的应用

讲解 MyBatis 二级缓存的原理和配置之后，接下来，本节通过一个实例演示 MyBatis 二级缓存的应用。

1. 修改映射文件

修改映射文件 StudentMapper. xml，在映射文件的< mapper >元素中加入如下代码。

```
<cache></cache>
```

2. 修改 POJO 类

由于此时配置的二级缓存为可读写缓存，因此需要被操作的类实现序列化接口。修改 Student 类，使其实现 Serializable 接口，具体代码如下。

```
public class Student implements Serializable{
```

3. 编写测试类

在 com. qfedu. test 包下新建类 TestCache03，具体代码如例 5-5 所示。

【例 5-5】 TestCache03. java

```
package com.qfedu.test;
import java.io.*;
import org.apache.ibatis.io.Resources;
import org.apache.ibatis.session.*;
import com.qfedu.pojo.Student;
public class TestCache03 {
    public static void main(String[] args) {
```

```
        String resource = "mybatis-config.xml";
        try {
            InputStream in = Resources.getResourceAsStream(resource);
            SqlSessionFactory factory = new
                SqlSessionFactoryBuilder().build(in);
            SqlSession sqlSession01 = factory.openSession();
            SqlSession sqlSession02 = factory.openSession();
            Student student01 = new Student();
            student01. setSid (1);
            Student stu01 = sqlSession01.selectOne
                ("student.findStudentBySid", student01);
            System.out.println(stu01.toString());
            sqlSession01.close();
            Student stu02 = sqlSession02.selectOne
                ("student.findStudentBySid",student01);
            System.out.println(stu02.toString());
            sqlSession02.close();
            } catch (IOException e) {
                e.printStackTrace();
            }
    }
}
```

在以上代码中，先创建两个 SqlSession 对象，然后分别使用这两个 SqlSession 对象查询同一 Mapper 中的查询，由于 MyBatis 的二级缓存已开启，当第二次执行查询时，可以从缓存中直接获取数据。

执行 TestCache03 类，执行结果如图 5.6 所示。

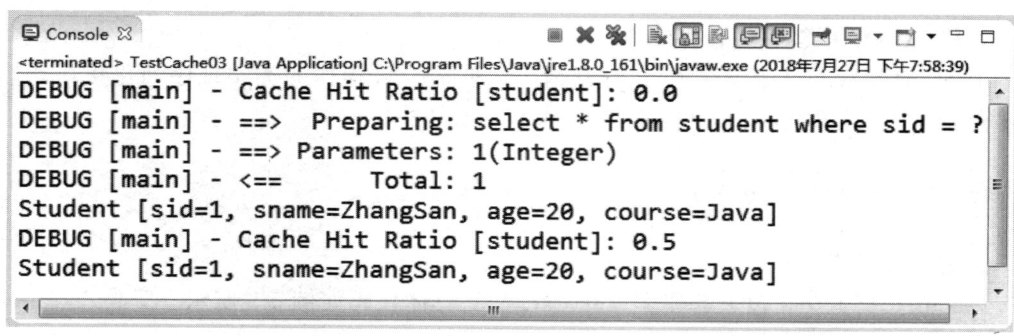

图 5.6 执行 TestCache03 类的结果

从执行结果可以看出，控制台输出了所有的 DEBUG 信息。通过分析 DEBUG 信息可以发现，当第一个 SqlSession 对象执行查询时，Cache Hit Ratio(命中率)为 0，程序发送了 SQL 语句；当第二个 SqlSession 对象执行相同的查询时，Cache Hit Ratio(命中率)为 0.5，程序没有发出 SQL 语句，这就说明，程序直接从二级缓存中获取了数据。

在 com.qfedu.test 包下新建类 TestCache04，具体代码如例 5-6 所示。

【例 5-6】 TestCache04. java

```
package com.qfedu.test;
import java.io.*;
import org.apache.ibatis.io.Resources;
import org.apache.ibatis.session.*;
import com.qfedu.pojo.Student;
public class TestCache04 {
    public static void main(String[] args) {
        String resource = "mybatis-config.xml";
        try {
            InputStream in = Resources.getResourceAsStream(resource);
            SqlSessionFactory factory = new
                SqlSessionFactoryBuilder().build(in);
            SqlSession sqlSession01 = factory.openSession();
            SqlSession sqlSession02 = factory.openSession();
            SqlSession sqlSession03 = factory.openSession();
            Student student01 = new Student();
            student01. setSid(1);
            Student stu01 = sqlSession01.selectOne
                ("student.findStudentBySid", student01);
            System.out.println(stu01.toString());
            sqlSession01.close();
            Student student02 = new Student();
            student02. setSid(3);
            student02.setSname("ZhaoLiu");
            student02.setAge("19");
            sqlSession02.update("student.updateStudent",student02);
            sqlSession02.commit();
            sqlSession02.close();
            Student stu02 = sqlSession03.selectOne
                ("student.findStudentBySid",student01);
            System.out.println(stu02.toString());
            sqlSession03.close();
            } catch (IOException e) {
                e.printStackTrace();
            }
    }
}
```

在以上代码中，程序通过 sqlSession01 查询 sid 为 1 的学生信息，接下来通过 sqlSession02 更新 sid 为 3 的学生信息，然后通过 sqlSession03 再次查询 sid 为 1 的学生信息。由于程序执行了 DML 操作，当再次查询 sid 为 1 的学生信息时，只能从数据库中获取。

执行 TestCache04 类，执行结果如图 5.7 所示。

从执行结果可以看出，控制台输出了所有的 DEBUG 信息。通过分析 DEBUG 信息可以发现，当程序第二次查询 sid 为 1 的学生信息时，Cache Hit Ratio(命中率)为 0，程序向数据库发送了 1 条 SQL 语句，由此可见，MyBatis 的二级缓存被清空，程序通过向数据库发送 SQL 语句来获取数据。

```
<terminated> TestCache04 [Java Application] C:\Program Files\Java\jre1.8.0_161\bin\javaw.exe (2018年7月30日 下午1:20:46)
DEBUG [main] - Cache Hit Ratio [student]: 0.0
DEBUG [main] - ==>  Preparing: select * from student where sid = ?
DEBUG [main] - ==> Parameters: 1(Integer)
DEBUG [main] - <==      Total: 1
Student [sid=1, sname=ZhangSan, age=20, course=Java]
DEBUG [main] - ==>  Preparing: update student set sname = ?,age = ? where sid = ?
DEBUG [main] - ==> Parameters: ZhaoLiu(String), 19(String), 3(Integer)
DEBUG [main] - <==    Updates: 1
DEBUG [main] - Cache Hit Ratio [student]: 0.0
DEBUG [main] - ==>  Preparing: select * from student where sid = ?
DEBUG [main] - ==> Parameters: 1(Integer)
DEBUG [main] - <==      Total: 1
Student [sid=1, sname=ZhangSan, age=20, course=Java]
```

图 5.7　执行 TestCache04 类的结果

5.4　整合 EhCache 缓存

5.4.1　EhCache 简介

在实际开发中，很多项目采用分布式的系统架构。分布式系统架构是将项目拆分成若干个子项目并使它们协同发挥功能的解决方案。在分布式系统架构下，为了提升系统性能，通常会采用分布式缓存对缓存数据进行集中管理。由于 MyBatis 自身无法实现分布式缓存，需要整合其他分布式缓存框架，如 EhCache 缓存。

EhCache 是一种获得广泛应用的开源分布式缓存框架，具有快速、简单、能够提供多种缓存策略等特点，主要面向通用缓存、Java EE 和轻量级容器等。

EhCache 通常采用名称为 ehcache.xml 的配置文件实现功能配置，EhCache 的配置文件有其自身特有的层次结构，具体如下所示。

```
<ehcache>
<! -- 磁盘存储 -->
<diskStore path = "" />
<! -- 默认的缓存配置信息 -->
        <defaultCache maxElementsInMemory = "" eternal = ""
        timeToIdleSeconds = "" timeToLiveSeconds = "" overflowToDisk = "" />
        <! -- 根据需要自定义的缓存配置信息 -->
        <cache name = "" maxElementsInMemory = "" eternal = ""
        timeToIdleSeconds = "" timeToLiveSeconds = "" overflowToDisk = "" />
    </ehcache>
```

在以上配置信息中，<diskStore>元素用于指定将缓存中的数据转移到磁盘，其 path 属性指定缓存数据在磁盘中的存储位置；<defaultCache>元素用于指定默认的缓存配置信息，如果没有其他配置项，则所有对象按此配置项处理；<cache>元素用于指定自定义的配置信息。

<defaultCache>元素包含了一些属性用于定义配置信息，具体如表 5.2 所示。

表 5.2 < defaultCache >元素的属性

属性名称	描 述
maxElementsInMemory	指定内存中最大缓存对象数
maxElementsOnDisk	指定磁盘中最大缓存对象数，若是 0，则表示无穷大
eternal	指定缓存的 elements 是否永远不过期。如果为 true，则缓存的数据始终有效，如果为 false，则根据 timeToIdleSeconds、timeToLiveSeconds 判断
overflowToDisk	指定当内存缓存溢出的时候是否将过期的 element 缓存到磁盘上
timeToIdleSeconds	指定 EhCache 中的数据前后两次被访问的时间间隔，如果超过，这些数据便会删除，默认值为 0，即可闲置时间无穷大
timeToLiveSeconds	指定缓存 element 的有效生命期，默认值为 0，即 element 存活时间无穷大
diskSpoolBufferSizeMB	指定磁盘缓存的缓存区大小，默认值为 30MB
diskPersistent	在 VM 重启时是否启用磁盘保存 EhCache 中的数据，默认值为 false
memoryStoreEvictionPolicy	当内存缓存达到最大，有新的 element 加入的时候，移除缓存中 element 的策略。默认值为 LRU

表 5.2 列举出了 EhCache 配置文件中< defaultCache >元素的属性，开发人员可以根据需要调整 EhCache 缓存的功能。

5.4.2 EhCache 下载

由于 EhCache 的诸多优势，MyBatis 提供了通过 EhCache 实现其二级缓存的方法，开发人员在使用时导入相应的 jar 包即可。接下来演示相关 jar 包的下载，具体如下。

(1) 打开浏览器，访问 https://github.com/mybatis/ehcache-cache/releases，浏览器显示的页面如图 5.8 所示。

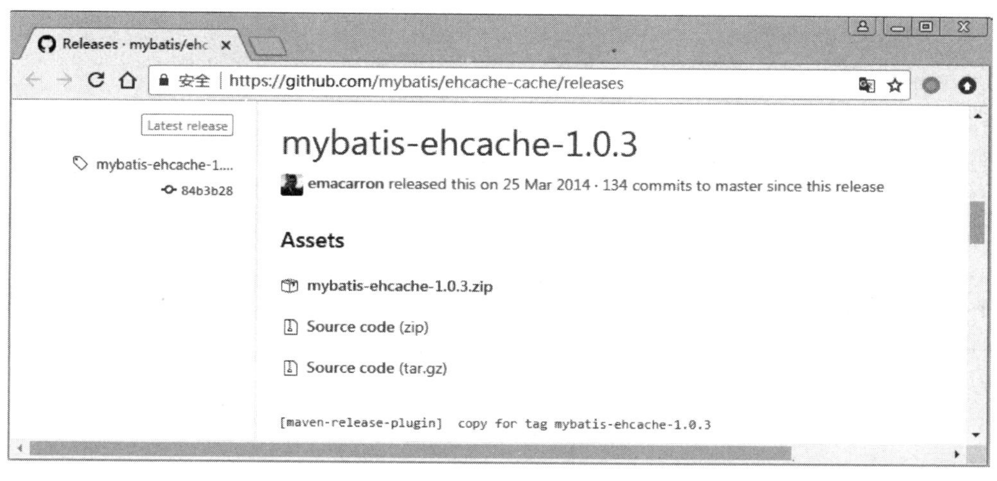

图 5.8 下载页面

(2) 单击 Assets 栏目下的 mybatis-ehcache-1.0.3.zip 超链接，将文件下载到本地。

(3) 下载完成后，将得到名称为 mybatis-ehcache-1.0.3.zip 的文件，将该文件解压，打开解压后的文件夹，可以找到 mybatis-ehcache-1.0.3.jar 文件，打开 lib 目录，可以找到 ehcache-core-2.6.8.jar 文件，这两个文件即为 MyBatis 整合 EhCache 缓存所需 jar 包。

5.4.3 MyBatis 整合 EhCache 缓存

完成 jar 包的下载后，接下来，本节通过一个实例演示 MyBatis 整合 EhCache 缓存。

1. 引入 EhCache 的 jar 包

将 mybatis-ehcache-1.0.3.jar 文件、lib 目录下的 ehcache-core-2.6.8.jar 文件复制到工程 chapter05 的 lib 目录下，完成导入包。

2. 配置 EhCache 的 type 属性

在 StudentMapper.xml 映射文件中，配置< cache >标签的 type 属性为 EhCache 对 cache 接口的实现类的完全限定名，具体代码如下所示。

```
<cache type = "org.mybatis.caches.ehcache.EhcacheCache"></cache>
```

3. 配置 EhCache

在工程 chapter05 的 src 目录下新建 EhCache 的配置文件 ehcache.xml，具体代码如例 5-7 所示。

【例 5-7】 ehcache.xml

```
<?xml version = "1.0" encoding = "UTF - 8" ?>
<ehcache xmlns:xsi = "http://www.w3.org/2001/XMLSchema - instance"
    xsi:noNamespaceSchemaLocation = "../config/ehcache.xsd" >
    <!-- 缓存数据要存放的磁盘地址 -->
    <diskStore  path = "D:/ehcache"/>
    <!-- 默认缓存 -->
    <defaultCache maxElementsInMemory = "10"
        maxElementsOnDisk = "10000000" eternal = "false" overflowToDisk = "true"
        timeToIdleSeconds = "120" timeToLiveSeconds = "120"
        diskExpiryThreadIntervalSeconds = "120"
        memoryStoreEvictionPolicy = "LRU">
    </defaultCache>
</ehcache>
```

在以上代码中，< diskStore >的 path 属性设置缓存地址为 D 盘下的 ehcache 目录，maxElementsInMemory 指定在内存中缓存元素的最大数目为 10，由于本例要演示缓存溢出后数据会存入磁盘的效果，此处将该值设置得过低，实际开发中根据具体需求设置；overflowToDisk 属性值为 true，当内存缓存溢出时，过期的 element 会被缓存到磁盘上。

执行 TestCache03 类，执行结果如图 5.9 所示。

从执行结果可以看出，和使用 MyBatis 默认的二级缓存相同，当第二个 SqlSession 对象执行相同的查询时，Cache Hit Ratio(命中率)为 0.5，程序没有发出 SQL 语句，这就说明，程序从 EhCache 缓存中获取了数据。

打开 D 盘，可以发现 D 盘中出现了 ehcache 目录，打开 D:\ehcache 目录，该目录中存有 EhCache 缓存的信息，如图 5.10 所示。

```
Console ☒
<terminated> TestCache03 [Java Application] C:\Program Files\Java\jre1.8.0_161\bin\javaw.exe (2018年7月30日 下午8:03:22)
DEBUG [main] - Cache Hit Ratio [student]: 0.0
DEBUG [main] - ==>  Preparing: select * from student where sid = ?
DEBUG [main] - ==> Parameters: 1(Integer)
DEBUG [main] - <==      Total: 1
Student [sid=1, sname=ZhangSan, age=20, course=Java]
DEBUG [main] - Cache Hit Ratio [student]: 0.5
Student [sid=1, sname=ZhangSan, age=20, course=Java]
```

图 5.9　重新执行 TestCache03 类的结果

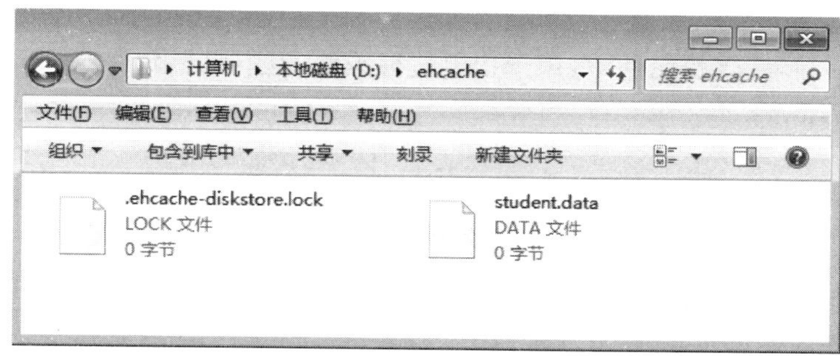

图 5.10　缓存信息

5.5　本章小结

本章首先介绍了 MyBatis 的缓存机制，接下来讲解了 MyBatis 中的一级缓存，包括一级缓存的原理、应用等，然后讲解了 MyBatis 的二级缓存，包括二级缓存的原理、配置及应用，最后讲解了 MyBatis 与 EhCache 缓存的整合。通过本章知识的学习，大家应该能理解 MyBatis 的缓存机制，重点掌握二级缓存的应用和 MyBatis 整合 EhCache 缓存的方法，能够使用缓存提升 MyBatis 程序的查询效率、优化系统整体性能。

5.6　习　　题

1. 填空题

(1) MyBatis 支持的缓存分为________和________。

(2) 在 MyBatis 中，________是 SqlSession 级别的缓存。

(3) 在 MyBatis 中，________是 Mapper 级别的缓存。

(4) 使用一级缓存时，SqlSession 对象引入________对象作为存储数据的区域。

(5) 在 MyBatis 中，一级缓存是默认开启的，二级缓存需要________。

2. 选择题

(1) 关于 MyBatis 的一级缓存，下列选项错误的是(　　)。

A. 当程序对数据库执行了 DML 操作，MyBatis 会删除一级缓存中的相关内容

B. 当一级缓存开启时，如果 SqlSession 第一次执行某查询语句，MyBatis 会将结果写入一级缓存

C. 当一级缓存开启时，如果 SqlSession 第二次执行某查询语句，MyBatis 一定可以从一级缓存中读取数据

D. MyBatis 一级缓存的作用域是 SqlSession

(2) 关于 MyBatis 的二级缓存，下列选项错误的是(　　)。

A. MyBatis 二级缓存的作用域是跨 Mapper 的

B. 在使用二级缓存时，MyBatis 以 namespace 区分 Mapper

C. MyBatis 二级缓存默认是关闭的

D. 如果 MyBatis 二级缓存为可读写缓存，则被操作的 POJO 类需实现序列化接口

(3) 在<cache>元素的属性中，用于指定收回策略的是(　　)。

A. eviction　　B. flushInterval

C. size　　D. readOnly

(4) 关于 MyBatis 整合 EhCache 缓存，下列说法错误的是(　　)。

A. MyBatis 的二级缓存可以自定义缓存源

B. EhCache 是一种获得广泛应用的开源分布式缓存框架

C. MyBatis 自身可以实现分布式缓存

D. EhCache 通常采用名称为 ehcache. xml 的配置文件实现功能配置

(5) 在 EhCache 缓存的配置文件中，用于指定磁盘存储的是(　　)。

A. <diskStore>　　B. <defaultCache>

C. <cache>　　D. <ehcache>

3. 思考题

(1) 简述 MyBatis 一级缓存的工作原理。

(2) 简述 MyBatis 二级缓存的工作原理。

4. 编程题

通过 MyBatis 查询数据表 student 中的所有学生信息，将查询结果存入 EhCache 缓存，注意 MyBatis 整合 EhCache 缓存的方法。

第6章 Spring 基础

本章学习目标

- 理解 Spring 的概念和优势
- 理解 Spring 的体系结构
- 掌握 Spring 的核心容器
- 掌握 Spring 入门程序的编写

作为 Java 生态圈影响最为久远的优秀框架之一，Spring 从诞生之日起就引起了开发人员的关注。Spring 具有良好的设计和分层架构，它改变了传统重量型框架臃肿、低效的劣势，大大降低项目开发中的技术复杂性，帮助开发人员将更多精力投入项目业务中。接下来，本章将对 Spring 的基础知识做详细讲解。

6.1 Spring 概述

6.1.1 Spring 简介

在 Spring 兴起之前，Java 企业级开发主要通过 EJB (Enterprise JavaBean)完成。EJB 是服务器端的组件模型，由于它过于依靠 EJB 容器并存在缓慢、复杂等问题，因此逐渐被 Spring 取代。

Spring 是一个分层的 Java SE/EE 应用的一站式轻量级开源框架，它最为核心的理念是 IOC(控制反转)和 AOP(面向切面编程)，其中，IOC 是 Spring 的基础，它支撑着 Spring 对 JavaBean 的管理功能，AOP 是 Spring 的重要特性。

Spring 提供了展现层 Spring MVC、持久层 Spring JDBC 以及事务管理等一站式的企业级应用技术，除此之外，Spring 目前已融入 Java EE 开发的各个领域，它可以整合 Java 生态圈里众多的第三方框架和类库，为企业开发提供全面的基础功能支持。

Spring 是一个易于开发、便于测试且功能齐全的开发框架，它的核心功能适用于任何 Java 应用，在较长一段时期内仍将被广泛采用。

6.1.2 Spring 的优势

与其他开发框架相比，Spring 具有无可比拟的优势，这主要表现在以下几个方面。

1. 降低耦合度，方便开发

通过 IOC 容器，Spring 可以管理对象的生命周期、控制对象之间的依赖关系，如此一来，因硬编码造成的程序过度耦合得以避免。

2. 支持 AOP 编程

通过 AOP,Spring 可以对程序进行权限拦截、安全监控等操作,这可以减少通过传统 OOP 方法带来的代码冗余和繁杂。

3. 支持声明式事务

在 Spring 中,可以直接通过 Spring 配置文件管理数据库事务,省去了手动编程的烦琐,提升了开发效率。

4. 方便程序测试

Spring 中集成了 Junit ,开发人员可以通过 Junit 进行单元测试。

5. 方便集成各种优秀框架

Spring 提供了一个广阔的基础平台,它不排斥各种优秀的开源框架,其内部提供了对各种优秀框架(如 MyBatis、Hibernate、Quartz 等)的直接支持。

6. 降低 Java EE API 的使用难度

Spring 封装了 Java EE 中使用难度较大的 API,经过封装,这些 API 更容易被开发人员理解和调用。

6.1.3 Spring 功能体系

Spring 框架采用分层和模块的架构方式,由核心容器、数据访问及集成、Web、AOP 等功能构成,具体如图 6.1 所示。

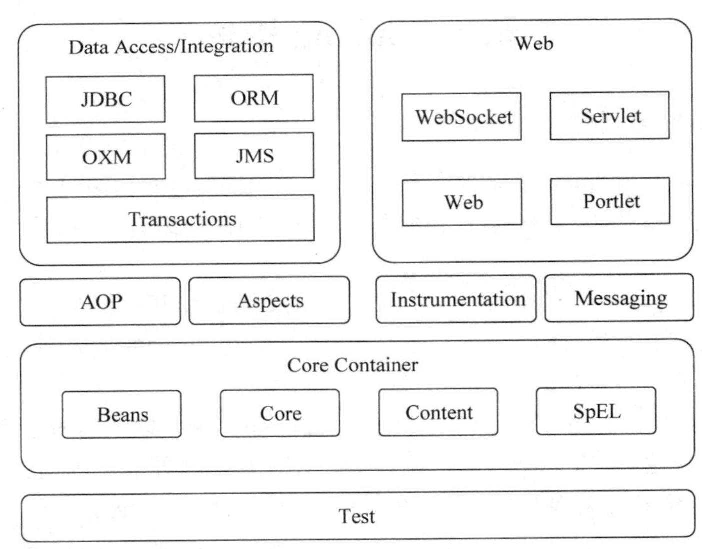

图 6.1 Spring 的功能体系

图 6.1 展示了 Spring 的功能体系,这些功能通常由 Spring 中的一个或多个模块联合实现。在实际开发中,开发人员可根据业务需求自主选择所需模块,同时也可集成其他第三方框架,提升开发的针对性和效率。

1. 核心容器(Core Container)

核心容器(Core Container)在 Spring 的功能体系中起到支撑性作用,是其他模块实现功能的基石。核心容器包含 Beans、Core、Context、SpEL 四个模块,具体如下。

1）Beans 和 Core 模块

Beans 和 Core 模块规定了创建、配置和管理 Bean 的方式，提供了 IOC 和 DI 功能，BeanFactory 类是 Beans 和 Core 模块的关键。

2）Context 模块

Context 模块在 Beans 和 Core 模块的基础之上扩展了功能，通过 ApplicationContext 接口提供上下文信息。

3）SpEL 模块

SpEL 模块提供了一个强大的表达式语言，该语言用于在 Spring 运行时查询和操纵对象。

2. 数据访问及集成（Data Access/Integration）

数据访问及集成（Data Access/Integration）主要用于访问和操作数据中的数据，它主要包含 JDBC、ORM、OXM、JMS 和 Transactions 模块。

1）JDBC 模块

JDBC 模块提供了 JDBC 抽象层，它消除了冗长的 JDBC 编码并解析数据库供应商特有的错误代码。

2）ORM 模块

ORM 模块为主流的对象关系映射 API 提供了集成层，这些主流的对象关系映射包括 MyBatis、Hibernate 和 JDO 等，除此之外，该模块可将对象关系映射框架与 Spring 提供的特性组合使用。

3）OXM 模块

OXM 模块提供了对 OXM 实现的支持，例如，JAXB、Castor、XML Beans、JiBX 和 XStream 等。

4）JMS 模块

JMS 模块包含生产（produce）和消费（consume）消息的功能。从 4.1 版本开始，Spring 集成了 spring-messaging 模块。

5）Transactions 模块

Transactions 模块的主要功能是事务管理，Spring 支持手写 beginTransaction()、commit()、rollback()的编程式事务，同时，它也支持通过注解或配置后由 Spring 自动处理的声明式事务。

3. Web

Web 功能的实现基于 ApplicationContext 基础之上，它提供了 Web 应用的各种工具类，包含 Web、Servlet、WebSocket 和 Protlet 模块。

1）Web 模块

Web 模块提供了基本的面向 Web 的集成功能，如大部分文件上传功能，使用 Servlet 侦听器的 IoC 容器的初始化以及面向 Web 的应用程序上下文。它还包含一个 HTTP 客户端和 Spring 远程处理支持的 Web 相关部分。

2）Servlet 模块

Servlet 模块包含 Spring 的模型、视图、控制器以及 Web 应用程序的 REST Web 服务实现。

3）WebSocket 模块

WebSocket 模块是 Spring 4 版本加入的一个模块，该模块用于适配不同的 WebSocket

引擎，并能够全面支持 WebSocket，它与 Java WebSocket API 标准保持一致，同时提供了额外的服务。

4）Protlet 模块

Protlet 模块提供了一个在 Protlet 环境中使用的 MVC 实现，相当于镜像了 Servlet 模块的功能。

4. AOP、Aspects 和 Instrumentation

AOP 提供了面向切面的编程实现，允许定义方法拦截器和切入点对代码解耦。Aspects 主要用于集成 AspectJ，AspectJ 是一个功能强大且成熟的 AOP 框架，为面向切面编程提供多种实现方法。Instrumentation 提供了类 instrumentation 的支持和类加载器的实现，通常在特定的服务器使用。

5. TEST

TEST 模块支持使用 JUnit 或 TestNG 对 Spring 组件进行单元测试和集成测试，它提供了一致加载 ApplicationContext 和这些 ApplicationContext 的缓存。除此之外，它还提供了可用于独立测试代码的模拟对象。

6. MESSAGE

MESSAGE 在 Spring 4 版本时开始引入 Spring 的功能体系中，主要用于为 Spring 框架集成一些基础的消息传送应用。

6.1.4 Spring 子项目

从狭义层面理解，Spring 通常是指 Spring 框架，但从广义层面理解，Spring 还包含有除 Spring 框架本身之外的很多子项目，这些子项目共同构建起一个企业级开发解决方案的生态系统。

Spring 官网提供了 Spring 的相关子项目信息，打开浏览器，访问 https://spring.io/projects，可以看到显示 Spring 子项目信息的页面，如图 6.2 所示。

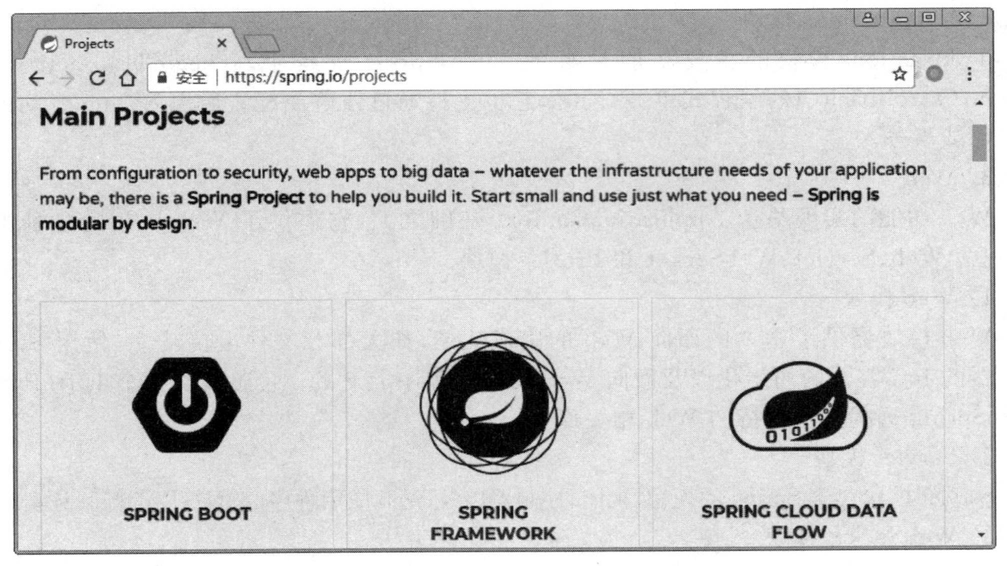

图 6.2 显示 Spring 子项目信息的页面

关于 Spring 主要子项目的详细介绍如表 6.1 所示。

表 6.1　Spring 的主要子项目

项目名称	描述
Spring Boot	是用来简化 Spring 应用的初始搭建以及开发过程的框架,简化了配置文件,使用嵌入式 Web 服务器,含有诸多开箱即用的微服务功能,可以和 Spring Cloud 联合部署
Spring Cloud	相当于微服务工具包,可以集成一系列框架。它利用 Spring Boot 的开发便利性巧妙地简化了分布式系统基础设施的开发,如服务发现注册、配置中心、消息总线、负载均衡等
Spring Data	是一个用于简化数据库访问,并支持云服务的开源框架,封装了很多种数据及数据库的访问相关技术。其主要目标是使得对数据的访问变得方便快捷,它支持 map-reduce 框架和云计算数据服务
Spring Security	是一个强大的、可高度定制化的认证和访问控制的框架,它为基于 Spring 的应用提供了约定俗成的安全标准
Spring Batch	一个轻量级的、综合的批处理框架,用于构建健壮的用于企业日常任务调度的批处理程序
Spring Integration	面向企业应用集成(EAI/ESB)的编程框架,支持的通信方式包括 HTTP、FTP、TCP/UDP、JMS、RabbitMQ、E-mail 等
Spring AMQP	操作消息队列的工具包,主要封装了对 RabbitMQ 的操作
Spring HATEOAS	是一个用于支持实现超文本驱动的 REST Web 服务的开发库
Spring Session	session 管理的开发工具包,可以把 session 保存到 redis 等,进行集群化 session 管理
Spring Web Services	基于 Spring 的 Web 服务框架,提供 SOAP 服务开发,允许通过多种方式创建 Web 服务

表 6.1 列举了 Spring 的主要子项目,开发者在构建应用程序的基础服务时,可以根据需要选择对应的 Spring 子项目。

6.1.5　Spring 5 新特性

Spring 5 是 Spring 的最新版本,与历史版本相比,Spring 5 增强了一些功能和特性以满足新的技术生态的需要,主要体现在以下几个方面。

1. JDK 基线更新

由于 Spring 5 代码库运行于 Java 8 之上,因此 Spring 5 对环境的最低要求是 Java 8,这可以促进 Spring 的使用者积极运用 Java 8 新特性。

2. 核心框架修订

Spring 5 利用 Java 8 的新特性对自身功能进行了修订,主要包括以下几个方面。

(1) 基于 Java 8 的反射增强,Spring 5 中的方法参数可以更加高效地进行访问。

(2) 核心的 Spring 接口提供基于 Java 8 的默认方法构建的选择性声明。

(3) 用@Nullable 和@NotNull 注解来表明可为空的参数以及返回值,如此一来,可以在编译时处理空值而不是在运行时抛出 NullPointerExceptions。

(4) 在日志记录方面,Spring 5 提供了 Commons Logging 桥接模块的封装,它被叫作

spring-jcl 而不是标准的 Commons Logging。

(5) 新加入 Resourse 抽象提供的 isFile 指示器以及 getFile 方法,防御式编程方法也得到进一步的推动。

3. 核心容器更新

Spring 5 支持候选组件索引作为类路径扫描的替代方案。从索引读取实体会使加载组件索引开销更低,因此,Spring 程序的启动时间将会缩减。

4. 支持响应式编程

响应式编程是另外一种编程风格,它专注于构建对事件做出响应的应用程序。Spring 5 包含响应流和 Reactor(Reactive Stream 的 Java 实现),响应流和 Reactor 支撑了 Spring 自身的功能及相关 API。

5. 支持函数式 Web 框架

Spring 5 提供了一个函数式 Web 框架。该框架使用函数式编程风格来定义端点,它引入了两个基本组件: HandlerFunction 和 RouterFunction。HandlerFunction 表示处理接收到的请求并生成响应的函数; RouterFunction 替代了@RequestMapping 注解,它用于将接收到的请求转发到处理函数。

6. 支持 Kotlin

Spring 5 引入对 Kotlin 语言的支持。Kotlin 是一种支持函数式编程风格的面向对象语言,它运行在 JVM 之上,可以让代码更具有表现力、简洁性和可读性。有了对 Kotlin 的支持,开发人员可以进行深度的函数式 Spring 编程,这拓宽了 Spring 的应用领域。

7. 提升测试功能

Spring 5 完全支持 JUnit 5 Jupiter,所以可以使用 JUnit 5 来编写测试以及扩展。除此之外,Spring 5 还提供了在 Spring TestContext Framework 中进行并行测试的扩展。针对响应式编程模型,Spring 5 引入了支持 Spring WebFlux 的 WebTestClient 集成测试。

6.2 Spring 的下载及使用

Spring 是一个不依赖于任何 Web 服务器或容器的框架,它既可以应用于 Java SE 项目,也可以应用于 Java EE 项目,在使用 Spring 之前要获取它的 jar 包,这些 jar 包可以从 Spring 的官网下载。本书编写时 Spring 的最新稳定版本为 5.0.8,因此书中基于该版本展开讲解。接下来,本节将讲解 Spring 的相关 jar 包的下载方法,具体步骤如下。

(1) 打开浏览器,访问 http://repo.spring.io/simple/libs-release-local/org/springframework/spring/5.0.8.RELEASE/,浏览器跳转到下载页面,如图 6.3 所示。

(2) 单击页面中 spring-framework-5.0.8.RELEASE-dist.zip 超链接,将文件下载到指定目录。

(3) 解压下载到本地的 zip 文件,此时获得名称为 spring-framework-5.0.8.RELEASE-dist 的文件夹,打开该文件夹,可以看到 Spring 的目录结构,具体如图 6.4 所示。

其中,docs 文件夹存放 Spring 的相关开发文档,libs 文件夹存放 Spring 的 jar 包和源代码,schema 文件夹存放规定配置文件结构的约束文件。

(4) 由于 Spring 的日志功能需要依靠 Commons Logging,因此,还需下载 Commons

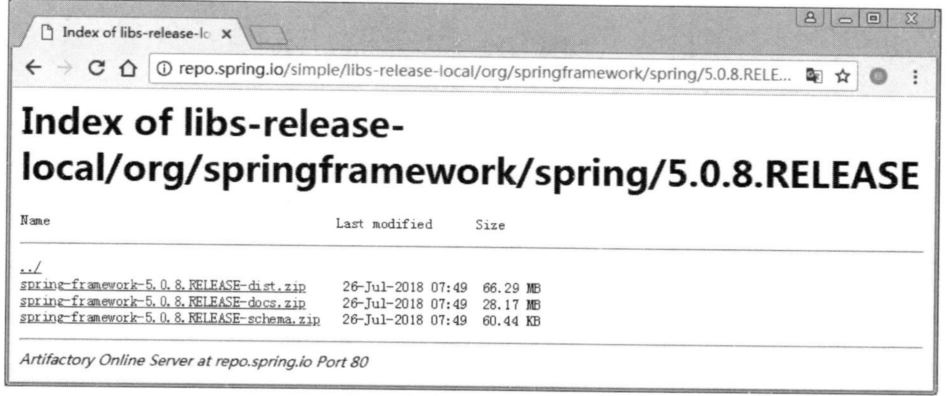

图 6.3　Spring 的下载页面

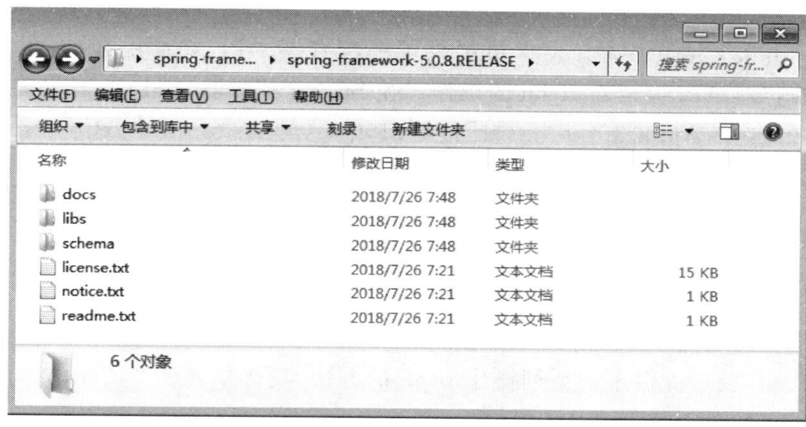

图 6.4　Spring 的目录结构

Logging 的 jar 包。打开浏览器，访问 http://commons.apache.org/proper/commons-logging/download_logging.cgi，浏览器跳转到下载页面，如图 6.5 所示。

图 6.5　Commons Logging 的下载页面

(5) 单击页面中 commons-logging-1.2-bin.zip 超链接，将文件下载到指定目录。

(6) 解压下载到本地的 zip 文件，此时获得名称为 commons-logging-1.2 的文件夹，打开 commons-logging-1.2-bin 文件夹，找到 lib 目录下的 commons-logging-1.2.jar 文件，该文件即为使用 commons-logging 所要导入的文件。

6.3 Spring 的容器机制

6.3.1 容器机制简介

容器是 Spring 框架实现功能的基础，Spring 容器类似一家超级工厂，当 Spring 启动时，所有被配置过的类都会被纳入 Spring 容器的管理之中。

Spring 把它管理的类称为 Bean，通常情况下，与 Java Bean 相比，Spring 并没有要求 Bean 必须遵循一定的规范，即使是普通的 Java 类，只要被配置到容器中，Spring 就可以管理它并把它作为 Bean 处理。

Spring 可以通过 XML 文件或注解获取配置信息，进而通过容器对象来管理 Bean。Spring 对 Bean 的管理体现在它负责创建 Bean 并管理 Bean 的生命周期。Bean 运行在 Spring 容器中，它只需发挥自己功能，而无须过多关注 Spring 容器的情况。

为了便于开发，Spring 为开发人员提供了一套容器 API，如此一来，开发人员可使用 Spring 提供的容器 API 完成对 Bean 的操作。在 Spring 的容器 API 中，开发人员接触最多的是 BeanFactory 和 ApplicaitonContext 接口，其中，ApplicaitonContext 是 BeanFactory 的子接口。接下来，本节将对这两个接口做详细讲解。

6.3.2 BeanFactory 接口

BeanFactory 是 IOC 最基本的容器接口，它定义了创建和管理 Bean 的方法，为其他容器提供了最基本的规范。

BeanFactory 接口提供了一系列操作 Bean 的方法，具体如表 6.2 所示。

表 6.2 BeanFactory 接口的方法

方法名称	描述
getBean(String name)	根据名称获取 Bean
getBean(String name, Class<T> type)	根据名称、类型获取 Bean
<T> T getBean(Class<T> requiredType)	根据类型获取 Bean
Object getBean(String name, Object... args)	根据名称获取 Bean
isSingleton(String name)	是否为单实例
isPrototype(String name)	是否为多实例
isTypeMatch(String name, ResolvableTypetype)	名称、类型是否匹配
isTypeMatch(String name, Class<?> type)	名称、类型是否匹配
Class<?> getType(String name)	根据名称获取类型
String[]getAliases(String name)	根据实例的名字获取实例的别名数组
boolean containsBean(String name)	根据 bean 的名称判断是否含有指定的 Bean

表 6.2 列举了 BeanFactory 接口的方法，开发者调用这些 API 即可完成对 Bean 的操作，无须关注 Bean 的实例化过程。

Spring 中提供了几种 BeanFactory 的实现类，其中最常用的是 XmlBeanFactory，它可以读取 XML 文件并根据 XML 文件中的配置信息来管理 Bean。

6.3.3 ApplicaitonContext 接口

ApplicaitonContext 接口的功能建立在 BeanFactory 接口的基础之上，它增强了 BeanFactory 的特性，增加了更多企业级的功能。

ApplicaitonContext 接口为应用提供国际化访问功能，提供资源(包括 URL 和文件系统)的访问支持，可同时加载多个配置文件，引入事件机制，让容器在上下文中提供了对应用事件的支持，以声明式方式启动并创建 Spring 容器。除此之外，ApplicaitonContext 接口可以为单例的 Bean 实行预初始化，并根据< property >元素执行 setter 方法，这决定了此时单例的 Bean 可以被直接使用，提升了程序获取 Bean 实例时的性能，因此，实际开发中使用 ApplicaitonContext 接口更多。

在实际开发中，如果想要获取 ApplicaitonContext 的实例，可以通过自定义一个实现 ApplicationContextAware 接口的工具类来完成，并且这个工具类也要配置到 Spring 容器中。ApplicationContextAware 接口有一个 setApplicationContext(ApplicationContext context)方法，该方法由 Spring 调用并传入 ApplicationContext 实例，工具类可通过该参数获取实例。

为了便于开发，Spring 提供了几种常用的 ApplicaitonContext 接口实现类，具体如表 6.3 所示。

表 6.3 ApplicaitonContext 接口的实现类

类　名　称	描　　述
ClassPathXmlApplicationContext	从类路径加载配置文件，创建 ApplicaitonContext 实例
FileSystemXmlApplicationContext	从文件系统加载配置文件，创建 ApplicaitonContext 实例
AnnotationConfigApplicationContext	从注解中加载配置文件，创建 ApplicaitonContext 实例
WebApplicationContext	在 Web 应用中使用，从相对于 Web 根目录的路径中加配置文件，创建 ApplicaitonContext 实例
ConfigurableWebApplicationContext	扩展了 WebApplicationContext，它允许通过配置的方式实例化 WebApplicationContext

表 6.3 列举了几种常用的 ApplicaitonContext 接口实现类，开发人员可根据具体需求酌情调用。

6.3.4 容器的启动过程

Spring 容器的底层原理相对复杂，因此，初学者无须在探究实现细节上耗费过多精力。在使用 Spring 编写入门程序之前，先大体了解 Spring 容器的启动过程即可。

Spring 容器的启动过程包括三个基本步骤：BeanDifinition 的 Resource 定位、BeanDifinition 的载入与解析、BeanDifinition 在 Spring 容器中的注册。

1. BeanDifinition 的 Resource 定位

在获取配置信息时，Spring 容器首先需要找到具体的 Resource。在实际开发中，Resource 可以是 XML 文件，也可以是注解，Resource 定位由 ResourceLoader 通过统一的 Resource 接口来完成。

2. BeanDifinition 的载入

读取配置信息，将< bean >配置信息转换为 Spring 容器内部的数据结构，这个数据结构就是 BeanDifinition。通过 BeanDifinition，Spring 能够方便地对 Bean 进行管理。

3. BeanDifinition 的注册

在完成 BeanDifinition 的载入后，需要将 BeanDifinition 注册到 Spring 的容器中。Spring 容器通过一个 HashMap 对象持有 BeanDifinition 数据。注册 BeanDifinition 就是将 BeanDifinition 数据置入 HashMap 对象中，这个过程通过调用 BeanDifinitionRegistry 接口实现。

随着 Spring 容器的启动，Bean 完成了在 Spring 容器中的定义。根据 Spring 容器的原理，在默认情况下，Bean 的实例化将在 Spring 容器的启动过程中完成。如果用户想要在第一次向 Spring 容器索要 Bean 时完成实例化，可以通过配置信息中 Bean 的 lazy-init 属性来实现。Bean 的 lazy-init 属性有三个可选值：default、false、true，其中，default 和 false 功能相同，均默认在 Spring 容器的启动过程中完成实例化，只有当 lazy-init 的属性值为 true 时，Bean 的实例化才会在用户第一次索要时执行。

6.4 Spring 的简单应用

6.4.1 环境准备

前面的小节讲解了 Spring 的基础知识，接下来，本节通过一个实例演示 Spring 的简单应用。

在 Eclipse 中新建 Web 工程 chapter06，将 Spring 核心容器的 jar 包和 Spring 依赖 jar 包 commons-logging-1.2.jar 导入到工程 chapter06 的 lib 目录下，导入的 jar 包如图 6.6 所示。

```
WebContent
  META-INF
  WEB-INF
    lib
      commons-logging-1.2.jar
      spring-beans-5.0.8.RELEASE.jar
      spring-context-5.0.8.RELEASE.jar
      spring-core-5.0.8.RELEASE.jar
      spring-expression-5.0.8.RELEASE.jar
```

图 6.6 要导入的 jar 包

6.4.2 创建 Bean

在工程 chapter06 的 src 目录下新建包 com.qfedu.bean，在该包下新建类 Student，如例 6-1 所示。

【例 6-1】 Student.Java

```
package com.qfedu.bean;
public class Student {
    //无参构造
    public Student() {
        super();
        System.out.println("Student 对象创建了");
    }
    //创建一个成员变量,并给出 get/set 方法
    private String msg = null;
    public void setMsg(String msg){
```

```java
        this.msg = msg;
    }
    public String getMsg() {
        return msg;
    }
    //study()方法中引用成员变量
    public void study(){
        System.out.println("学生在学习" + msg);
    }
}
```

在以上代码中，当 Student 类的无参构造被调用时，控制台将输出“Student 对象创建了”的提示信息，除此之外，Student 类声明了一个成员变量 msg，Student 类的 study()方法引用了成员变量 msg。

在包 com.qfedu.bean 下新建类 Teacher，如例 6-2 所示。

【例 6-2】 Teacher.Java

```java
package com.qfedu.bean;
public class Teacher {
    public Teacher() {
        super();
        System.out.println("Teacher 对象创建了");
    }
    private String msg  = null;
    public void setMsg(String msg){
        this.msg = msg;
    }
    public String getMsg() {
        return msg;
    }
    public void teach(){
        System.out.println("老师讲解" + msg + "知识");
    }
}
```

在以上代码中，当 Teacher 类的无参构造被调用时，控制台将输出“Teacher 对象创建了”的提示信息，除此之外，Teacher 类声明了一个成员变量 msg，Teacher 类的 teach()方法引用了成员变量 msg。

6.4.3 创建配置文件

在工程 chapter06 的 src 目录下新建配置文件 applicationContext.xml，Spring 通过该文件获取 Bean 的配置信息，如例 6-3 所示。

【例 6-3】 applicationContext.xml

```xml
<?xml version = "1.0" encoding = "UTF - 8"?>
<beans xmlns = "http://www.springframework.org/schema/beans"
```

```
    xmlns:xsi = "http://www.w3.org/2001/XMLSchema - instance"
    xsi:schemaLocation = "http://www.springframework.org/schema/beans
        http://www.springframework.org/schema/beans/spring - beans.xsd">
    <! -- 将指定的类配置给 Spring -- >
    < bean id = "student" class = "com.qfedu.bean.Student">
        < property name = "msg">
            < value > Java </value >
        </property >
    </bean >
</beans >
```

在以上代码中,< bean >元素指定了一个将由 Spring 管理的类。< bean >元素的 class 属性指定该类的完全限定名,id 属性指定该类的唯一 id 值,< property >元素的 name 属性指定该类的成员变量名称,< property >的子元素< value >给对应成员变量注入值。除此之外,由于该配置文件的约束信息较长,开发人员可以从 Spring 下载包中包含的官方文档中获取。打开下载包的 docs\spring-framework-reference 目录,打开 core.html 文件,找到配置文件的约束模板,直接复制到本地工程中即可。

6.4.4 测试功能

在工程 chapter06 的 src 目录下新建 com.qfedu.test 包,在该包下新建测试类 TestSpring,如例 6-4 所示。

【例 6-4】 TestSpring.java

```
package com.qfedu.test;
import org.springframework.context.ApplicationContext;
import org.springframework.context.support.
    ClassPathXmlApplicationContext;
import com.qfedu.bean.Student;
public class TestSpring {
    public static void main(String[ ] args) throws Exception  {
        //通过读取配置信息获取 ApplicationContext 对象
        ApplicationContext  applicationContext = new
                ClassPathXmlApplicationContext("applicationContext.xml");
        //通过 id 值获取 Student 对象
        Student student  = applicationContext.getBean
                ("student",Student.class);
        //调用 Student 对象的方法
        student.study();
    }
}
```

在以上代码中,程序通过 Spring 获取 Student 对象,进而调用 Student 对象的 study()方法。由于 Spring 通过配置文件获得了 bean 的配置信息,当调用 Student 对象的 study()方法时,控制台将输出相应的提示信息。

执行 TestSpring 类,执行结果如图 6.7 所示。

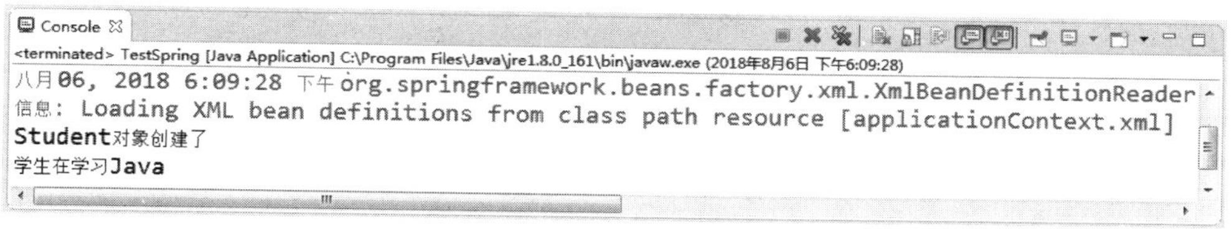

图 6.7　执行 TestSpring 类的结果

从图 6.7 中可以发现，控制台输出 Student 类的无参构造方法和 study()方法被执行的信息。这就说明，Spring 成功完成 Student 对象的创建及成员方法调用，除此之外，Student 类的成员变量 msg 的值被成功注入。

修改配置文件 applicationContext.xml，将第 9 行代码中< value >的属性值由“Java”修改为“Spring”。重新执行 TestSpring 类，执行结果如图 6.8 所示。

```
Console
<terminated> TestSpring [Java Application] C:\Program Files\Java\jre1.8.0_161\bin\javaw.exe (2018年8月6日 下午6:22:26)
八月 06, 2018 6:22:26 下午 org.springframework.beans.factory.xml.XmlBeanDefinitionReader
信息: Loading XML bean definitions from class path resource [applicationContext.xml]
Student对象创建了
学生在学习Spring
```

图 6.8　重新执行的结果

从图 6.8 中可以发现，控制台输出的提示信息发生变化，Student 类的成员变量 msg 的值被修改为“Spring”，这与配置文件的修改是一致的。

修改配置文件 applicationContext.xml，在< bean >元素中添加配置信息，具体代码如下。

```
<bean id = "teacher" class = "com.qfedu.bean.Teacher">
    <property name = "msg">
        <value > Spring </value >
    </property >
</bean >
```

在以上代码中，< bean >元素再次为 Spring 配置了一个类，具体配置项与例 6-3 类似，此处不再赘述。

修改测试类 TestSpring，在 main()方法中添加代码，具体如下。

```
Teacher teacher = applicationContext.getBean("teacher",Teacher.class);
 teacher.teach();
```

在以上代码中，程序获取 Teacher 对象并调用 Teacher 对象的 teach()方法。

第三次执行 TestSpring 类，执行结果如图 6.9 所示。

从图 6.9 中可以发现，控制台输出了 Student 类和 Teacher 类的相关信息，这就说明，多个 Bean 同时可以被 Spring 容器管理。除此之外，虽然获取 Teacher 对象的方法在调用 Student 类的 study()方法之后执行，但 Teacher 类的无参构造方法却在 Student 类的 study()

```
Console
<terminated> TestSpring [Java Application] C:\Program Files\Java\jre1.8.0_161\bin\javaw.exe (2018年8月6日 下午6:42:35)
八月 06, 2018 6:42:35 下午 org.springframework.beans.factory.xml.XmlBeanDefinitionReader
信息: Loading XML bean definitions from class path resource [applicationContext.xml]
Student对象创建了
Teacher对象创建了
学生在学习Spring
老师在讲解Spring
```

图 6.9　第三次执行的结果

方法之前执行，这就说明，Teacher 类在容器启动时完成了实例化，因此，它的实例化时间要比调用 Student 对象的时间早。

修改配置文件 applicationContext. xml，修改 Teacher 类的配置信息，加入 lazy-init 的属性，具体代码如下。

```
<bean id = "teacher" class = "com.qfedu.bean.Teacher"   lazy - init = "true">
    <property name = "msg">
        <value> Spring </value>
    </property>
</bean>
```

在以上代码中，< bean >元素的 lazy-init 属性为 true，这意味着 Teacher 类将在用户第一次索取时才能被实例化。

第四次执行 TestSpring 类，执行结果如图 6.10 所示。

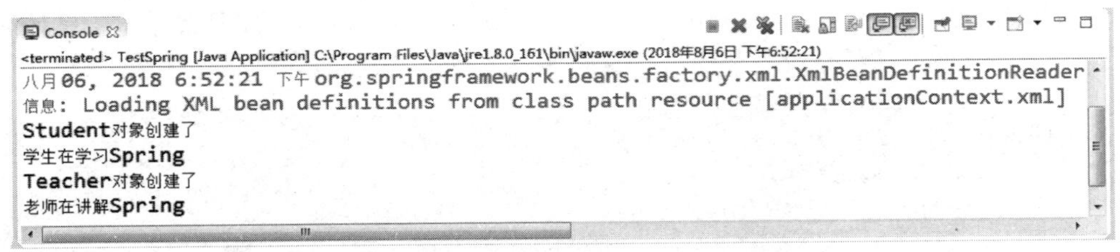

图 6.10　第四次执行的结果

从图 6.10 中可以发现，Teacher 类的无参构造方法在 Student 类的 study()方法之后执行，由此可见，Teacher 类在容器启动时并没有完成实例化，而是在第一次索取时才完成实例化，因此，它的实例化时间要比调用 Student 对象的时间晚。

6.5　本 章 小 结

本章首先介绍了 Spring 的基础知识，包括 Spring 的基本概念、优势、功能体系、子项目等，接下来详细讲解了 Spring 的下载及使用，然后重点讲解了 Spring 的容器功能，包括容器的基本概念、两大容器接口 BeanFactory 和 ApplicationContext、容器的启动过程，最后通过一个实例讲解了 Spring 的简单应用。通过本章知识的学习，大家应该能理解 Spring 的基础概念和生态圈，掌握两大容器接口的用法，能够通过 Spring 完成 Bean 的配置和 Bean 实例的获取，进而通过 Spring 编写简单的应用程序。

6.6 习 题

1. 填空题

(1) Spring 最为核心的理念是________________和________________。

(2) 在 Spring 中，________和________模块规定了创建、配置和管理 Bean 的方式。

(3) 在 Spring 中，________模块可以通过 ApplicationContext 接口提供上下文信息。

(4) Spring 的最新版本是________，它对 Java 版本的最低要求是________。

(5) Spring 最基本的容器接口是____________________。

2. 选择题

(1) 关于 Spring，下列选项错误的是(　　)。

A. Spring 是一个分层的 Java SE/EE 应用的一站式轻量级框架

B. Spring 尚未将它的源代码公开

C. Spring 提供了展现层 SpringMVC、持久层 Spring JDBC 以及事务管理等技术

D. Spring 可以整合 Java 生态圈里众多的第三方框架和类库

(2) 在下列选项中，不属于 Spring 优势的是(　　)。

A. 降低程序的耦合度

B. 支持 AOP 编程，对程序进行权限拦截、安全监控等操作

C. 降低 Java API 的使用难度，方便开发

D. Spring 自身提供了 AOP 方式的日志系统

(3) 在 BeanFactory 接口的方法中，用于判断是否为单实例的是(　　)。

A. isSingleton(String name)　　B. isPrototype(String name)

C. String[]getAliases(String name)　　D. boolean containsBean(String name)

(4) 在 ApplicaitonContext 接口的实现类中，用于通过注解加载配置信息的是(　　)。

A. ClassPathXmlApplicationContext

B. FileSystemXmlApplicationContext

C. AnnotationConfigApplicationContext

D. WebApplicationContext

(5) Spring 的子项目，可被作为微服务工具包使用的是(　　)。

A. Spring Cloud　　B. Spring Data

C. Spring Security　　D. Spring Batch

3. 思考题

(1) 简述 Spring 的概念及优势。

(2) 简述 Spring 的容器功能。

4. 编程题

编写一个 Person 类，Person 类中封装有成员变量 age、成员方法 say()，将 Person 类配置到 Spring 中，通过 Spring 的配置文件为 Person 类的成员变量 age 注入值，然后在 Spring 程序中调用成员方法 say()输出成员变量 age 的值。

第7章 使用 Spring 管理 Bean

本章学习目标

- 理解 IOC 和 DI 的概念
- 掌握依赖注入常用的两种方法
- 掌握 Bean 的配置方法
- 掌握 Bean 的生命周期管理方法
- 掌握 Spring 入门程序的编写

容器功能是 Spring 的基础功能，当 Bean 被配置到 Spring 之后，对象的生命周期及依赖关系就会纳入 Spring 的统一管理。在实际开发中，基于容器功能，Spring 可有效降低类与类之间的耦合，同时对类的统一管理也可以简化开发，提升开发效率。因此，对于开发人员来说，深入理解 Spring 管理 Bean 的技术细节显得尤为重要。接下来，本章将详细讲解使用 Spring 管理 Bean 的相关知识。

7.1 IOC 和 DI

7.1.1 简介

IOC 是 Inversion Of Control(控制反转)的缩写，它是一种设计思想，是指将对象的控制权由程序代码反转给外部容器。

在 Spring 中，控制反转是实现 Spring 容器的指导思想。有了 Spring 容器，开发人员无须编写管理对象生命周期和依赖关系的代码，此项工作将由 Spring 容器根据配置自动完成，如此一来，对象的控制权由程序代码反转给 Spring 容器。

DI 是 Dependency Injection(依赖注入)的缩写，它是控制反转的另一种说法，同时也为控制反转提供了实现方法。依赖注入是指调用类对其他类的依赖关系由容器注入，这就避免了调用类对其他类的过度依赖，降低了类与类之间的耦合。

接下来以一个实例来说明依赖注入。现有一个学生类和一个图书类，学生类对图书类存在有依赖关系，如图 7.1 所示。

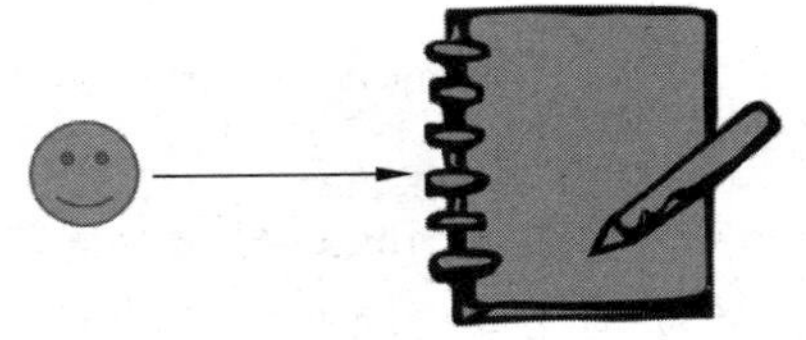

图 7.1 学生类依赖图书类

如果需要在学生类中调用图书类实现功能，按照原始的做法，在学生类中使用 new 关键字创建一个图书类的对象，然后再调用图书类的对象完成下一步操作。但是，在实际开发中，这种做法会导致调用者与被依赖者的硬编码耦合，各个类之间的职责不明确，对项目后期的升级维护非常不利，因而不建议采用该方法。

此时如果引入 Spring 框架并采用依赖注入的做法，那么上述问题就可以避免。使用依赖注入的做法后，学生类无须主动创建图书类对象，只须被动等待 Spring 容器注入即可，Spring 容器如此一来，学生类和图书类之间高度解耦，程序的可扩展性增强。

7.1.2 依赖注入的方式

依赖注入有三种方式，它们分别是构造器注入、属性注入和接口注入，其中，构造器注入和属性注入是 Spring 常用的方式，接下来对这两种注入方式做详细讲解。

1. 构造器注入

构造器注入指的是在被注入的类中声明一个构造方法，而构造方法可以是有参或无参的。Spring 在读取配置信息后，会通过反射方式调用构造方法，如果是有参构造方法，可以在构造方法中传入所需的参数值，进而创建类对象。

接下来通过一个实例演示构造器注入，具体步骤如下。

(1) 在 Eclipse 中新建 Web 工程 chapter07，将 Spring 核心容器的 jar 包和 Spring 依赖的 commons-logging 的 jar 包导入到工程 chapter07 的 lib 目录下，由于第 6 章相应章节中已给出相关 jar 包，此处不再赘述。

(2) 在工程 chapter07 的 src 目录下新建包 com. qfedu. bean，在该包下新建类 Student，如例 7-1 所示。

【例 7-1】 Student. Java

```
package com.qfedu.bean;
public class Student {
    private int sid;
    private String name;
    private String age;
    private String course;
    public Student(int sid, String name, String age, String course) {
        super();
        this.sid = sid;
        this.name = name;
        this.age = age;
        this.course = course;
    }
    @Override
    public String toString() {
        return "Student [sid=" + sid + ", name=" + name + ", age=" + age +
         ", course=" + course + "]";
    }
}
```

在以上代码中，Student 类提供了有参构造方法和 toString()方法，当 Spring 通过构造器注入相应的值以后，通过 toString()方法可以获取这些注入的值。

(3) 在工程 chapter07 的 src 目录下新建配置文件 applicationContext. xml，Spring 通过该文件获取 Bean 的配置信息，如例 7-2 所示。

【例 7-2】 applicationContext. xml

```
<?xml version = "1.0" encoding = "UTF - 8"?>
<beans xmlns = "http://www.springframework.org/schema/beans"
    xmlns:xsi = "http://www.w3.org/2001/XMLSchema - instance"
    xsi:schemaLocation = "http://www.springframework.org/schema/beans
        http://www.springframework.org/schema/beans/spring - beans.xsd">
    <!-- 通过构造器注入成员变量的值 -->
    <bean id = "student" class = "com.qfedu.bean.Student">
    <constructor - arg name = "sid" value = "1"></constructor - arg>
    <constructor - arg name = "name" value = "ZhangSan"></constructor - arg>
    <constructor - arg name = "age" value = "20"></constructor - arg>
    <constructor - arg name = "course" value = "Java"></constructor - arg>
    </bean>
</beans>
```

在以上代码中，< constructor-arg >元素用于给类构造方法的参数注入值，Spring 通过构造器注入获取这些值，最终这些值将会赋给 Spring 创建的 Student 对象。

（4）在工程 chapter07 的 src 目录下新建 com. qfedu. test 包，在该包下新建测试类 TestSpring01，如例 7-3 所示。

【例 7-3】 TestSpring01. java

```
package com.qfedu.test;
import org.springframework.context.ApplicationContext;
import org.springframework.context.support
    .ClassPathXmlApplicationContext;
import com.qfedu.bean.Student;
public class TestSpring01 {
    public static void main(String[] args) throws Exception  {
        ApplicationContext  applicationContext  = new
            ClassPathXmlApplicationContext("applicationContext.xml");
        Student student  = applicationContext.getBean
            ("student",Student.class);
        //输出 Student 类对象信息
        System.out.println(student);
    }
}
```

在以上代码中，程序通过 Spring 获取 Student 对象，进而输出 Student 对象的信息。执行测试类 TestSpring01，执行结果如图 7.2 所示。

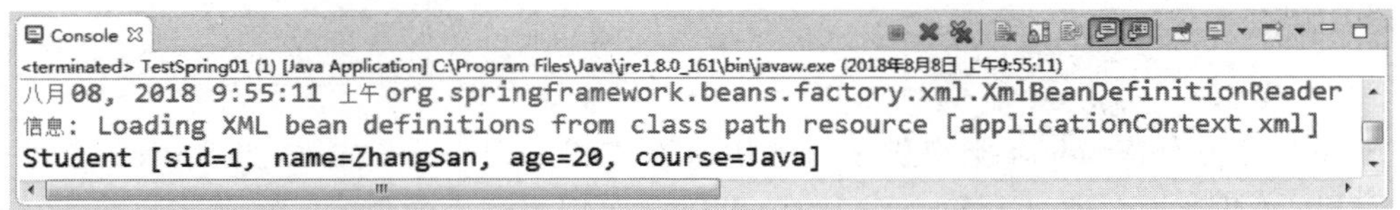

图 7.2 执行 TestSpring01 类的结果

从以上执行结果可以看出,控制台输出 Student 对象的信息,其中成员变量的值和配置文件< constructor-arg >元素的 value 值相同,由此可见,Spring 实现了构造器注入。

2. 属性注入

属性注入是 Spring 中最主流的注入方法,这种注入方法简单、直观,它是在被注入的类中声明一个 Set 方法,并通过该 Set 方法的参数注入对应的值。

接下来通过一个实例演示属性注入,具体步骤如下。

(1) 在 com. qfedu. bean 包下新建类 Teacher,如例 7-4 所示。

【例 7-4】 Teacher. Java

```
package com.qfedu.bean;
public class Teacher {
    private int tid;
    private String name;
    public void setTid(int tid) {
        this.tid = tid;
    }
    public void setName(String name) {
        this.name = name;
    }
    @Override
    public String toString() {
        return "Teacher [tid=" + tid + ", name=" + name + "]";
    }
}
```

在以上代码中,Teacher 类提供了成员变量、set 方法和 toString()方法,当 Spring 通过属性注入相应的值以后,通过 toString()方法可以获取这些注入的值。

(2) 修改配置文件 applicationContext. xml,在 applicationContext. xml 文件的 bean 元素中添加配置信息,具体代码如下所示。

```
<bean id="teacher" class="com.qfedu.bean.Teacher">
    <property name="tid" value="1"></property>
    <property name="name" value="LiSi"></property>
</bean>
```

在以上代码中,< property >元素的 name 属性指定该类的成员变量名称,value 属性给对应的成员变量注入值。

(3) 在 com. qfedu. test 包下新建测试类 TestSpring02,如例 7-5 所示。

【例 7-5】 TestSpring02. java

```
package com.qfedu.test;
import org.springframework.context.ApplicationContext;
import org.springframework.context.support
    .ClassPathXmlApplicationContext;
import com.qfedu.bean.Teacher;
```

```
public class TestSpring02 {
    public static void main(String[] args) throws Exception  {
        ApplicationContext  applicationContext = new
            ClassPathXmlApplicationContext("applicationContext.xml");
        Teacher teacher = applicationContext.getBean
            ("teacher",Teacher.class);
        //输出 Teacher 类对象信息
        System.out.println(teacher);
    }
}
```

执行测试类 TestSpring02，执行结果如图 7.3 所示。

```
Console
<terminated> TestSpring02 [Java Application] C:\Program Files\Java\jre1.8.0_161\bin\javaw.exe (2018年8月8日 上午11:51:10)
八月 08, 2018 11:51:10 上午 org.springframework.beans.factory.xml.XmlBeanDefinitionReader
信息: Loading XML bean definitions from class path resource [applicationContext.xml]
Teacher [tid=1, name=LiSi]
```

图 7.3　执行 TestSpring02 类的结果

从以上执行结果可以看出，控制台输出 Student 对象的信息，其中成员变量的值和配置文件< property >元素的 value 值相同，由此可见，Spring 实现了属性注入。

7.2　Bean 的配置

7.2.1　Bean 的定义

在本书前面的讲解中，通常使用 XML 文件来实现 Bean 的配置。XML 配置文件的根元素是< beans >，< beans >元素中有很多< bean >子元素，每个< bean >子元素都定义了一个 Bean，并且描述了该 Bean 应该被如何配置到 Spring 容器中。

< bean >元素提供了一系列属性用于描述 Bean 的信息，其中常用的如表 7.1 所示。

表 7.1　< bean >元素的属性

属　　性	描　　述
Class	指定 Bean 对应类的全路径
Name	指定 Bean 对应对象的一个标识
Scope	指定 Bean 对象的作用域
id	id 是 Bean 对象的唯一标识，Spring 通常通过 id 属性完成对 Bean 的配置、管理
lazy-init	是否延时加载，默认值为 false
init-method	对象初始化方法
destory	对象销毁方法

表 7.1 列举了< bean >元素的常用属性，由于本书前面已介绍过< bean >元素的具体用法，并且对 class、name、id 等属性也有讲解，因此本节不再赘述。

除了以上所列举的属性之外，< bean >元素还包含有一些配置依赖信息的子元素，其中，<property>子元素用于通过属性注入的方式设置值，<constructor-arg>子元素用于自动寻找 Bean 的构造函数，并在初始化时将设置的参数注入。

7.2.2 注入集合

在实际开发中，根据业务需求，有时需要将集合注入到 Bean，例如，List、Set、Map、Array 等，此时可以通过<list>、<set>、<map>、<array>等集合元素实现。

接下来，本节通过一个实例演示集合的注入，具体步骤如下。

(1) 在 com.qfedu.bean 包下新建类 Mix，如例 7-6 所示。

【例 7-6】 Mix.Java

```
package com.qfedu.bean;
import java.util.*;
public class Mix {
    private List<String> myList;
    private Map<String,String> myMap;
    private String[] myArray;
    public void setMyList(List<String> myList) {
        this.myList = myList;
    }
    public void setMyMap(Map<String, String> myMap) {
        this.myMap = myMap;
    }
    public void setMyArray(String[] myArray) {
        this.myArray = myArray;
    }
    @Override
    public String toString() {
        return "Mix [myList=" + myList + ", myMap=" + myMap + ", myArray=" +
                Arrays.toString(myArray) + "]";
    }
}
```

在以上代码中，Mix 类封装了 List、Map、Array 三种类型的成员变量并提供了 set 方法和 toString()方法，当 Spring 通过属性注入相应的值以后，通过 toString()方法可以获取这些注入的值。

(2) 修改配置文件 applicationContext.xml，在 applicationContext.xml 文件的<bean>元素中添加配置信息，具体代码如下所示。

```
<bean id="mix" class="com.qfedu.bean.Mix">
    <!-- 注入 List -->
    <property name="myList">
        <list>
            <value>list01</value>
            <value>list02</value>
        </list>
    </property>
```

```
    <!-- 注入 Map -->
    <property name="myMap">
        <map>
            <entry key="key01" value="map01"></entry>
            <entry key="key02" value="map02"></entry>
        </map>
    </property>
    <!-- 注入 array -->
    <property name="myArray">
        <array>
            <value>array01</value>
            <value>array02</value>
        </array>
    </property>
</bean>
```

在以上代码中，<bean>元素下共有三个<property>元素，这三个<property>元素分别包含了<list>、<map>和<array>元素，其中，<list>元素用于注入 List 集合，<map>元素用于注入 Map 集合，<array>元素用于注入 Array。

（3）在 com.qfedu.test 包下新建测试类 TestSpring03，如例 7-7 所示。

【例 7-7】 TestSpring03.java

```
package com.qfedu.test;
import org.springframework.context.ApplicationContext;
import org.springframework.context.support
    .ClassPathXmlApplicationContext;
import com.qfedu.bean.Mix;
public class TestSpring03 {
    public static void main(String[] args) throws Exception  {
        ApplicationContext  applicationContext = new
            ClassPathXmlApplicationContext("applicationContext.xml");
        Mix mix = applicationContext.getBean("mix",Mix.class);
        //输出 Mix 类对象信息
        System.out.println(mix);
    }
}
```

执行测试类 TestSpring03，执行结果如图 7.4 所示。

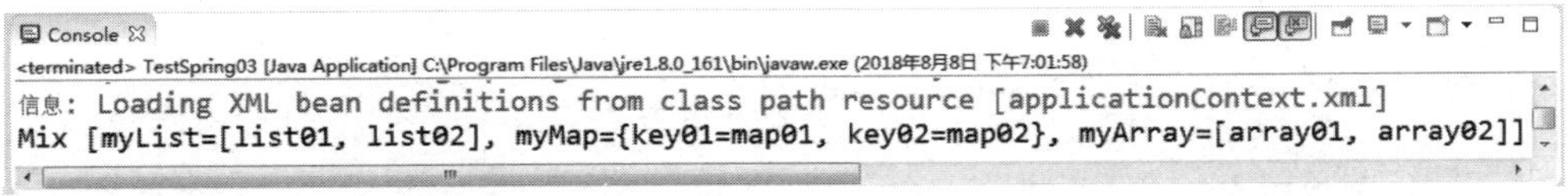

```
Console
<terminated> TestSpring03 [Java Application] C:\Program Files\Java\jre1.8.0_161\bin\javaw.exe (2018年8月8日 下午7:01:58)
信息: Loading XML bean definitions from class path resource [applicationContext.xml]
Mix [myList=[list01, list02], myMap={key01=map01, key02=map02}, myArray=[array01, array02]]
```

图 7.4 执行 TestSpring03 类的结果

从以上执行结果可以看出，控制台输出 Mix 对象的信息，其中成员变量的值和配置文件<property>元素配置的值相同，由此可见，Spring 实现了集合类型的注入。

7.2.3 注入其他 Bean

在 Spring 容器中，一个 Bean 的属性值可能是另外一个 Bean 的实例，即当前 Bean 对其他 Bean 存在有依赖关系，此时就需要将其他 Bean 的实例配置为当前 Bean 的属性，这可以通过< ref >元素或< property >元素的 ref 属性来实现。

接下来，本节通过一个实例演示在当前 Bean 中注入其他 Bean，具体步骤如下。

(1) 在 com. qfedu. bean 包下新建类 School，如例 7-8 所示。

【例 7-8】 School. Java

```
package com.qfedu.bean;
public class School {
    private int sid;
    private Student stu;
     public int getSid() {
        return sid;
    }
    public void setSid(int sid) {
        this.sid = sid;
    }
    public void setStu(Student stu) {
        this.stu = stu;
    }
    public Student getStu() {
        return stu;
    }
}
```

在以上代码中，School 类封装了 sid 和 stu 两个成员变量并提供了它们的 getter 和 setter 方法，当 Spring 通过属性注入相应的值以后，可以通过 School 对象获取这些注入的值。

(2) 修改配置文件 applicationContext. xml，在 applicationContext. xml 文件的< bean >元素中添加配置信息，具体代码如下所示。

```
<bean id = "school" class = "com.qfedu.bean.School">
    <property name = "stu">
        <ref bean = "student"/>
    </property>
</bean>
```

在以上代码中，< property >元素下包含一个< ref >元素，< ref >元素用于将 Student 类的实例配置给 Shcool 类作为属性。

(3) 在 com. qfedu. test 包下新建测试类 TestSpring04，如例 7-9 所示。

【例 7-9】 TestSpring04. java

```
package com.qfedu.test;
import org.springframework.context.ApplicationContext;
```

```
import org.springframework.context.support
      .ClassPathXmlApplicationContext;
import com.qfedu.bean.School;
import com.qfedu.bean.Student;
public class TestSpring04 {
    public static void main(String[] args) throws Exception  {
        ApplicationContext  applicationContext = new
            ClassPathXmlApplicationContext("applicationContext.xml");
         School school  = applicationContext.getBean("school",School.class);
        //通过 School 对象获取 Student 对象,并输出 Student 对象的信息
        Student stu  =  school.getStu();
        System.out.println(stu);
    }
}
```

执行测试类 TestSpring04,执行结果如图 7.5 所示。

```
Console ☒
<terminated> TestSpring04 [Java Application] C:\Program Files\Java\jre1.8.0_161\bin\javaw.exe (2018年8月9日 下午3:02:33)
八月 09, 2018 3:02:33 下午 org.springframework.beans.factory.xml.XmlBeanDefinitionReader
信息: Loading XML bean definitions from class path resource [applicationContext.xml]
Student [sid=1, name=ZhangSan, age=20, course=Java]
```

图 7.5　执行 TestSpring04 类的结果

从以上执行结果可以看出,控制台输出 Student 对象的信息。程序通过调用 School 对象获取 Student 对象,而且 Student 对象的属性值和配置文件中 Student 类的配置信息是匹配的,由此可见,Spring 实现了为当前 Bean 的属性注入其他 Bean 的实例。

除了< ref >元素外,以上功能还可以通过< property >元素的 ref 属性实现。替换配置文件 applicationContext.xml 中 School 类的配置信息,替换后的代码如下所示。

```
<bean id = "school" class = "com.qfedu.bean.School">
    <property name = "stu"  ref = "student"></property>
</bean>
```

重新执行测试类 TestSpring04,执行结果如图 7.6 所示。

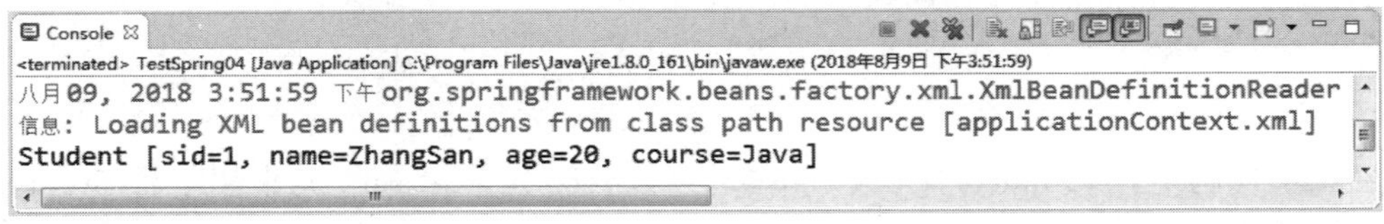

图 7.6　重新执行的结果

从以上执行结果可以看出,控制台输出相同的执行结果。这就说明,通过< property >元素的 ref 属性可以实现与< ref >元素同样的功能。

7.2.4 使用 P:命名空间注入

前面的章节讲过，Spring 的配置文件一般基于< property >元素来配置 Bean 的属性，但是当 Bean 实例的属性足够多时，使用大量的< property >元素配置属性就显得冗余，为了解决这个问题，Spring 引入了 P:命名空间。

P:命名空间也是通过属性注入实现的，和使用< property >元素不同的是，当导入 P:命名空间后，可以通过 P:属性完成属性注入，而不再依靠< property >元素。

接下来，本节以 School 类为例演示如何使用 P:命名空间完成注入，具体步骤如下。

(1) 引入 P:命名空间，修改配置文件 applicationContext.xml 中< beans >元素的代码，具体如下。

```
1 <beans xmlns = "http://www.springframework.org/schema/beans"
2     xmlns:xsi = "http://www.w3.org/2001/XMLSchema-instance"
3     xmlns:p = "http://www.springframework.org/schema/p"
4     xsi:schemaLocation = "http://www.springframework.org/schema/beans
5         http://www.springframework.org/schema/beans/spring-beans.xsd">
```

以上代码的第 3 行用于引入 P:命名空间，在使用 P:命名空间时，开发人员要首先添加此行代码。

(2) 修改 School 类的配置信息，修改后的代码如下。

```
<bean id = "school" class = "com.qfedu.bean.School" p:sid = "1"
p:stu-ref = "student"></bean>
```

在以上代码中，p:sid、p:stu-ref 用于设置 School 类的属性，它们可以实现和< property >元素同样的功能。

(3) 在 com.qfedu.test 包下新建测试类 TestSpring05，如例 7-10 所示。

【例 7-10】 TestSpring05.java

```
package com.qfedu.test;
import org.springframework.context.ApplicationContext;
import org.springframework.context.support
      .ClassPathXmlApplicationContext;
import com.qfedu.bean.School;
import com.qfedu.bean.Student;
public class TestSpring05 {
    public static void main(String[] args) throws Exception  {
        ApplicationContext  applicationContext = new
            ClassPathXmlApplicationContext("applicationContext.xml");
        School school = applicationContext.getBean("school",School.class);
        int sid = school.getSid();
        System.out.println("sid:" + sid);
        Student stu = school.getStu();
        System.out.println(stu);
    }
}
```

执行测试类 TestSpring05，执行结果如图 7.7 所示。

```
Console ☒
<terminated> TestSpring05 [Java Application] C:\Program Files\Java\jre1.8.0_161\bin\javaw.exe (2018年8月9日 下午6:10:59)
八月 09, 2018 6:11:00 下午 org.springframework.beans.factory.xml.XmlBeanDefinitionReader
信息: Loading XML bean definitions from class path resource [applicationContext.xml]
sid:1
Student [sid=1, name=ZhangSan, age=20, course=Java]
```

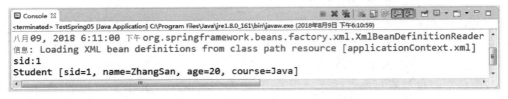

图 7.7　执行 TestSpring05 类的结果

从以上执行结果可以看出，控制台输出了 sid 值和 Student 对象的信息。经过分析发现，程序输出的信息和配置文件中 P:属性设置的值是匹配的，因此，程序实现了使用 P:命名空间注入值。

7.2.5　使用 SpEL 注入

SpEL 是 Spring Expression Language(Spring 表达式语言)的缩写，它是一种支持查询和操作运行时对象导航图功能的表达式语言，可以简化开发，能够减少代码逻辑和配置信息的编写。

SpEL 的语法与 EL 类似，它以 #{...}为定界符，能够为 Bean 的属性进行动态赋值，它可以通过 Bean 的 id 引用 Bean，调用对象的方法或引用对象的属性、计算表达式的值、匹配正则表达式等。

接下来，本节将通过一个实例演示如何使用 SpEL 完成注入，具体步骤如下。

(1) 在 com.qfedu.bean 包下新建类 Employee，如例 7-11 所示。

【例 7-11】　Employee.Java

```
package com.qfedu.bean;
public class Employee {
    private int eid;
    private String name;
    private String department;
    public int getEid() {
        return eid;
    }
    public void setEid(int eid) {
        this.eid = eid;
    }
    public String getName() {
        return name;
    }
    public void setName(String name) {
        this.name = name;
    }
    public String getDepartment() {
        return department;
    }
    public void setDepartment(String department) {
```

```
        this.department = department;
    }
    @Override
    public String toString() {
        return "Employee [eid=" + eid + ", name=" + name + ", department="
         + department + "]";
    }
}
```

在以上代码中，Employee 类封装了 eid、name、department 三种类型的成员变量并提供了 getter、setter 方法以及 toString()方法，此处需要注意的是，由于要使用 SpEL 表达式获取 Employee 的成员变量的值，必须要为 Employee 类声明 getter 方法。

（2）在 com.qfedu.bean 包下新建类 Person，如例 7-12 所示。

【例 7-12】 Person.Java

```
package com.qfedu.bean;
public class Person {
    private int pid;
    private String name;
    public void setPid(int pid) {
        this.pid = pid;
    }
    public void setName(String name) {
        this.name = name;
    }
    @Override
    public String toString() {
        return "Person [pid=" + pid + ", name=" + name + "]";
    }
}
```

（3）修改配置文件 applicationContext.xml，在 applicationContext.xml 文件的< bean >元素中添加配置信息，具体代码如下所示。

```
<bean id="employee" class="com.qfedu.bean.Employee">
    <property name="eid" value="1"></property>
    <property name="name" value="WangWu"></property>
    <property name="department" value="研发部"></property>
</bean>
<bean id="person" class="com.qfedu.bean.Person">
    <property name="pid" value="#{employee.eid}"></property>
    <property name="name" value="#{employee.name}"></property>
</bean>
```

以上代码中的第二个< bean >元素下共有两个< property >元素，这两个< property >元素分别将 Employee 类的 eid 属性、name 属性注入为 Person 类的 pid 属性、name 属性。

（4）在 com.qfedu.test 包下新建测试类 TestSpring06，如例 7-13 所示。

【例 7-13】 TestSpring06.Java

```
package com.qfedu.test;
import org.springframework.context.ApplicationContext;
import org.springframework.context.support
        .ClassPathXmlApplicationContext;
import com.qfedu.bean.Person;
public class TestSpring06 {
    public static void main(String[] args) throws Exception  {
        ApplicationContext  applicationContext = new
             ClassPathXmlApplicationContext("applicationContext.xml");
        Person person = applicationContext.getBean("person",Person.class);
        System.out.println(person);
    }
}
```

执行测试类 TestSpring06,执行结果如图 7.8 所示。

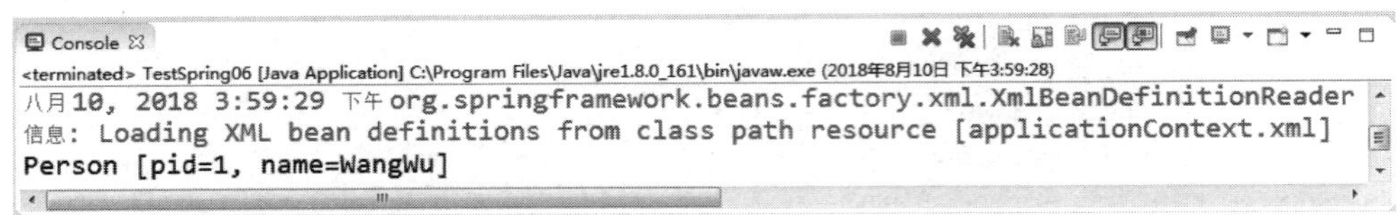

图 7.8 执行 TestSpring06 类的结果

从以上执行结果可以看出,控制台输出了 Person 对象的信息。经过分析发现,Person 对象的属性信息和 Employee 对象的属性信息是匹配的,因此,程序实现了使用 SpEL 注入值。

7.2.6 Bean 的作用域

当 Spring 容器创建一个 Bean 实例时,不仅可以完成 Bean 的实例化,还可以为该 Bean 指定作用域。在 Spring 容器中,Bean 的作用域是指 Bean 实例相对于其他 Bean 实例的请求可见范围。

Spring 支持五种作用域,具体如表 7.2 所示。

表 7.2 Spring 支持的作用域

属 性	描 述
Singleton	单例模式,作用域为 singleton 的 Bean 在 Spring 容器中只会存在一个共享的 Bean 实例,所有对 Bean 的请求只要 id 与 Bean 的定义相匹配,则只会返回 Bean 的同一实例
Prototype	每次从容器中调用 Bean 时,都会产生一个新的 Bean 实例
Request	一个 HTTP 请求会产生一个 Bean 对象,也就是说,每一个 HTTP 请求都有自己的 Bean 实例,只在基于 Web 的 Spring ApplicationContext 中可用
Session	限定一个 Bean 的作用域为 HTTPsession 的生命周期,只有基于 Web 的 Spring ApplicationContext 才能使用
global session	限定一个 Bean 的作用域为全局 HTTPsession 的生命周期,只有在 Web 应用中使用 Spring 时,该作用域才有效

表 7.2 列举了 Spring 支持的五种作用域，其中，singleton 和 prototype 是两种较为常用的作用域，如果不指定 Bean 的作用域，Spring 会默认该 Bean 的作用域是 singleton。

当 Bean 的作用域是 singleton 时，Spring 容器只为该 Bean 创建一个实例，并且该实例可以被重复使用。Spring 容器管理着 Bean 的生命周期，可以控制 Bean 的创建、初始化、销毁。由于创建和销毁 Bean 实例会带来一定的系统开销，因此，singleton 作用域的 Bean 避免了反复创建和销毁实例造成的资源消耗。

当 Bean 的作用域是 prototype 时，每次调用 Bean 时 Spring 容器都会返回一个新的不同的实例，此时，Spring 容器只负责创建 Bean 实例而不再跟踪其生命周期。

接下来，本节将通过一个实例演示 Bean 的 singleton 和 prototype 作用域，具体步骤如下。

（1）修改配置文件 applicationContext. xml，在 applicationContext. xml 文件的< bean >元素中添加配置信息，具体代码如下所示。

```
<bean id = "person01" class = "com.qfedu.bean.Person"/>
<bean id = "person02" class = "com.qfedu.bean.Person" scope = "prototype"/>
```

以上代码中通过< bean >元素配置了两个 Bean，其中，第 1 个 Bean 的作用域默认为 singleton，第 2 个 Bean 的作用域为 prototype 。

（2）在 com. qfedu. test 包下新建测试类 TestSpring07，如例 7-14 所示。

【例 7-14】 TestSpring07. Java

```
package com.qfedu.test;
import org.springframework.context.ApplicationContext;
import org.springframework.context.support.
    ClassPathXmlApplicationContext;
import com.qfedu.bean.Person;
public class TestSpring07 {
    public static void main(String[ ] args) throws Exception  {
        ApplicationContext  applicationContext  = new
             ClassPathXmlApplicationContext("applicationContext.xml");
        //获取 id 为 person01 的 Person 实例
         Person person01_1  =  applicationContext.getBean
            ("person01",Person.class);
        Person person01_2  =  applicationContext.getBean
            ("person01",Person.class);
        //获取 id 为 person02 的 Person 实例
        Person person02_1  =  applicationContext.getBean
            ("person02",Person.class);
        Person person02_2  =  applicationContext.getBean
            ("person02",Person.class);
        //判断两次获取的 id 为 person01 的 Person 实例是否相等
        System.out.println(person01_1 == person01_2);
        //判断两次获取的 id 为 person02 的 Person 实例是否相等
        System.out.println(person02_1 == person02_2);
    }
}
```

在以上代码中，程序分两次获取了 person01 和 person02 的实例，并比较所获取的实例是否相同。如果相同，控制台将输出 true，如果不相同，控制台将输出 false。

执行测试类 TestSpring07，执行结果如图 7.9 所示。

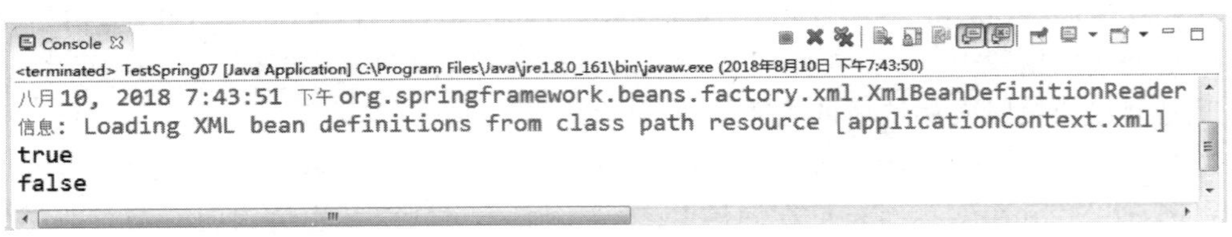

图 7.9　执行 TestSpring07 类的结果

从以上执行结果可以看出，对于 singleton 作用域的 Bean，程序每次请求 Bean 都会返回一个共享的实例，因此两次获取 Bean 的实例是相同的，对于作用域为 prototype 的 Bean，程序每次请求 Bean 都会返回一个新的实例，因此，两次获取 Bean 的实例是不相同的。

7.2.7　Bean 的生命周期

Bean 的生命周期是指 Bean 实例被创建、初始化和销毁的过程。前面的章节讲到过，Spring 容器可以管理作用域为 singleton 的 Bean 的生命周期。对于作用域为 prototype 的 Bean，Spring 容器只负责创建实例而不负责跟踪其生命周期。

在 Bean 的生命周期中，有两个时间节点尤为重要，这两个时间节点分别是 Bean 实例初始化以后和 Bean 实例销毁之前，实际开发中，有时需要在这两个时间节点完成一些指定操作，例如，在 Bean 实例初始化之后申请某些资源、在 Bean 实例销毁之前回收某些资源等。

为了便于监控 Bean 生命周期中的特殊节点，Spring 提供了相关的 API，当需要在 Bean 实例初始化后执行指定行为时，可以通过使用 init-method 属性或实现 initializingBean 接口的方式；当需要在 Bean 实例销毁前执行指定行为时，可以通过使用 destroy-method 属性或实现 DisposableBean 接口的方式。

1. Bean 实例初始化之后执行指定方法

采用 init-method 属性指定 Bean 实例初始化之后的方式相对简单，并且这种方式不会将代码和 Spring 的接口耦合，接下来通过一个实例演示 init-method 属性的使用，具体步骤如下。

(1) 在 com. qfedu. bean 包下新建类 Bean01，如例 7-15 所示。

【例 7-15】　Bean01. Java

```
package com.qfedu.bean;
public class Bean01 {
    private String bid;
    private String name;
    public Bean01() {
    }
    public String getBid() {
        return bid;
    }
```

```
    public void setBid(String bid) {
        this.bid = bid;
    }
    public String getName() {
        return name;
    }
    public void setName(String name) {
        this.name = name;
    }
    @Override
    public String toString() {
        return "Bean01 [bid=" + bid + ", name=" + name + "]";
    }
    public void init(){
        System.out.println("Bean的初始化完成,调用init()方法");
        System.out.println(this.toString());
    }
}
```

在以上代码中,Bean01 类提供了 init()方法,如果该方法的名称被指定为配置文件中 init-method 属性,当对应的 Bean 完成初始化时,该方法将被调用。

(2) 修改配置文件 applicationContext.xml,在 applicationContext.xml 文件的< bean >元素中添加配置信息,具体代码如下所示。

```
<bean id="bean01" class="com.qfedu.bean.Bean01" init-method="init">
    <property name="bid" value="1"></property>
    <property name="name" value="xiaoqian"></property>
</bean>
```

以上代码中通过< bean >元素的 init-method 属性指定 Bean 完成初始化时要调用 init()方法。

(3) 在 com.qfedu.test 包下新建测试类 TestSpring08,如例 7-16 所示。

【例 7-16】 TestSpring08.Java

```
package com.qfedu.test;
import org.springframework.context.ApplicationContext;
import org.springframework.context.support
    .ClassPathXmlApplicationContext;
public class TestSpring08 {
    public static void main(String[] args) throws Exception {
        ApplicationContext applicationContext = new
            ClassPathXmlApplicationContext("applicationContext.xml");
    }
}
```

执行测试类 TestSpring08,执行结果如图 7.10 所示。

从以上执行结果可以看出,当 Spring 容器启动且 Bean 完成初始化时,程序自动调用了

```
Console
<terminated> TestSpring08 [Java Application] C:\Program Files\Java\jre1.8.0_161\bin\javaw.exe (2018年8月28日 上午10:45:53)
八月 28, 2018 10:45:53 上午 org.springframework.beans.factory.xml.XmlBeanDefinitionReade
信息: Loading XML bean definitions from class path resource [applicationContext.xml]
Bean的初始化完成，调用init()方法
Bean01 [bid=1, name=xiaoqian]
```

图 7.10　执行 TestSpring08 类的结果

Bean01 类中的 init()方法。

除了 init-method 属性之外，还可以通过实现 initializingBean 接口的方式实现以上功能，initializingBean 接口中定义了一个 afterPropertiesSet()方法，如果某个 Bean 实现了 initializingBean 接口，那么该 Bean 在完成初始化时，它的 afterPropertiesSet()方法将被执行。

接下来通过一个实例演示 initializingBean 接口的使用。

（1）在 com. qfedu. bean 包下新建类 Bean02，如例 7-17 所示。

【例 7-17】　Bean02. Java

```
package com.qfedu.bean;
import org.springframework.beans.factory.InitializingBean;
public class Bean02 implements InitializingBean {
    private String bid;
    private String name;
    public Bean02() {
    }
    public String getBid() {
        return bid;
    }
    public void setBid(String bid) {
        this.bid = bid;
    }
    public String getName() {
        return name;
    }
    public void setName(String name) {
        this.name = name;
    }
    @Override
    public String toString() {
        return "Bean02 [bid=" + bid + ", name=" + name + "]";
    }
    @Override
    public void afterPropertiesSet() throws Exception {
        System.out.println("Bean的初始化完成,调用 afterPropertiesSet()方法");
        System.out.println(this.toString());
    }
}
```

在以上代码中，Bean02 类实现了 InitializingBean 接口并提供了 afterPropertiesSet()方

法，当 Bean02 类完成初始化时，它的 afterPropertiesSet()方法将被调用。

（2）修改配置文件 applicationContext.xml，在 applicationContext.xml 文件的<bean>元素中添加配置信息，具体代码如下所示。

```
<bean id = "bean02" class = "com.qfedu.bean.Bean02">
    <property name = "bid" value = "1"></property>
    <property name = "name" value = "xiaoqian"></property>
</bean>
```

为了避免 Bean01 类干扰执行结果，此时需要删除或注释掉 applicationContext.xml 文件中 Bean01 类的配置信息。

执行测试类 TestSpring08，执行结果如图 7.11 所示。

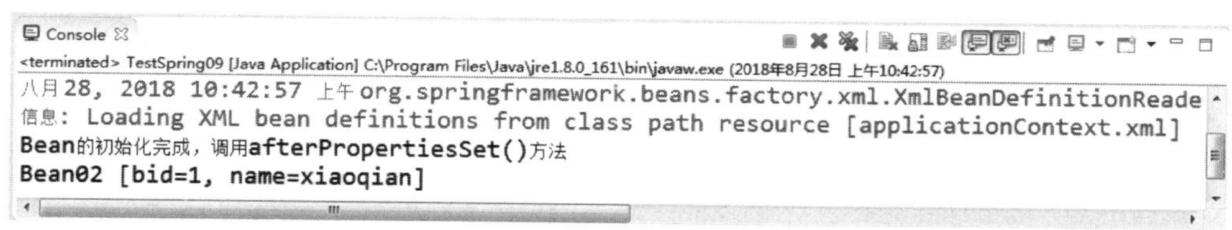

图 7.11　执行 TestSpring08 类的结果

从以上执行结果可以看出，当 Spring 容器启动且 Bean 完成初始化时，程序自动调用了 Bean02 类中的 afterPropertiesSet()方法。

2. Bean 实例销毁之前执行指定方法

如果要在 Bean 实例销毁之前执行指定方法，可通过 destroy-method 属性或实现 DisposableBean 接口的方式。接下来通过一个实例演示 destroy-method 属性的使用，具体步骤如下。

（1）在 com.qfedu.bean 包下新建类 Bean03，如例 7-18 所示。

【例 7-18】　Bean03.Java

```
package com.qfedu.bean;
public class Bean03 {
    private String bid;
    private String name;
    public Bean03() {
    }
    public String getBid() {
        return bid;
    }
    public void setBid(String bid) {
        this.bid = bid;
    }
    public String getName() {
        return name;
    }
    public void setName(String name) {
```

```
        this.name = name;
    }
    @Override
    public String toString() {
        return "Bean03 [bid=" + bid + ", name=" + name + "]";
    }
    public void close(){
        System.out.println("Bean实例即将销毁,调用close()方法");
        System.out.println(this.toString());
    }
}
```

在以上代码中,Bean03类提供了close()方法,如果该方法的名称被指定为配置文件中的destroy-method属性,那么对应的Bean实例在销毁之前,该方法将被调用。

(2) 修改配置文件applicationContext.xml,在applicationContext.xml文件的<bean>元素中添加配置信息,具体代码如下所示。

```
<bean id="bean03" class="com.qfedu.bean.Bean03" destroy-method="close">
    <property name="bid" value="1"></property>
    <property name="name" value="xiaoqian"></property>
</bean>
```

以上代码中通过<bean>元素的destroy-method属性指定Bean实例在销毁之前要调用close()方法。为了避免Bean02类干扰执行结果,此时需要删除或注释掉applicationContext.xml文件中Bean02类的配置信息。

(3) 在com.qfedu.test包下新建测试类TestSpring09,如例7-19所示。

【例7-19】 TestSpring09.Java

```
package com.qfedu.test;
import org.springframework.context.ApplicationContext;
import org.springframework.context.support.*;
public class TestSpring09 {
    public static void main(String[] args) throws Exception {
        ApplicationContext applicationContext = new
            ClassPathXmlApplicationContext("applicationContext.xml");
        //关闭容器,此时Bean实例将被销毁
        AbstractApplicationContext ac = (AbstractApplicationContext)
            applicationContext;
        ac.registerShutdownHook();
    }
}
```

在以上代码中,程序通过AbstractApplicationContext提供的registerShutdownHook()方法关闭Spring容器,此时Bean实例将被销毁。

执行测试类TestSpring09,执行结果如图7.12所示。

从以上执行结果可以看出,当Spring容器关闭且Bean实例将要销毁时,程序自动调用

```
Console ☒
<terminated> TestSpring10 [Java Application] C:\Program Files\Java\jre1.8.0_161\bin\javaw.exe (2018年8月28日 上午11:37:49)
八月 28, 2018 11:37:50 上午 org.springframework.context.support.ClassPathXmlApplicationContext
信息: Closing org.springframework.context.support.ClassPathXmlApplicationContext@20fa23c1:
Bean实例即将销毁，调用close()方法
Bean03 [bid=1, name=xiaoqian]
```

图 7.12　执行 TestSpring09 类的结果

了 Bean03 类中的 close()方法。

除了 destroy-method 属性之外，还可以通过实现 DisposableBean 接口的方式实现以上功能，DisposableBean 接口中定义了一个 destroy() 方法，如果某个 Bean 实现了 DisposableBean 接口，那么这个 Bean 的实例在要被销毁时，它的 destroy()方法将被执行。

接下来通过一个实例演示 DisposableBean 接口的使用。

(1) 在 com.qfedu.bean 包下新建类 Bean04，如例 7-20 所示。

【例 7-20】　Bean04.Java

```java
package com.qfedu.bean;
import org.springframework.beans.factory.DisposableBean;
public class Bean04 implements DisposableBean {
    private String bid;
    private String name;
    public Bean04() {
    }
    public String getBid() {
        return bid;
    }
    public void setBid(String bid) {
        this.bid = bid;
    }
    public String getName() {
        return name;
    }
    public void setName(String name) {
        this.name = name;
    }
    @Override
    public String toString() {
        return "Bean04 [bid=" + bid + ", name=" + name + "]";
    }
    @Override
    public void destroy() throws Exception {
        System.out.println("Bean 实例即将销毁,调用 destroy()方法");
        System.out.println(this.toString());
    }
}
```

在以上代码中，Bean04 类实现了 DisposableBean 接口并提供了 destroy() 方法，当 Bean04 实例在销毁之前，它的 destroy()方法将被调用。

(2) 修改配置文件 applicationContext. xml，在 applicationContext. xml 文件的 beans 元素中添加配置信息，具体代码如下所示。

```
<bean id="bean04" class="com.qfedu.bean.Bean04">
    <property name="bid" value="1"></property>
    <property name="name" value="xiaoqian"></property>
</bean>
```

为了避免 Bean03 类干扰执行结果，此时需要删除或注释 applicationContext. xml 文件中 Bean03 类的配置信息。

执行测试类 TestSpring09，执行结果如图 7.13 所示。

```
Console ☒
<terminated> TestSpring10 [Java Application] C:\Program Files\Java\jre1.8.0_161\bin\javaw.exe (2018年8月28日 上午11:34:23)
八月28, 2018 11:34:24 上午 org.springframework.context.support.ClassPathXmlApplicationContext
信息: Closing org.springframework.context.support.ClassPathXmlApplicationContext@20fa23c1:
Bean实例即将销毁，调用destroy()方法
Bean04 [bid=1, name=xiaoqian]
```

图 7.13　再次执行 TestSpring09 类的结果

从以上执行结果可以看出，当 Spring 容器关闭且 Bean 将要销毁时，程序自动调用了 Bean04 类中的 destroy()方法。

7.3　注　　解

7.3.1　Spring 支持的注解简介

除了 XML 文件，Spring 还支持通过注解实现 Bean 的管理。注解直接写在代码中，使用注解可以减少 XML 文件的配置内容，并且注解能够实现自动装配，提供的功能也更为强大。

Spring 支持的注解相对较多，其中常用的注解如表 7.3 所示。

表 7.3　常用的注解

注　　解	描　　述
@Component	指定一个普通的 Bean
@Controller	指定一个控制器组件 Bean，功能上等同于@Component
@Service	指定一个业务逻辑组件 Bean，功能上等同于@Component
@Repository	指定一个 DAO 组件 Bean，功能上等同于@Component
@Scope	指定 Bean 实例的作用域
@Value	指定 Bean 实例的注入值
@Autowired	指定要自动装配的对象
@Qualifier	指定要自动装配的对象名称，通常与@Autowired 联合使用
@Resource	指定要注入的对象
@PostConstruct	指定 Bean 实例完成初始化后调用的方法
@PreDestroy	指定 Bean 实例销毁之前调用的方法

以上列举出了 Spring 支持的常用注解，开发人员可根据具体需求选择使用。

7.3.2 注解的应用

上个小节介绍了 Spring 注解的基础知识，接下来通过一个实例演示注解的应用，具体步骤如下。

(1) 导入 spring-aop 的 jar 包。将 spring-aop-5.0.8.RELEASE.jar 文件复制到 chapter07 的 lib 目录下，完成导入包。

(2) 引入 Context 约束，修改配置文件 applicationContext.xml 中< beans >元素的代码，修改后的代码如下。

```
1 <?xml version = "1.0" encoding = "UTF - 8"?>
2 < beans xmlns = "http://www.springframework.org/schema/beans"
3     xmlns:xsi = "http://www.w3.org/2001/XMLSchema - instance"
4     xmlns:context = "http://www.springframework.org/schema/context"
5     xmlns:p = "http://www.springframework.org/schema/p"
6     xsi:schemaLocation = "http://www.springframework.org/schema/beans
7         http://www.springframework.org/schema/beans/spring - beans.xsd
8         http://www.springframework.org/schema/context
9         http://www.springframework.org/schema/context/spring - context.xsd">
```

以上代码的第 4 行、第 8 行和第 9 行用于引入 Context 约束。

(3) 配置注解扫描。在 applicationContext.xml 的< beans >元素中添加配置信息，具体代码如下。

```
< context:component - scan base - package = "com.qfedu.bean"/>
```

以上代码指定了程序将扫描 com.qfedu.bean 包中下所有类的注解。

(4) 在 com.qfedu.bean 包下新建类 Stu，如例 7-21 所示。

【例 7-21】 Stu.Java

```
package com.qfedu.bean;
import org.springframework.beans.factory.annotation.Value;
import org.springframework.stereotype.Component;
@Component("stu")
public class Stu {
    @Value("001")
    private int sid;
    @Value("ZhangSan")
    private String name;
    private String age;
    private String course;
    public int getSid() {
        return sid;
    }
    public void setSid(int sid) {
```

```
        this.sid = sid;
    }
    public String getName() {
        return name;
    }
    public void setName(String name) {
        this.name = name;
    }
    public String getAge() {
        return age;
    }
    public void setAge(String age) {
        this.age = age;
    }
    public String getCourse() {
        return course;
    }
    public void setCourse(String course) {
        this.course = course;
    }
    @Override
    public String toString() {
        return "Stu [sid=" + sid + ", name=" + name + ", age=" + age + ", "
                + "course=" + course + "]";
    }
}
```

在以上代码中,@Component 注解将 Stu 类注册为 Spring 容器中的 Bean,@value 注解为 Stu 类的 sid 属性、name 属性注入值。

(5) 在 com.qfedu.bean 包下新建类 Sch,如例 7-22 所示。

【例 7-22】 Sch.Java

```
package com.qfedu.bean;
import javax.annotation.*;
import org.springframework.beans.factory.annotation.*;
import org.springframework.context.annotation.Scope;
import org.springframework.stereotype.Component;
@Component("sch")
@Scope(scopeName="singleton")
public class Sch {
    @Value("005")
    private int sid;
    @Autowired
    @Qualifier("stu")
    private Stu stu;
    public int getSid() {
        return sid;
    }
```

```
    public void setSid(int sid) {
        this.sid = sid;
    }
    public Stu getStu() {
        return stu;
    }
    public void setStu(Stu stu) {
        this.stu = stu;
    }
    @Override
    public String toString() {
        return "Sch [sid=" + sid + ", stu=" + stu + "]";
    }
    @PostConstruct
    public  void init(){
        System.out.println("Bean的初始化完成,调用 init()方法");
    }
    @PreDestroy
        public void destroy(){
        System.out.println("Bean的初始化完成,调用 destroy()方法");
    }
}
```

在以上代码中,@Component 注解将 Sch 类注册为 Spring 容器中的 Bean,@Scope 注解指定 Sch 类的作用域,@Value 注解为 Sch 类的 sid 属性注入值,@Autowired 和 @Qualifier 注解为 Sch 类的 sid 属性注入另外一个 Bean 的实例。@PostConstruct 注解指定 Bean 实例完成初始化后调用的方法,@PreDestroy 注解指定 Bean 实例销毁之前调用的方法。

(6) 在 com.qfedu.bean 包下新建测试类 TestSpring10,如例 7-23 所示。

【例 7-23】 TestSpring10.Java

```
package com.qfedu.test;
import org.springframework.context.ApplicationContext;
import org.springframework.context.support.*;
import com.qfedu.bean.Sch;
public class TestSpring10 {
    public static void main(String[] args) throws Exception  {
        ApplicationContext  applicationContext = new
            ClassPathXmlApplicationContext("applicationContext.xml");
        Sch sch = applicationContext.getBean("sch",Sch.class);
        System.out.println(sch);
        AbstractApplicationContext  ac = (AbstractApplicationContext)
            applicationContext;
        ac.registerShutdownHook();
    }
}
```

执行 TestSpring10 类,执行结果如图 7.14 所示。

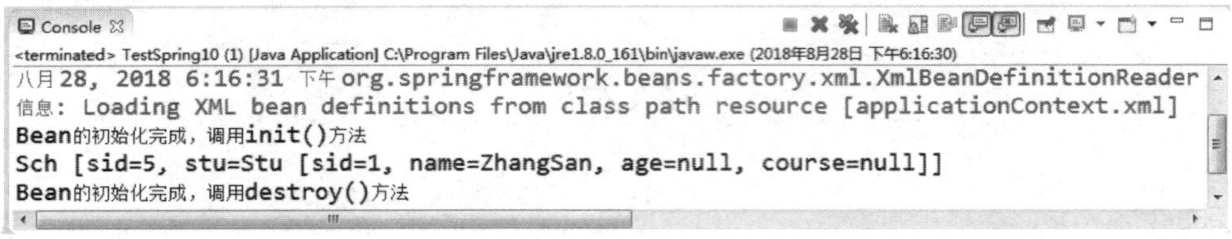

```
<terminated> TestSpring10 (1) [Java Application] C:\Program Files\Java\jre1.8.0_161\bin\javaw.exe (2018年8月28日 下午6:16:30)
八月28, 2018 6:16:31 下午 org.springframework.beans.factory.xml.XmlBeanDefinitionReader
信息: Loading XML bean definitions from class path resource [applicationContext.xml]
Bean的初始化完成，调用init()方法
Sch [sid=5, stu=Stu [sid=1, name=ZhangSan, age=null, course=null]]
Bean的初始化完成，调用destroy()方法
```

图 7.14　执行 TestSpring10 类的结果

从以上执行结果可以看出，程序输出了 Sch 类对象的信息，由此可见，使用注解可以实现和采用 XML 文件形式相同的效果。

7.4　本章小结

本章首先介绍了 Spring 的 IOC 和 DI 的概念，然后通过实例演示了依赖注入的方式，接着讲解了 Spring 中 Bean 的配置、作用域、生命周期等，最后详细讲解了 Spring 支持的注解。通过本章知识的学习，大家应该能理解 Spring 的依赖注入，掌握 Bean 的定义和配置，掌握 Spring 中注解的使用，能够使用 Spring 编写简单的应用程序。

7.5　习　　题

1. 填空题

(1) 依赖注入主要有两种方式，它们分别是________和________。

(2) 在 Spring 配置文件< bean >元素的属性中，________属性指定 Bean 对应类的全路径。

(3) 在 Spring 配置文件< bean >元素的属性中，________属性指定 Bean 对象的作用域。

(4) 在默认情况下，Spring 中 Bean 实例的作用域是________。

(5) 当需要在 Bean 实例初始化后执行指定行为时，可通过实现________接口来完成。

2. 选择题

(1) 下列选项中哪个不是依赖注入的注入方式(　　)。

A. 构造器注入

B. 属性注入

C. 接口注入

D. 命名空间注入

(2) 下列关于< bean >元素属性的描述中错误的是(　　)。

A. class：指定 Bean 对应类的全路径

B. name：指定 Bean 对应对象的一个标识

C. scope：指定 Bean 对象的作用域

D. init-method：是否延时加载，默认值为 false

(3) 下列关于 P：命名空间的说法中正确的是(　　)。

A. P:命名空间也是通过属性注入实现的

B. 引入了 P:命名空间不能够有效地解决元素冗余问题

C. 程序不能使用 P:命名空间实现值的注入

D. 以上说法都正确

(4) 下列关于 Bean 的生命周期说法中不正确的是(　　)。

A. Spring 容器可以管理作用域为 singleton 的 Bean 的生命周期

B. 通过 init-method 属性可以在 Bean 实例销毁之前执行指定方法

C. Bean 的生命周期是指 Bean 实例被创建、初始化和销毁的过程

D. 作用域为 prototype 的 Bean,Spring 容器只创建实例而不跟踪其生命周期

(5) 下列哪个注解可以指定要自动装配的对象(　　)。

A. @Autowired　　B. @Scope

C. @Repository　　D. @PostConstruct

3. 思考题

(1) 简述 Spring 中 Bean 的作用域。

(2) 简述 Spring 中 Bean 的生命周期。

4. 编程题

编写一个 Person 类和一个 Address 类,Person 类中封装有成员变量 name、age,Address 类中封装有成员变量 name、address。将 Person 类和 Address 类配置到 Spring 中,通过 Spring 的配置文件将 Address 对象注入到 Person 中,然后在 PersonServiceImpl 中,打印出 name、age 和 address 的内容。

第8章 Spring 的 AOP

本章学习目标

- 理解 AOP 的概念及基本术语
- 理解 Spring AOP 的实现机制
- 掌握 Spring AOP 的开发方法
- 掌握多个切面的优先级
- 掌握 Spring AOP 的应用

除了 IOC 之外,AOP 是 Spring 框架的另外一个重要特性。AOP 意为面向切面编程,它是对 OOP(面向对象编程)的完善和补充,已成为一种较为成熟的编程方式,它通过横向抽取的方式降低程序中的代码冗余和功能耦合。在实际开发中,事务管理、日志记录、性能监控、权限检查等功能通常会采用 AOP 的方式实现。接下来,本章将对 Spring AOP 涉及的相关知识进行详细讲解。

8.1 AOP 基础

8.1.1 AOP 简介

AOP 是 Aspect Oriented Programming(面向切面编程)的缩写,和 OOP 不同,它主张将程序中的相同业务逻辑进行横向隔离,并将重复的业务逻辑抽取到一个独立的模块中,最终实现提升程序可复用性和开发效率的目的。

在传统的 OOP 编程中,借助于面向对象的分析和设计,程序的功能通过对象与对象之间的协作来实现。OOP 引入抽象、封装、继承等概念,将具有相同属性或行为的对象纳入一个层次分明的类结构体系中,由于类可以继承,因此这种体系是纵向的。

随着软件规模的不断扩大,系统中出现了一些 OOP 难以彻底解决的问题。例如,系统的某个类中有若干个方法都包含事务管理的业务逻辑,如图 8.1 所示。

从图 8.1 中可以看出,查询用户信息、修改用户信息、删除用户信息的方法体中都包含事务管理的业务逻辑,这会带来一定数量的重复代码并使程序的维护成本增加。基于横向抽取机制,AOP 为此类问题提供了完美的解决方案,它将事务管理的业务逻辑从这三个方法体中抽取到一个可重用的模块,进而降低耦合,减少重复代码。

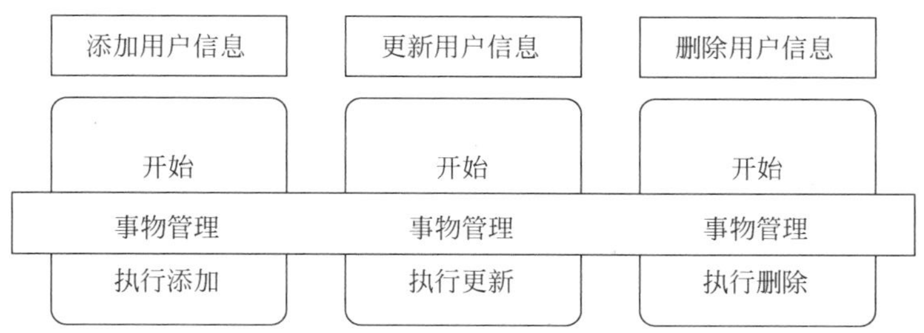

图 8.1 操作用户信息

8.1.2 AOP 的基本术语

上个小节讲解了 AOP 的基本概念，接下来对 AOP 涉及的基本术语进行详细讲解。

1. 连接点(Joinpoint)

程序执行过程中某个特定的节点，例如，某个类的初始化完成后、某个方法执行之前、程序处理异常时等。广义上讲，一个类或一段程序代码拥有一些具有边界性质的特定点都可以被作为连接点，但由于 Spring 仅仅支持方法连接点，因此，在 Spring AOP 中，一个连接点是指与方法执行相关的特定节点。

2. 通知(Advice)

通知是在目标类连接点上执行的一段代码，包括 around、before 和 after 等不同类型。在 Spring AOP 中，它主要描述围绕方法调用而注入的行为，相比之下，功能更加细化。Spring AOP 提供的通知类型，具体如表 8.1 所示。

表 8.1 Spring AOP 提供的通知类型

通 知 类 型	描 述
前置通知(Before)	在目标方法被调用之前调用通知
后置通知(After)	在目标方法被调用之后调用通知
返回通知(After-returning)	在目标方法成功执行之后调用通知
异常通知(After-throwing)	在目标方法抛出异常之后调用通知
环绕通知(Around)	通知包裹了被通知的方法，在被通知的方法调用前和调用后执行自定义的行为

表 8.1 列举了 Spring AOP 提供的通知类型，关于这些通知类型的使用方法，本书在后文中会有讲解，此处不再赘述。

3. 切点(Pointcut)

匹配连接点的断言，AOP 通过切点来定位特定的连接点。通知和一个切入点表达式关联，并在满足这个切入点的连接点上运行(例如，当执行某个特定名称的方法时)。切入点表达式如何和连接点匹配是 AOP 的核心。

4. 目标对象(Target)

通知所作用的目标业务类。如果缺少 AOP 的支持，那么目标业务类就要独立完成所有的业务逻辑，为了降低冗余，目标业务类可以借助 AOP 将重复代码抽取出来。

5. 引介(Introduction)

引介是一种特殊的通知,它为类添加一些属性和方法。如此一来,即使一个业务类原本没有实现某一个接口,通过 AOP 的引介功能,也可以动态地为该业务类添加接口的实现逻辑,让业务类成为这个接口的实现类。

6. 切面(Aspect)

对系统中的横切关注点逻辑进行模块化封装的 AOP 概念实体。关注点模块化之后,可能会横切多个对象。Spring AOP 是实施切面的具体方法,它将切面所定义的横切逻辑添加到切面所指定的连接点中。

7. 织入(Weaving)

织入是将通知添加到目标类具体连接点的过程,这些可以在编译时、类加载时和运行时完成,Spring 采用动态代理织入,而 AspectJ 采用编译期织入和类装载器织入。

8. 代理(Proxy)

目标类被 AOP 织入增强后产生的一个结果类,这个结果类融合了原类和增强的逻辑,根据不同的代理方式,代理类既可能是和原类具有相同接口的类,也可能就是原类的子类,所以可以采用与调用原类相同的方法调用代理类。

8.2 Spring AOP 的实现机制

8.2.1 JDK 动态代理

在 Spring 中,AOP 代理由 Spring 的 IOC 容器负责创建,其依赖关系也将由 IOC 容器负责管理,因此 Spring AOP 可以直接将 IOC 容器中的其他 Bean 实例作为目标对象。在默认情况下,Spring AOP 使用 JDK 动态代理,当目标对象是一个类并且这个类没有实现接口时,Spring 会切换为使用 CGLib 代理。

JDK 动态代理主要涉及两个 API:InvocationHandler 和 Proxy,它们位于 java. lang. reflect 包中,其中,InvocationHandler 是一个接口,代理类可以通过实现该接口定义横切逻辑,并将横切逻辑和业务逻辑编织在一起;Proxy 利用 InvocationHandler 动态生成目标类的代理对象。

接下来,本节通过一个案例演示 JDK 动态代理技术的代码实现。

(1) 在 Eclipse 中新建 Web 工程 chapter08,在 chapter08 的 src 目录下新建包 com. qfedu. demo01,在 com. qfedu. demo01 包下创建接口 LoginService,具体代码如例 8-1 所示。

【例 8-1】 LoginService. Java

```
package com.qfedu.demo01;
public interface LoginService {
    public void login();
}
```

(2) 在 com. qfedu. demo01 包下创建接口 LoginService 的实现类 LoginServiceImpl,具体代码如例 8-2 所示。

【例 8-2】 **LoginServiceImpl. Java**

```
package com.qfedu.demo01;
public class LoginServiceImpl implements LoginService {
    @Override
    public void login() {
        System.out.println("执行 login()方法");
    }
}
```

(3) 在 com. qfedu. demo01 包下创建类 PerformHandler,具体代码如例 8-3 所示。

【例 8-3】 **PerformHandler. Java**

```
1  package com.qfedu.demo01;
2  import java.lang.reflect.InvocationHandler;
3  import java.lang.reflect.Method;
4  public class PerformHandler implements InvocationHandler {
5      //目标对象
6      private Object target;
7      public PerformHandler(Object target){
8          this.target = target;
9      }
10     @Override
11     public Object invoke(Object proxy, Method method, Object[] args) throws
12         Throwable {
13         //增强的方法
14         System.out.println("方法开始执行");
15         //执行被代理类的原方法
16         Object invoke = method.invoke(target, args);
17         //增强的方法
18         System.out.println("方法执行完毕");
19         return invoke;
20     }
21 }
```

在以上代码中,程序通过构造方法传入被代理的目标对象,第 14 行和第 17 行的输出是需要增强的横切逻辑,第 16 行的 method. invoke()通过 Java 的反射机制间接调用被代理类的原方法。

(4) 在 com. qfedu. demo01 包下创建测试类 TestPerformHandler,具体代码如例 8-4 所示。

【例 8-4】 **TestPerformHandler. Java**

```
package com.qfedu.demo01;
import java.lang.reflect.Proxy;
public class TestPerformHandler {
    public static void main(String[] args) {
        LoginService loginService = new LoginServiceImpl();
```

```
        PerformHandler performHandler = new PerformHandler(loginService);
        //创建代理对象
        loginService = (LoginService)Proxy.newProxyInstance
            (loginService.getClass().getClassLoader(),
        loginService.getClass().getInterfaces(),performHandler);
        loginService.login();
    }
}
```

在以上代码中，程序通过 Proxy 类生成了代理对象，程序可以通过代理对象调用目标对象的方法。

执行 TestPerformHandler 类，执行结果如图 8.2 所示。

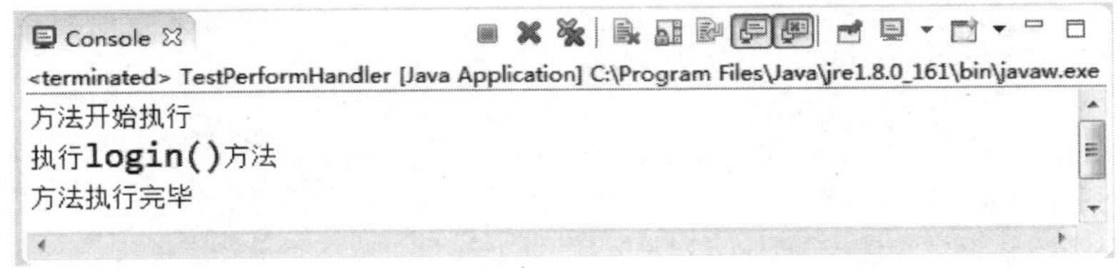

图 8.2　执行 TestPerformHandler 类的结果

从图 8.2 中可以看出，当执行代理对象的 login()方法时，原接口中的 login()方法和增强的方法都被执行。

8.2.2　CGLib 动态代理

JDK 动态代理存在缺陷，它只能为接口创建代理实例，当需要为类创建代理实例时，就要使用 CGLib 动态代理。CGLib 动态代理不要求目标对象实现接口，它采用底层的字节码技术，通过继承的方式动态创建代理对象。

CGLib 动态代理依靠 Enhancer 类创建代理实例，依靠 MethodInterceptor 接口织入要增强的方法，接下来，本节通过一个案例演示 CGLib 动态代理技术的代码实现。

(1) 在工程 chapter08 的 lib 目录下导入所需 jar 包，要导入的 jar 包如图 8.3 所示。

```
WebContent
  META-INF
  WEB-INF
    lib
      aopalliance-1.0.jar
      aspectjweaver-1.7.4.jar
      commons-logging-1.2.jar
      spring-aop-5.0.8.RELEASE.jar
      spring-aspects-5.0.8.RELEASE.jar
      spring-beans-5.0.8.RELEASE.jar
      spring-context-5.0.8.RELEASE.jar
      spring-core-5.0.8.RELEASE.jar
      spring-expression-5.0.8.RELEASE.jar
```

图 8.3　导入的 jar 包

(2) 在 com.qfedu.demo01 包下创建类 CglibProxy，具体代码如例 8-5 所示。

【例 8-5】　CglibProxy.Java

```
package com.qfedu.demo01;
import java.lang.reflect.Method;
import org.springframework.cglib.proxy.*;
public class CglibProxy implements MethodInterceptor{
```

```java
    private Enhancer enhancer = new Enhancer();
    //生成代理对象的方法
    public Object getProxy(Class clazz){
        enhancer.setSuperclass(clazz);
        enhancer.setCallback(this);
        return enhancer.create();
    }
    //回调方法
    @Override
    public Object intercept(Object obj, Method method,Object[]
        objects,MethodProxy methodProxy) throws Throwable {
        System.out.println("CGLig 代理之前");
        Object invoke = methodProxy.invokeSuper(obj,objects);
        System.out.println("CGLig 代理之后");
        return invoke;
    }
}
```

在以上代码中,CglibProxy 类提供了一个生成代理对象的方法 getProxy(),除此之外,CglibProxy 类实现了 MethodInterceptor 接口并提供了一个 intercept()方法,该方法将拦截目标类中所有方法的调用。

(3) 在 com.qfedu.demo01 包下创建类 TestCGLlib,具体代码如例 8-6 所示。

【例 8-6】 TestCGLlib.Java

```java
package com.qfedu.demo01;
public class TestCGLlib {
    public static void main(String[] args) {
        CglibProxy cglibProxy = new CglibProxy();
        //创建代理对象
        LoginServiceImpl userService = (LoginServiceImpl)
            cglibProxy.getProxy(LoginServiceImpl.class);
        userService.login();
    }
}
```

在以上代码中,程序通过 TestCGLlib 类生成了代理对象,程序可以通过代理对象调用目标对象的方法。

执行 TestCGLlib 类,执行结果如图 8.4 所示。

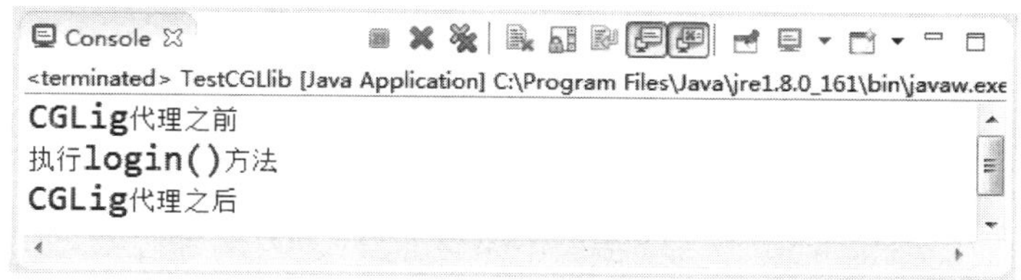

图 8.4 执行 TestCGLib 类的结果

从图 8.4 中可以看出，与 JDK 动态代理的结果相同，当执行代理对象的 login()方法时，原接口中的 login()方法和增强的方法都被执行。

8.3 Spring AOP 的开发方法

8.3.1 基于 XML 开发 Spring AOP

上个小节介绍了 Spring AOP 的实现机制，接下来讲解 Spring AOP 的开发方法。Spring AOP 的常用开发方法有两种，它们分别是基于 XML 文件和基于注解，本节首先对基于 XML 的 Spring AOP 开发做详细讲解。

由于 Spring AOP 中的代理对象由 IOC 容器自动生成，因此，开发者无须过多关注代理过程的实现过程，只需完成选择连接点、创建切面、定义切点并在 XML 文件中添加配置信息即可。

Spring 提供了一系列配置 AOP 的 XML 元素，具体如表 8.2 所示。

表 8.2　XML 配置 Spring AOP 的元素

元素名称	描　述
<aop:config>	AOP 配置的根元素
<aop:aspect>	指定切面
<aop:advisor>	指定通知器
<aop:pointcut>	指定切点
<aop:before>	指定前置通知
<aop:after>	指定后置通知
<aop:around>	指定环绕方式
<aop:after-returning>	指定返回通知
<aop:after-throwing>	指定异常通知

表 8.2 列举了 XML 配置文件中有关 Spring AOP 的元素，关于这些元素的具体使用，本书在后文中有详细讲解。

接下来通过一个案例演示通过 XML 方式开发 Spring AOP。

(1) 在 chapter08 的 src 目录下创建 applicationContext.xml，此时引入 AOP 的命名空间，具体代码如下所示。

```
<?xml version = "1.0" encoding = "UTF - 8"?>
<beans xmlns = "http://www.springframework.org/schema/beans"
    xmlns:xsi = "http://www.w3.org/2001/XMLSchema - instance"
    xmlns:context = "http://www.springframework.org/schema/context"
    xmlns:aop = "http://www.springframework.org/schema/aop"
    xsi:schemaLocation = "http://www.springframework.org/schema/beans
    http://www.springframework.org/schema/beans/spring - beans.xsd
    http://www.springframework.org/schema/context
    http://www.springframework.org/schema/context/spring - context.xsd
    http://www.springframework.org/schema/aop
```

```
11        http://www.springframework.org/schema/aop/spring-aop.xsd">
12 </beans>
```

在以上代码中,第 5 行、第 10 行和第 11 行用于引入 AOP 的命名空间。

(2) 在 chapter08 的 src 目录下创建 com. qfedu. demo02 包,在该包下创建接口 UserService,具体代码如例 8-7 所示。

【例 8-7】 UserService. Java

```
package com.qfedu.demo02;
public interface UserService {
    void insert();
    void delete();
    void update();
    void select();
}
```

(3) 在 com. qfedu. demo02 包下创建 UserService 接口的实现类 UserServiceImpl,具体代码如例 8-8 所示。

【例 8-8】 UserServiceImpl. Java

```
package com.qfedu.demo02;
public class UserServiceImpl implements UserService {
    @Override
    public void insert() {
        System.out.println("添加用户信息");
    }
    @Override
    public void delete() {
        System.out.println("删除用户信息");
    }
    @Override
    public void update() {
        System.out.println("更新用户信息");
    }
    @Override
    public void select() {
        System.out.println("查询用户信息");
    }
}
```

(4) 在 com. qfedu. demo02 包下创建类 XmlAdvice,该类用于定义通知,具体代码如例 8-9 所示。

【例 8-9】 XmlAdvice. Java

```
package com.qfedu.demo02;
import org.aspectj.lang.ProceedingJoinPoint;
public class XmlAdvice {
```

```
    //前置通知
    public void before(){
        System.out.println("这是前置通知");
    }
    //后置通知
    public void afterReturning(){
        System.out.println("这是后置通知(方法不出现异常时调用)");
    }
    //环绕通知
    public Object around(ProceedingJoinPoint point) throws Throwable {
        System.out.println("这是环绕通知之前的部分!!");
        //调用目标方法
        Object object = point.proceed();
        System.out.println("这是环绕通知之后的部分!!");
        return object;
    }
    //异常通知
    public void afterException(){
        System.out.println("异常通知!!");
    }
    //后置通知
    public void after(){
        System.out.println("这是后置通知!!");
    }
}
```

以上代码中分别定义了五种通知,这些通知可以被植入对应的切点,进而增强目标对象的方法。

(5) 在 applicationContext.xml 文件的< bean >元素中添加 Spring AOP 的配置信息,具体代码如下所示。

```
<!-- 注册 Bean -->
<bean name = "userService" class = "com.qfedu.demo02.UserServiceImpl"/>
<bean name = "xmlAdvice"  class = "com.qfedu.demo02.XmlAdvice"/>
<!-- 配置 Spring AOP -->
<aop:config>
        <!-- 指定切点   -->
      <aop:pointcut  id = "pointcut"
          expression = "execution( * com.qfedu.demo02.UserServiceImpl.*(..))"/>
        <!-- 指定切面   -->
          <aop:aspect ref = "xmlAdvice">
              <!-- 指定前置通知 -->
              <aop:before method = "before" pointcut-ref = "pointcut"/>
              <!-- 指定返回通知   -->
              <aop:after-returning method = "afterReturning"
                    pointcut-ref = "pointcut"/>
              <!-- 指定环绕方式 -->
              <aop:around method = "around" pointcut-ref = "pointcut"/>
```

```
                <! -- 指定异常通知  -->
                <aop:after - throwing  method = "afterException"
                       pointcut - ref = "pointcut"/>
                <! -- 指定后置通知  -->
                <aop:after method = "after" pointcut - ref = "pointcut"/>
            </aop:aspect>
</aop:config>
```

在以上代码中，<aop:pointcut>元素指定了一个切点，<aop:aspect>元素指定了一个切面，<aop:aspect>元素内部可以指定多个通知，在<aop:before>、<aop:after-returning>、<aop:around>、<aop:after-throwing>、<aop:after>元素中，method 属性指定通知的方法，pointcut-ref 属性指定要匹配的切点。

(6) 在 com. qfedu. demo02 包下创建测试类 TestXml，具体代码如例 8-10 所示。

【例 8-10】 TestXml. Java

```
package com.qfedu.demo02;
import org.springframework.context.ApplicationContext;
import org.springframework.context.support
    .ClassPathXmlApplicationContext;
public class TestXml {
    public static void main(String[] args) {
        ApplicationContext context = new
             ClassPathXmlApplicationContext("applicationContext.xml");
        UserService userService = context.getBean
              ("userService",UserService.class);
        userService.delete();
    }
}
```

执行 TestXml 类，执行结果如图 8.5 所示。

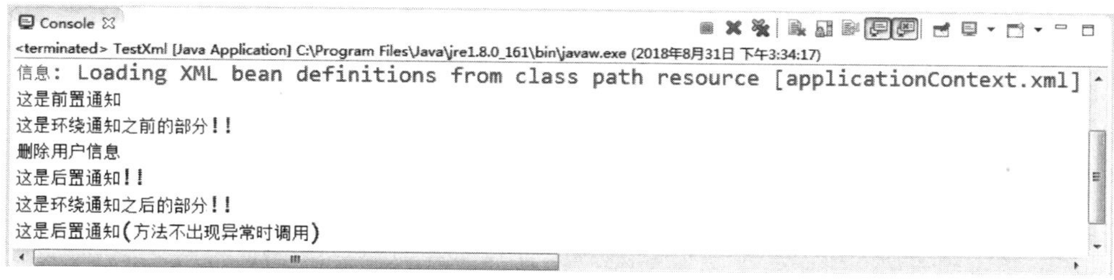

图 8.5　执行 TestXml 类的结果

从图 8.5 中可以看出，程序执行了 XmlAdvice 类中要增强的方法，由此可见，Spring AOP 实现了对目标对象的方法增强。

8.3.2　基于注解开发 Spring AOP

上个小节介绍了基于 XML 形式的 Spring AOP 开发，接下来讲解基于注解完成 Spring AOP 开发的具体方法。相比 XML 的配置方式，基于注解的 Spring AOP 开发是将配置信息编写到代码中。

为了便于开发，Spring AOP 支持一系列注解的使用，具体如表 8.3 所示。

表 8.3 Spring AOP 支持的注解

注解名称	描　述
@Aspect	指定切面
@Pointcut	指定切点
@Before	指定前置通知
@After	指定后置通知
@Around	指定环绕方式
@AfterReturning	指定返回通知
@AfterThrowing	指定异常通知

表 8.3 列举了 Spring AOP 支持的注解，开发人员可根据具体情况选择使用。

接下来通过注解的方式实现以上案例。

(1) 在 com.qfedu.demo 包下新建 AnnoAdvice 类，修改后的代码如例 8-11 所示。

【例 8-11】 AnnoAdvice.Java

```
package com.qfedu.demo02;
import org.aspectj.lang.ProceedingJoinPoint;
import org.aspectj.lang.annotation.*;
@Aspect
public class AnnoAdvice {
    //切点
    @Pointcut("execution(* com.qfedu.demo02.UserServiceImpl.*(..))")
    public void pointcut(){
    }
    //前置通知
    @Before("pointcut()")
    public  void before(){
        System.out.println("这是前置通知");
    }
    //后置通知
    @AfterReturning("pointcut()")
    public  void afterReturning(){
        System.out.println("这是后置通知(方法不出现异常时调用)");
    }
    //环绕通知
    @Around("pointcut()")
    public  Object around(ProceedingJoinPoint point) throws Throwable {
        System.out.println("这是环绕通知之前的部分!!");
        //调用目标方法
        Object object = point.proceed();
        System.out.println("这是环绕通知之后的部分!!");
        return object;
    }
    //异常通知
```

```
    @AfterThrowing("pointcut()")
    public void afterException(){
        System.out.println("异常通知!!");
    }
    //后置通知
    @After("pointcut()")
    public void after(){
        System.out.println("这是后置通知!!");
    }
}
```

在以上代码中，通过@Pointcut 注解指定了一个切点，然后通过其他注解指定了相应的通知，这些通知将织入到目标代理对象的相应方法中。

（2）修改配置文件 applicationContext. xml，将< bean >元素中的的配置代码替换为如下所示。

```
<!-- 注册 Bean -->
<bean name = "userService" class = "com.qfedu.demo02.UserServiceImpl"/>
<bean name = "annoAdvice" class = "com.qfedu.demo02.AnnoAdvice"/>
<!-- 开启@aspectj 的自动代理支持 -->
<aop:aspectj-autoproxy/>
```

以上代码将 UserServiceImpl 和 AnnoAdvice 注册为 Spring 容器中的 Bean，并且开启了@ aspectj 的自动代理支持。

（3）在 com. qfedu. demo02 包下新建测试类 TestAnnotation，具体代码如例 8-12 所示。

【例 8-12】 TestAnnotation. Java

```
package com.qfedu.demo02;
import org.springframework.context.ApplicationContext;
import org.springframework.context.support
    .ClassPathXmlApplicationContext;
public class TestAnnotation {
    public static void main(String[] args) {
        ApplicationContext context = new
            ClassPathXmlApplicationContext("applicationContext.xml");
         UserService userService =
             context.getBean("userService",UserService.class);
        userService.insert();
    }
}
```

执行 TestAnnotation 类，执行结果如图 8.6 所示。

从图 8.6 中可以看出，程序执行了 AnnoAdvice 类中要增强的方法，由此可见，采用注解方式实现了和采用 XML 配置文件方式同样的效果。

```
Console
<terminated> TestAnnotation (1) [Java Application] C:\Program Files\Java\jre1.8.0_161\bin\javaw.exe (2018年9月6日 下午8:25:26)
信息: Loading XML bean definitions from class path resource [applicationContext.xml]
这是环绕通知之前的部分!!
这是前置通知
添加用户信息
这是环绕通知之后的部分!!
这是后置通知!!
这是后置通知(方法不出现异常时调用)
```

图 8.6　执行 TestAnnotation 类的结果

8.4　多个切面的优先级

8.4.1　基于注解配置

上个小节介绍了基于注解形式的 Spring AOP 开发，实际开发中可能需要在多个切面中定义多个通知织入同一个切点，这就要考虑多个切面的优先级。切面的优先级有三种配置方式，可以通过 XML 文件、@Order 注解或切面类实现 Ordered 接口的形式来实现。优先级越高的切面，它的通知方法优先执行并最后退出。

接下来演示通过注解的形式实现 Aspect 的优先级，具体步骤如下。

(1) 在 com.qfedu.demo02 包下新建类 Aspect01，具体代码如例 8-13 所示。

【例 8-13】　Aspect01.Java

```
package com.qfedu.demo02;
import org.aspectj.lang.annotation.*;
import org.springframework.core.annotation.Order;
@Aspect
@Order(1)
public class Aspect01{
    //切点
    @Pointcut("execution(* com.qfedu.demo02.UserServiceImpl.*(..))")
    public void pointcut(){
    }
    //前置通知
    @Before("pointcut()")
    public  void before(){
        System.out.println("这是 Aspect01 的前置通知!!");
    }
    //后置通知
    @After("pointcut()")
    public void after(){
        System.out.println("这是 Aspect01 的后置通知!!");
    }
}
```

在以上代码中，@Aspect 注解指定了切面，@Order 注解指定了优先级，其中值越小优先级越高，此处为 1，@Pointcut 注解指定了一个切点，@Before 注解指定了前置通知，@After 注解指定了后置通知。

(2) 在 com.qfedu.demo02 包下新建类 Aspect02，具体代码如例 8-14 所示。

【例 8-14】 Aspect02.Java

```
package com.qfedu.demo02;
import org.aspectj.lang.annotation.*;
import org.springframework.core.Ordered;
import org.springframework.core.annotation.Order;
@Aspect
@Order(0)
public class Aspect02 {
    //切点
    @Pointcut("execution(* com.qfedu.demo02.UserServiceImpl.*(..))")
    public void pointcut(){
    }
    //前置通知
    @Before("pointcut()")
    public  void before(){
        System.out.println("这是 Aspect02 的前置通知!!");
    }
    //后置通知
    @After("pointcut()")
    public void after(){
        System.out.println("这是 Aspect02 的后置通知!!");
    }
}
```

在以上代码中，@Order 注解指定了优先级，其中值越小优先级越高，此处为 0，显而易见，Aspect02 的优先级高于 Aspect01。

(3) 修改配置文件 applicationContext.xml，注释掉 AnnoAdvice 类的配置信息，将 Aspect01 和 Aspect02 的配置信息添加到<bean>元素中，修改后的代码如下所示。

```
<!-- 注册 Bean -->
<bean name="userService" class="com.qfedu.demo02.UserServiceImpl"/>
<!-- <bean name="annoAdvice"  class="com.qfedu.demo02.AnnoAdvice"/> -->
<bean name="aspect01"  class="com.qfedu.demo02.Aspect01"/>
<bean name="aspect02"  class="com.qfedu.demo02.Aspect02"/>
<!-- 扫描包设置 -->
<aop:aspectj-autoproxy/>
```

以上代码将 Aspect01 类和 Aspect02 类注册为 Spring 容器中的 Bean，并且开启了@aspectj 的自动代理支持。

(4) 在 com.qfedu.demo02 包下新建一个类 TestAspect，具体代码如例 8-15 所示。

【例 8-15】 TestAspect. Java

```
package com.qfedu.demo02;
import org.springframework.context.ApplicationContext;
import org.springframework.context.support
     .ClassPathXmlApplicationContext;
public class TestAspect {
    public static void main(String[] args) {
        ApplicationContext context = new
            ClassPathXmlApplicationContext("applicationContext.xml");
        UserService userService =
            context.getBean("userService",UserService.class);
        userService.select();
    }
}
```

执行 TestAspect 类,执行结果如图 8.7 所示。

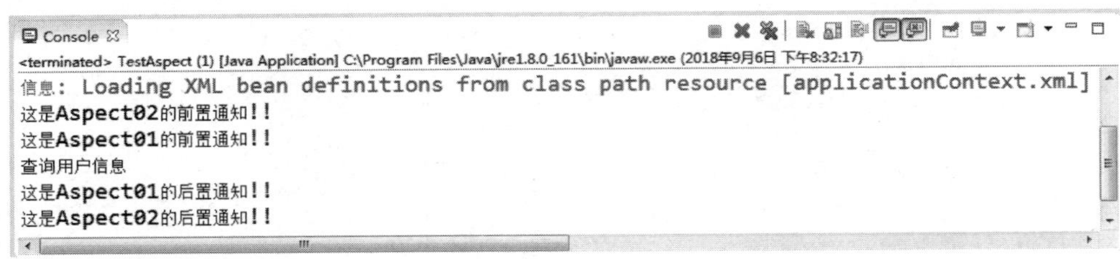

图 8.7 执行 TestAspect 类的结果

从图 8.7 可以看出,程序依次执行了 Aspect02 类和 Aspect01 类中的多个方法,通过分析发现,优先级高的切面,它包含的通知先执行后退出。

修改 Aspect02 类中@Order 注解的值,将该值修改为 2,此时,Aspect02 的优先级低于 Aspect01,重新执行 TestAspect 类,执行结果如图 8.8 所示。

```
Console
<terminated> TestAspect (1) [Java Application] C:\Program Files\Java\jre1.8.0_161\bin\javaw.exe (2018年9月6日 下午8:33:29)
信息: Loading XML bean definitions from class path resource [applicationContext.xml]
这是Aspect01的前置通知!!
这是Aspect02的前置通知!!
查询用户信息
这是Aspect02的后置通知!!
这是Aspect01的后置通知!!
```

图 8.8 重新执行 TestAspect 类的结果

从图 8.8 可以看出,Aspect01 类中的方法先执行后退出,由此可见,@Order 注解实现了切面的优先级配置。

8.4.2 基于 Ordered 接口配置

除了以注解方式配置切面优先级之外,还可以通过实现 Ordered 接口的方式配置切面优先级。Ordered 接口位于 org.springframework.core 包中,它提供了一个 getOrder()方法，该方法的返回值可以指定当前切面类的优先级。接下来演示使用 Ordered 接口配置切面优先级，具体步骤如下。

(1) 在 com.qfedu.demo02 包下新建类 Aspect03，具体代码如例 8-16 所示。

【例 8-16】 Aspect03.Java

```
package com.qfedu.demo02;
import org.aspectj.lang.annotation.*;
import org.springframework.core.Ordered;
@Aspect
public class Aspect03 implements Ordered{
    @Pointcut("execution(* com.qfedu.demo02.UserServiceImpl.*(..))")
    public void pointcut(){
    }
    @Before("pointcut()")
    public void before(){
        System.out.println("这是 Aspect03 的前置通知!!");
    }
    @After("pointcut()")
    public void after(){
        System.out.println("这是 Aspect03 的后置通知!!");
    }
    //该方法的返回值指定切面优先级
    @Override
    public int getOrder() {
        return 1;
    }
}
```

在以上代码中，Aspect03 类实现了 Ordered 接口并在 getOrder()方法中返回数值 1，这相当于使用注解@Order(1)。

(2) 在 com.qfedu.demo02 包下新建类 Aspect04，具体代码如例 8-17 所示。

【例 8-17】 Aspect04.Java

```
package com.qfedu.demo02;
import org.aspectj.lang.annotation.*;
import org.springframework.core.Ordered;
@Aspect
public class Aspect04 implements Ordered{
    @Pointcut("execution(* com.qfedu.demo02.UserServiceImpl.*(..))")
    public void pointcut(){
    }
    @Before("pointcut()")
    public void before(){
        System.out.println("这是 Aspect04 的前置通知!!");
    }
    @After("pointcut()")
    public void after(){
```

```
            System.out.println("这是 Aspect04 的后置通知!!");
        }
        //该方法的返回值指定切面优先级
        @Override
        public int getOrder() {
            return 0;
        }
}
```

在以上代码中，getOrder()方法返回的数值为 0，由于数值越低优先级越高，因此 Aspect04 的优先级高于 Aspect03 的优先级。

(3) 修改配置文件 applicationContext.xml，注释掉 Aspect01 类、Aspect02 类的配置信息，将 Aspect03 和 Aspect04 的配置信息添加到< bean >元素中，修改后的代码如下所示。

```
<!-- <bean name = "aspect01"  class = "com.qfedu.demo02.Aspect01"/>  -->
<!-- <bean name = "aspect02"  class = "com.qfedu.demo02.Aspect02"/>  -->
<bean name = "aspect03"  class = "com.qfedu.demo02.Aspect03"/>
<bean name = "aspect04"  class = "com.qfedu.demo02.Aspect04"/>
```

重新执行 TestAspect 类，执行结果如图 8.9 所示。

```
Console ☒
<terminated> TestAspect (1) [Java Application] C:\Program Files\Java\jre1.8.0_161\bin\javaw.exe (2018年9月6日 下午8:36:31)
信息: Loading XML bean definitions from class path resource [applicationContext.xml]
这是Aspect04的前置通知!!
这是Aspect03的前置通知!!
查询用户信息
这是Aspect03的后置通知!!
这是Aspect04的后置通知!!
```

图 8.9　执行 TestAspect 类的结果

从图 8.9 可以看出，程序依次执行了 Aspect04 类和 Aspect03 类中的多个方法，通过分析发现，优先级高的切面，它包含的通知先执行后退出。

修改 Aspect04 类中 getOrder()方法的返回值，将该值修改为 2，此时，Aspect04 的优先级低于 Aspect03，重新执行 TestAspect 类，执行结果如图 8.10 所示。

```
Console ☒
<terminated> TestAspect (1) [Java Application] C:\Program Files\Java\jre1.8.0_161\bin\javaw.exe (2018年9月6日 下午8:37:33)
信息: Loading XML bean definitions from class path resource [applicationContext.xml]
这是Aspect03的前置通知!!
这是Aspect04的前置通知!!
查询用户信息
这是Aspect04的后置通知!!
这是Aspect03的后置通知!!
```

图 8.10　重新执行 TestAspect 类的结果

从图 8.10 可以看出，Aspect03 类中的方法先执行后退出，由此可见，通过实现 Ordered 接口可以完成切面的优先级配置。

8.4.3 基于 XML 配置

XML 文件是配置切面优先级的第三种方式，主要通过< aop:aspect >元素的 order 元素实现。接下来以一个案例演示使用 XML 元素配置切面优先级。

(1) 在 com. qfedu. demo02 包下新建类 Aspect05，具体代码如例 8-18 所示。

【例 8-18】 Aspect05. Java

```
package com.qfedu.demo02;
public class Aspect05 {
    public void before(){
        System.out.println("这是 Aspect05 的前置通知!!");
    }
    public void after(){
        System.out.println("这是 Aspect05 的后置通知!!");
    }
}
```

(2) 在 com. qfedu. demo02 包下新建类 Aspect06，具体代码如例 8-19 所示。

【例 8-19】 Aspect06. Java

```
package com.qfedu.demo02;
public class Aspect06 {
    public void before(){
        System.out.println("这是 Aspect06 的前置通知!!");
    }
    public void after(){
        System.out.println("这是 Aspect06 的后置通知!!");
    }
}
```

(3) 在 src 目录下新建配置文件 applicationContext_aop. xml，具体代码参照 8. 3. 1 小节中 applicationContext. xml 文件，此处不再赘述。

(4) 在 applicationContext_aop. xml 文件的< bean >元素中添加配置信息，具体代码如下所示。

```
<bean name = "userService" class = "com.qfedu.demo02.UserServiceImpl"/>
<bean name = "aspect05"   class = "com.qfedu.demo02.Aspect05"/>
<bean name = "aspect06"   class = "com.qfedu.demo02.Aspect06"/>
<aop:config>
<aop:pointcut id = "pointcut"expression =
      "execution( * com.qfedu.demo02.UserServiceImpl. * (..))" />
   <! -- 配置切面 -->
<aop:aspect ref = "aspect05" order = "1" >
      <aop:before method = "before" pointcut - ref = "pointcut"/>
      <aop:after method = "after"  pointcut - ref = "pointcut"/>
   </aop:aspect>
<! -- 配置切面 -->
```

```
        <aop:aspect ref = "aspect06" order = "0">
            <aop:before method = "before"   pointcut - ref = "pointcut"/>
            <aop:after method = "after"   pointcut - ref = "pointcut"/>
        </aop:aspect>
    </aop:config>
```

在以上代码中,< aop:config >元素配置了两个切面,它们分别为 aspect05 和 aspect06,其中 aspect05 的 order 值为 1,aspect06 的 order 值为 0,显而易见,aspect06 的优先级高于 aspect05。

(5) 修改 TestAspect 类的代码,将其第 7 行、第 8 行代码修改为如下所示。

```
ApplicationContextcontext = new
ClassPathXmlApplicationContext("applicationContext_aop.xml");
```

以上代码将 ApplicationContext 获取配置信息的文件更改为 applicationContext_aop.xml。

执行 TestAspect 类,执行结果如图 8.11 所示。

```
Console ☒
<terminated> TestAspect (1) [Java Application] C:\Program Files\Java\jre1.8.0_161\bin\javaw.exe (2018年9月6日 下午8:43:06)
信息: Loading XML bean definitions from class path resource [applicationContext_aop.xml]
这是Aspect06的前置通知!!
这是Aspect05的前置通知!!
查询用户信息
这是Aspect05的后置通知!!
这是Aspect06的后置通知!!
```

图 8.11 执行 TestAspect 类的结果

从图 8.11 可以看出,程序依次执行了 Aspect06 类和 Aspect05 类中的多个方法,通过分析发现,优先级高的切面,它包含的通知先执行后退出。

修改 applicationContext_aop.xml 文件中的配置信息,将切面 Aspect06 的 order 值修改为 2,此时,Aspect06 的优先级低于 Aspect05,重新执行 TestAspect 类,执行结果如图 8.12 所示。

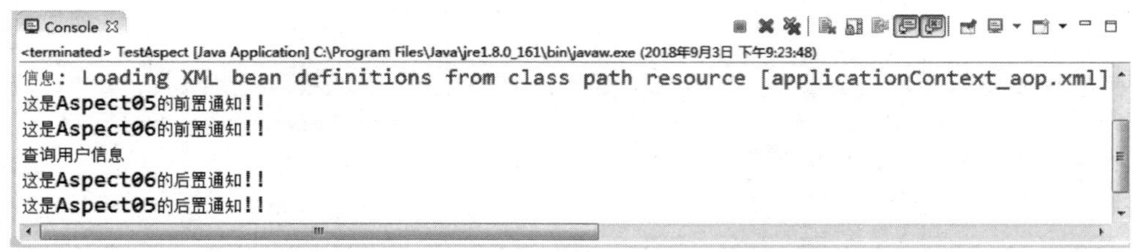

图 8.12 重新执行 TestAspect 类的结果

从图 8.12 中可以看出,Aspect05 类中的方法先执行后退出,由此可见,通过 XML 配置实现了切面的优先级配置。

8.5 Spring AOP 的应用

8.5.1 性能监控

性能监控是一个成熟系统必须具备的功能，它主要通过监测运行状况并识别瓶颈来提高应用程序的性能。通常情况下，程序针对各个组件的性能和参数进行对比，并通过这些对比发现触发问题的事件和条件，为系统排错提供支持。

在实际开发中，系统中的重要模块都要编写监控性能的代码，而且部分代码是非常细粒度的，这就会带来代码冗余的问题。除此之外，随着程序的变动，如果要在各个模块中分别修改监控性能的代码，必然会增加系统维护的工作量。为了避免这些问题，可以使用 AOP 的形式来实现性能监控。

接下来通过 Spring AOP 模拟性能监控的实现，具体步骤如下。

(1) 在 src 目录下新建 com. qfedu. demo03 包，在 com. qfedu. demo03 包下新建类 Service01，具体代码如例 8-20 所示。

【例 8-20】 Service01. Java

```
package com.qfedu.demo03;
public class Service01 {
    public void service() throws Exception{
        System.out.println("执行 service()方法");
        Thread.sleep(1000);
    }
}
```

在以上代码中，Thread 的 sleep()方法用于让当前线程停止执行一段时间，减缓当前线程的执行，其目的是间接拉长 service()方法的执行时间，让测试结果更加直观。

(2) 在 com. qfedu. demo03 包下新建类 Record，该类用于记录程序的执行信息，具体代码如例 8-21 所示。

【例 8-21】 Record. Java

```
package com.qfedu.demo03;
import java.util.Date;
public class Record {
    private String className;      //类名称
    private String methodName;     //方法名称
    private Date recordTime;       //记录时间
    private Long expendTime;       //程序执行耗费的时间
    public String getClassName() {
        return className;
    }
    public void setClassName(String className) {
        this.className = className;
    }
```

```
    public String getMethodName() {
        return methodName;
    }
    public void setMethodName(String methodName) {
        this.methodName = methodName;
    }
    public Date getRecordTime() {
        return recordTime;
    }
    public void setRecordTime(Date recordTime) {
        this.recordTime = recordTime;
    }
    public Long getExpendTime() {
        return expendTime;
    }
    public void setExpendTime(Long expendTime) {
        this.expendTime = expendTime;
    }
    @Override
    public String toString() {
        return "Record [className=" + className + ", methodName="
                + methodName + ", recordTime=" + recordTime + ",
                expendTime=" + expendTime + "]";
    }
}
```

(3) 在 com.qfedu.demo03 包下新建类 RecordAspect,该类用于定义一个切面,具体代码如例 8-22 所示。

【例 8-22】 RecordAspect.Java

```
package com.qfedu.demo03;
import java.util.Date;
import org.aspectj.lang.ProceedingJoinPoint;
import org.aspectj.lang.annotation.*;
@Aspect
public class RecordAspect {
    @Pointcut("execution(* com.qfedu.demo03.Service*.*(..))")
    public void record() {
    }
    @Around("record()")
    public Object recordTimer(ProceedingJoinPoint thisJoinPoint) throws
          Throwable {
            //获取目标对象的类名称
            String clazzName =
                thisJoinPoint.getTarget().getClass().getName();
            //获取目标对象的方法名称
            String methodName = thisJoinPoint.getSignature().getName();
            //计算目标对象对应方法执行所耗时间
```

```
        long startTime = System.currentTimeMillis();
        Object result = thisJoinPoint.proceed();
        long time = System.currentTimeMillis() - startTime;
        Record record = new Record();
        record.setExpendTime(time);                    //记录执行所耗时间
        record.setClassName(clazzName);                //记录类名称
        record.setMethodName(methodName);              //记录对应方法名称
        record.setRecordTime(new Date());              //记录时间
        System.out.println(record.toString());
        return result;
    }
}
```

在以上代码中，程序首先通过 ProceedingJoinPoint 获取目标对象的类名称和方法名称，然后执行目标对象中的对应方法并计算出执行时间，最后将包括目标对象的类名称、方法名称、方法执行时间等在内的信息存储到 Record 类对象中。在实际开发中，需要将 Record 类对象封装的信息提交到监控系统，由于当前是模拟程序，因此直接将 Record 类对象封装的信息输出到控制台。

（4）在 src 目录下新建配置文件 applicationContext_record.xml，具体代码参照 8.3.1 小节中 applicationContext.xml 文件，此处不再赘述。

（5）在 applicationContext_record.xml 文件的< bean >元素中添加配置信息，具体代码如下所示。

```
<bean name = "service01" class = "com.qfedu.demo03.Service01"/>
<bean name = "recordAspect"  class = "com.qfedu.demo03.RecordAspect"/>
<aop:aspectj-autoproxy/>
```

（6）在 com.qfedu.demo03 包下创建类 TestRecord，该类用于测试系统监控功能，具体代码如例 8-23 所示。

【例 8-23】 TestRecord.Java

```
package com.qfedu.demo03;
import org.springframework.context.ApplicationContext;
import org.springframework.context.support
    .ClassPathXmlApplicationContext;
import com.qfedu.demo03.Service01;
public class TestRecord {
    public static void main(String[] args) throws Exception {
        ApplicationContext context = new ClassPathXmlApplicationContext
            ("applicationContext_record.xml");
        Service01 service01 = context.getBean("service01",Service01.class);
        service01.service();
    }
}
```

执行 TestRecord 类,执行结果如图 8.13 所示。

```
Console ☒
<terminated> TestRecord [Java Application] C:\Program Files\Java\jre1.8.0_161\bin\javaw.exe (2018年9月6日 下午8:50:10)
loadBeanDefinitions
信息: Loading XML bean definitions from class path resource
[applicationContext_record.xml]
执行service()方法
Record [className=com.qfedu.demo03.Service01, methodName=service, recordTime=Thu Sep
06 20:50:12 CST 2018, expendTime=1019]
```

图 8.13　执行 TestRecord 类的结果

从图 8.13 中可以看出,程序向控制台输出 service()方法被执行的详细信息,包括类名、方法名、记录时间、执行方法所耗时间等,由此可见,程序通过 Spring AOP 实现了性能监控。

8.5.2　异常监控

在软件系统的评测中,处理异常的策略及成效也是一项重要指标,这些因素影响着软件的健壮性和可靠性。为保证系统程序的稳定运行、减少异常对程序运转造成的干扰,系统中通常会有自身的异常监控机制。

在企业项目中,异常监控通常不会与业务逻辑耦合在一起,它通常会被抽取出来并成为一项独立的模块,这个模块的主要功能是统一监控并处理项目中的异常。根据业务性质的特点,异常监控会有不同的具体做法,但将业务逻辑与异常监控解耦是最基本的思路,因此可以使用 Spring AOP 来实现项目中的异常监控。

接下来通过 Spring AOP 来模拟异常监控的实现,具体步骤如下。

(1) 在 com.qfedu.demo03 包下新建类 MyException,该类用于自定义异常,具体代码如例 8-24 所示。

【例 8-24】　MyException.Java

```
package com.qfedu.demo03;
public class MyException extends Exception {
    private static final long serialVersionUID = 1L;
    private String msg;
    public MyException(String msg) {
        super();
        this.msg = msg;
    }
    public String getMsg() {
        return msg;
    }
}
```

在以上代码中,MyException 提供了一个有参构造方法,可以通过该方法传入具体的异常信息。

(2) 在 com.qfedu.demo03 包下新建类 Service02,具体代码如例 8-25 所示。

【例 8-25】 **Service02. Java**

```
package com.qfedu.demo03;
public class Service02 {
    public void service() throws Exception{
        System.out.println("执行 service()方法");
        int num = 105;
        if(num > 100 || num < 0) {
            throw new MyException("您输入的不正确,请输入 0~100 之间的数字");
            } else {
              System.out.println("您输入的数字是" + num);
            }
    }
}
```

在以上代码中,首先声明变量 num 的值为 105,根据条件判断,当 num 的值大于 100 或小于 0 时,程序将抛出异常。

(3) 在 com. qfedu. demo03 包下新建类 Message,具体代码如例 8-26 所示。

【例 8-26】 **Message. Java**

```
package com.qfedu.demo03;
import java.util.Date;
public class Message {
    private String className;
    private String methodName;
    private Date recordTime;             //异常记录时间
    private String exceptionMsg;         //异常信息
    public String getClassName() {
        return className;
    }
    public void setClassName(String className) {
        this.className = className;
    }
    public String getMethodName() {
        return methodName;
    }
    public void setMethodName(String methodName) {
        this.methodName = methodName;
    }
    public Date getRecordTime() {
        return recordTime;
    }
    public void setRecordTime(Date recordTime) {
        this.recordTime = recordTime;
    }
    public String getExceptionMsg() {
        return exceptionMsg;
    }
    public void setExceptionMsg(String exceptionMsg) {
```

```
        this.exceptionMsg = exceptionMsg;
    }
    @Override
    public String toString() {
        return "Message [className = " + className + ", "
                + "methodName = " + methodName + ", recordTime = " + recordTime
                    + ", exceptionMsg = " + exceptionMsg + "]";
    }
}
```

在以上代码中，Message 类用于封装包括异常所在的类名称、方法名称、异常记录时间、异常信息在内的信息。

（4）在 com.qfedu.demo03 包下新建类 MessageAspect，具体代码如例 8-27 所示。

【例 8-27】 MessageAspect.Java

```
package com.qfedu.demo03;
import java.util.Date;
import org.aspectj.lang.ProceedingJoinPoint;
import org.aspectj.lang.annotation.*;
@Aspect
public class MessageAspect {
    @Pointcut("execution( * com.qfedu.demo03.Service*.*(..))")
    public void exceptionMsg() {
    }
    @Around("exceptionMsg()")
    public Object msgMethod(ProceedingJoinPoint thisJoinPoint) throws
          Throwable {
            //获取目标对象的类名称
            String clazzName =
                 thisJoinPoint.getTarget().getClass().getName();
            //获取目标对象的方法名称
            String methodName = thisJoinPoint.getSignature().getName();
            try {
                return thisJoinPoint.proceed();
            } catch (MyException e) {
            Message msg = new Message();
            //封装异常信息
            msg.setClassName(thisJoinPoint.getTarget()
                .getClass().getName());
            msg.setMethodName(thisJoinPoint.getSignature().getName());
            msg.setRecordTime(new Date());
                msg.setExceptionMsg(e.getMsg());
                System.out.println(msg.toString());
                return null;
            }
    }
}
```

在以上代码中，程序首先通过 ProceedingJoinPoint 获取目标对象的类名称和方法名称，然后执行目标对象中的对应方法并获取异常，最后将包括发生异常的类名称、方法名称、异常信息等在内的信息存储到 Message 类对象中。在实际开发中，需要将 Message 类对象封装的信息提交到监控系统，由于当前是模拟程序，因此直接将 Message 类对象封装的信息输出到控制台。

（5）在 src 目录下新建配置文件 applicationContext_msg. xml，具体代码参照 8.3.1 小节中 applicationContext. xml 文件，此处不再赘述。

（6）在 applicationContext_msg. xml 文件的< bean >元素中添加配置信息，具体代码如下所示。

```
<bean name = "service02" class = "com.qfedu.demo03.Service02"/>
<bean name = "messageAspect"   class = "com.qfedu.demo03.MessageAspect"/>
<aop:aspectj-autoproxy/>
```

（7）在 com. qfedu. demo03 包下创建类 TestMessage，该类用于测试异常监控功能，具体代码如例 8-28 所示。

【例 8-28】 TestMessage. Java

```
package com.qfedu.demo03;
import org.springframework.context.ApplicationContext;
import org.springframework.context.support
    .ClassPathXmlApplicationContext;
public class TestMessage {
    public static void main(String[] args) throws Exception {
        ApplicationContext context = new
        ClassPathXmlApplicationContext("applicationContext_msg.xml");
        Service02 service02 = context.getBean("service02",Service02.class);
        service02.service();
    }
}
```

执行 TestMessage 类，执行结果如图 8.14 所示。

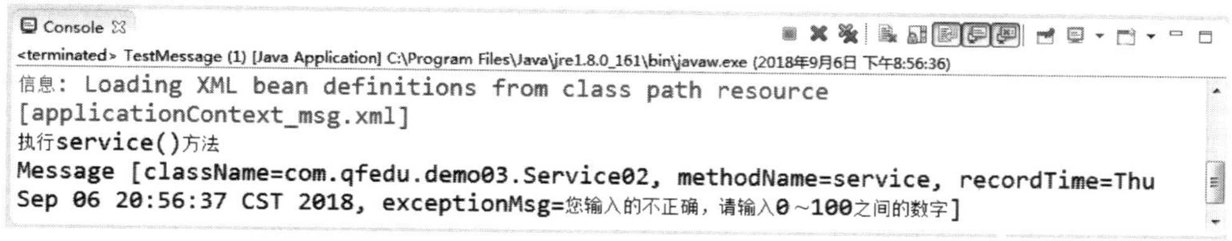

图 8.14 执行 TestMessage 类的结果

从图 8.14 中可以看出，程序向控制台输出 service()方法被执行的详细信息，由于 Service02 类的 service()方法抛出了一个异常，因此控制台窗口显示了发生异常的类名称、方法名称、异常信息等在内的内容。

8.6 本章小结

本章首先介绍了 Spring AOP 的基础知识，包括 AOP 简介、AOP 的基本术语等；其次通过实例演示了 Spring AOP 的实现机制，包括 JDK 动态代理机制和 CGLib 动态代理机制；接下来讲解了 Spring AOP 的开发方法，Spring AOP 的开发方法主要有两种，包括基于 XML 或基于注解；接着介绍了 Spring AOP 中多个切面的优先级，切面的优先级有三种配置方式，包括基于注解配置、基于 Ordered 接口配置和基于 XML 配置；最后通过具体案例讲解了 Spring AOP 的应用。通过本章知识的学习，大家应该能理解 AOP 的概念和基本术语，能理解 Spring AOP 的实现机制，掌握 Spring AOP 的多种开发方法，能够通过 Spring AOP 实现一些常用的功能。

8.7 习　　题

1. 填空题

(1) AOP 指的是____________________。

(2) Spring AOP 支持的通知类型共有________种。

(3) Spring AOP 的实现机制有两种，它们分别是________和________。

(4) 在默认情况下，Spring AOP 的实现方式是________。

(5) 当目标对象是一个没有实现任何接口的类时，Spring AOP 通过________实现。

2. 选择题

(1) 关于 AOP，下列选项错误的是(　　)。

A. AOP 是 Spring 框架的一个重要特性

B. AOP 引入了抽象、封装、继承的概念

C. AOP 主张将程序中的相同业务逻辑进行横向隔离

D. AOP 能够降低耦合，提升程序的可维护性

(2) 关于 AOP 的基本术语，下列选项错误的是(　　)。

A. 通知(Advice)是在目标类连接点上执行的一段代码

B. 目标对象(Target)是指通知所作用的目标业务类

C. 代理(Proxy)是指通知所作用的目标业务类

D. 织入(Weaving)是将通知添加到目标类具体链接点上的过程

(3) 关于 JDK 动态代理，下列选项错误的是(　　)。

A. JDK 动态代理主要通过 InvocationHandler 和 Proxy 这两个 API

B. InvocationHandler 是一个接口，可以通过实现该接口来定义横切逻辑

C. Proxy 利用 InvocationHandler 动态生成目标类的代理对象

D. 当目标对象为一个类时，Spring AOP 通常使用 JDK 动态代理来实现

(4) 关于 Spring AOP 配置文件(XML 文件)中的元素，下列选项错误的是(　　)。

A. <aop:config>元素是 AOP 配置的根元素

B. <aop:aspect>元素用于指定切面

C. <aop:pointcut>元素用于指定切点

D. <aop:after-returning>用于指定异常通知

(5) 在下列注解中,可以指定切面优先级的是(　　)。

A. @Aspect　　B. @Pointcut

C. @Around　　D. @Order

3. 思考题

(1) 简述 AOP 的概念。

(2) 简述 Spring AOP 的实现机制。

4. 编程题

现有一个 Calculation 类,Calculation 类中提供有 add()、subtract()、multiply()和 divide()共四个方法,请通过 Spring AOP 完成以下步骤:

(1) 当上述方法执行之前,控制台输出提示信息"方法开始执行"。

(2) 当上述方法执行之后,控制台输出提示信息"方法执行完毕"。

(3) 以注解或 XML 形式均可。

Calculation. Java

```
public class Calculation {
      public void add(){
           System.out.println("add()执行中……");
      }
      public void subtract(){
           System.out.println("subtract()执行中……");
      }
      public void multiply(){
           System.out.println("multiply()执行中……");
      }
      public void divide(){
           System.out.println("divide()执行中……");
      }
}
```

第 9 章　Spring 的 JDBC

本章学习目标

- 理解 Spring 中的数据库开发
- 理解 JDBCTemplate 类的常用 API
- 掌握 JDBCTemplate 操作数据库的方法
- 掌握通过 JDBCTemplate 封装 Dao 的方法

数据库用于持久化业务处理产生的数据，正因为此，操作数据库成为应用程序在运行过程中经常完成的动作。通常情况下，操作数据库的功能由持久层来实现。作为扩展性较强的一站式开发框架，Spring 提供了持久层 Spring JDBC 功能，Spring JDBC 可以管理数据库的连接资源，简化传统 JDBC 的操作，进而提升程序操作数据库的效率。接下来，本章将对 Spring JDBC 涉及的相关知识进行详细讲解。

9.1　Spring JDBC 基础

9.1.1　Spring JDBC 简介

本书在前面介绍过，传统 JDBC 存在代码烦琐、表关系维护复杂、硬编码等缺陷。为了解决这些问题，Spring 提供了改善和增强 JDBC 的方案，即 Spring JDBC。

Spring 的 JDBC 模块由四个包组成，它们分别是 core（核心包）、object（对象包）、dataSource（数据源包）和 support（支持包），这些包相互协作，共同支撑了 Spring 的 JDBC 功能。

Spring 对 JDBC 的增强主要体现在它封装了传统 JDBC 并提供了 JDBCTemplate 类，JDBCTemplate 类是 Spring JDBC 的核心类。

9.1.2　JDBCTemplate 类

通常情况下，Spring JDBC 通过 JDBCTemplate 类来避免传统 JDBC 在实际应用中的各项问题，JDBCTemplate 类是一个模板类，同时，Spring JDBC 中的更高层次的抽象类均在 JDBCTemplate 类基础上构建。JDBCTemplate 类包含了所有操作数据库的基本方法，包括添加、删除、查询、更新等，除此之外，JDBCTemplate 类还省去了传统 JDBC 中的复杂步骤，这可以让开发人员将更多精力投入到业务逻辑中。

在使用 JDBCTemplate 类对象操作数据库之前，首先要为其提供数据源，数据源可以是第三方提供的，如 C3P0、DBCP 等。

为了便于操作数据库，JDBCTemplate 类提供了一系列的方法，其中常用的如表 9.1 所示，在实际开发中，开发者可根据具体需要选择使用。

表 9.1　JDBCTemplate 类的常用方法

方　法	描　述
int[]batchUpdate(String sql)	使用批处理在单个 JDBC 语句上发出多个 SQL 更新
int[]batchUpdate(String sql, BatchPreparedStatementSetter pss)	在单个 PreparedStatement 上发出多个更新语句，使用批处理更新和 BatchPreparedStatementSetter 来设置值
void execute(String sql)	发出单个 SQL 执行，通常是 DDL 语句
<T> T execute(String sql, PreparedStatementCallback<T> psc)	执行 JDBC 数据访问操作，实现 JDBC PreparedStatement 上的回调操作
<T> T execute(String callString, CallableStatementCallback<T> csc)	执行 JDBC 数据访问操作，实现为处理 JDBC CallableStatement 的回调操作
int getFetchSize()	返回为此 JABCTemplate 指定的获取大小
int getMaxRows()	返回为此 JABCTemplate 指定的最大行数
Boolean isIgnoreWarnings()	返回是否忽略 SQLWarnings
List query(String sql, Object[]args, RowMapper<T> rowMapper)	查询给定 SQL 以从 SQL 创建预准备语句以及绑定到查询的参数列表，通过 RowMapper 将每一行映射到 Java 对象
List query(String sql, PreparedStatementSetter pss, RowMapper < T > rowMapper)	查询给定 SQL 以从 SQL 创建预准备语句和 PreparedStatementSetter 实现，该实现将值绑定到查询，通过 RowMapper 将每一行映射到 Java 对象
List query (String sql, RowMapper <T> rowMapper)	执行给定静态 SQL 的查询，通过 RowMapper 将每一行映射到 Java 对象
List query (String sql, RowMapper <T> rowMapper, Object args)	查询给定 SQL 以从 SQL 创建预准备语句以及绑定到查询的参数列表，通过 RowMapper 将每一行映射到 Java 对象
queryForObject (String sql, Class <T> requiredType)	执行 SQL 语句，返回结果对象
queryForObject (String sql, Class <T> requiredType, Object args)	执行 SQL 语句，传入参数，返回结果对象
List queryForList(String sql)	执行 SQL 语句，返回包含有执行结果的 List 集合
List queryForList (String sql, Class <T> elementType)	执行 SQL 语句，返回包含有执行结果的 List 集合
List queryForList (String sql, Class <T> elementType, Object args)	查询给定 SQL 以从 SQL 创建预准备语句以及绑定到查询的参数列表，期望结果列表
queryForList(String sql, Object args)	查询给定 SQL 以从 SQL 创建预准备语句以及绑定到查询的参数列表，期望结果列表
void setFetchSize(int fetchSize)	设置此 JABCTemplate 的获取大小
void setIgnoreWarnings(boolean b)	设置是否要忽略 SQLWarnings
int update(String sql)	发出单个 DML 操作
int update(String sql, Object args)	发出单个 DML 操作，绑定给定的参数
int update(String sql, Object[]args, int[]argTypes)	发出单个 DML 操作，绑定给定的参数

9.1.3　使用 JDBCTemplate 类完成简单程序

上个小节介绍了 JDBCTemplate 类的常用方法，接下来通过一个实例演示

JDBCTemplate 类的简单应用。

(1) 在 MySQL 中创建数据库 chapter09 和数据表 student,SQL 语句如下所示。

```
DROP DATABASE IF EXISTS chapter09;
CREATE DATABASE chapter09;
USE chapter09;
CREATE TABLE student(
sid INT PRIMARY KEY AUTO_INCREMENT, #ID
sname VARCHAR(20),      #学生姓名
age VARCHAR(20),        #学生年龄
course VARCHAR(20)      #学科
);
```

(2) 向数据表 student 添加数据,SQL 语句如下所示。

```
INSERT INTO student(sname,age,course) VALUES ('ZhangSan','20','Java');
INSERT INTO student(sname,age,course) VALUES ('LiSi','21','Python');
```

(3) 通过 SQL 语句测试数据是否添加成功,运行结果如下所示。

```
mysql >  SELECT * FROM student;
+-----+----------+------+--------+
| sid | sname    | age  | course |
+-----+----------+------+--------+
|  1  | ZhangSan | 20   | Java   |
|  2  | LiSi     | 21   | Python |
+-----+----------+------+--------+
2 rows in set (0.00 sec)
```

从以上运行结果可以看出,数据添加成功。

(4) 在 Eclipse 中新建 Web 工程 chapter09,将相关 jar 包添加到 lib 目录下,完成导包。与第 8 章使用的 jar 包相比,本次需要新添加 Spring JDBC、C3P0 连接池以及 JDBC 的 jar 包,要导入的所有 jar 包如图 9.1 所示。

```
WebContent
  META-INF
  WEB-INF
    lib
      aopalliance-1.0.jar
      aspectjweaver-1.7.4.jar
      c3p0-0.9.1.2.jar
      commons-logging-1.2.jar
      mchange-commons-java-0.2.3.4.jar
      mysql-connector-java-5.1.10-bin.jar
      spring-aop-5.0.8.RELEASE.jar
      spring-aspects-5.0.8.RELEASE.jar
      spring-beans-5.0.8.RELEASE.jar
      spring-context-5.0.8.RELEASE.jar
      spring-core-5.0.8.RELEASE.jar
      spring-expression-5.0.8.RELEASE.jar
      spring-jdbc-5.0.8.RELEASE.jar
      spring-tx-5.0.8.RELEASE.jar
```

图 9.1　需要导入的 jar 包

(5) 在工程 chapter09 的 src 目录下新建 com. qfedu. test 包，在该包下新建类 TestJDBCTemplate01，具体代码如例 9-1 所示。

【例 9-1】 TestJDBCTemplate01. Java

```
package com.qfedu.test;
import org.springframework.jdbc.core.JdbcTemplate;
import com.mchange.v2.c3p0.ComboPooledDataSource;
public class TestJDBCTemplate01 {
    public static void main(String[] args) throws Exception {
        //创建数据源
        ComboPooledDataSource dataSource   = new ComboPooledDataSource();
        dataSource.setDriverClass("com.mysql.jdbc.Driver");
        dataSource.setJdbcUrl("jdbc:mysql://localhost:3306/chapter09");
        dataSource.setUser("root");
        dataSource.setPassword("root");
        //创建 JDBCTemplate 类对象
        JdbcTemplate jdbcTemplate = new JdbcTemplate(dataSource);
        String sql = "insert into student (sname ,age,course) value (? , ?,?)";
        //调用 JDBCTemplate 类对象的 update()方法
        int result = jdbcTemplate.update(sql,"WangWu","21","Java");
        if (result>0) {
            System.out.println("成功添加" + result + "条数据");
        }else {
            System.out.println("数据添加失败");
        }
    }
}
```

执行 TestJDBCTemplate01 类，执行结果如图 9.2 所示。

```
Console
<terminated> TestJDBCTemplate01 [Java Application] C:\Program Files\Java\jre1.8.0_161\bin\javaw.exe (2018年9月11日 下午1:51:34)
password=******}, propertyCycle -> 0, statementCacheNumDeferredCloseThreads -> 0,
testConnectionOnCheckin -> false, testConnectionOnCheckout -> false,
unreturnedConnectionTimeout -> 0, userOverrides -> {},
usesTraditionalReflectiveProxies -> false ]
成功添加1条数据
```

图 9.2 执行 TestJDBCTemplate01 类的结果

从以上运行结果可以看出，程序显示成功添加数据的提示信息。

9.1.4 在 Spring 中管理 JDBCTemplate 类

在上个小节的案例中，JDBCTemplate 类对象由开发者手动创建，除此之外，开发者还需手动为 JDBCTemplate 类对象配置数据源。为了充分发挥 Spring IOC 的优势，通常将 JDBCTemplate 类交给 Spring 容器直接管理，因此，在使用 JDBCTemplate 类之前要首先完成 JDBCTemplate 类在 Spring 容器中的配置。

将 JDBCTemplate 类配置到 Spring 容器中主要分两步，首先是在 Spring 容器中注册数据源，其次是在 Spring 容器中注册 JDBCTemplate 类，然后向 JDBCTemplate 类注入数据源，此时，Spring 容器可以直接管理 JDBCTemplate 类。

1. 在 Spring 容器中注册数据源

JDBCTemplate 类依赖的数据源既可以是 Spring 内部提供的，也可以是第三方连接池。根据数据源具体实现的不同，要配置的属性也略有差别，以 C3P0 连接池为例，要配置的属性如表 9.2 所示。

表 9.2　C3P0 连接池的属性

属　　性	描　　述
driverClass	数据库的驱动名称
jdbcUrl	连接数据库的路径
user	数据库用户名
password	数据库密码
maxPoolSize	最大连接数
minPoolSize	最小连接数
initialPoolSize	初始化连接数
maxIdleTime	连接的最大空闲时间

表 9.2 列举出了 C3P0 连接池要配置的属性，在实际使用中，driverClass、jdbcUrl、user、password 这四个属性是必须要配置的，其余属性为可选项，在没有被额外配置时，这些属性将直接采用默认值。

2. 在 Spring 容器中注册 JDBCTemplate 类

完成数据源的注册之后，可以在 Spring 容器中注册 JDBCTemplate 类。JDBCTemplate 类要依靠数据源实现功能，因此要将数据源注入到 JDBCTemplate 类中，完成注册后，Spring 容器可以直接管理 JDBCTemplate 类，至此，开发人员可以不用过多关注 JDBCTemplate 类对象的创建过程，这就可以提升开发效率和程序性能，也减少了代码的冗余。

接下来通过一个案例演示以上过程，具体如下。

(1) 在工程 chapter09 的 src 目录下新建文件 jdbc. properties，具体代码如例 9-2 所示。

【例 9-2】　jdbc. properties

```
jdbc.driverClass = com.mysql.jdbc.Driver
jdbc.jdbcUrl = jdbc:mysql://localhost:3306/chapter09
jdbc.user = root
jdbc.password = root
```

以上代码中定义了 C3P0 连接池实现功能所需的属性，它们分别是 driverClass、jdbcUrl、user 和 password。

(2) 在工程 chapter09 的 src 目录下新建配置文件 applicationContext. xml，具体代码参照 8.3.1 小节中 applicationContext. xml 文件，此处不再赘述。

(3) 在 applicationContext. xml 文件的< bean >元素中添加配置信息，具体如下。

```
1 <!-- 引入外部 properties 文件 -->
2 <context:property-placeholder location="classpath:jdbc.properties"/>
3 <!-- 注册数据源 -->
4    <bean name="dataSource"
5              class="com.mchange.v2.c3p0.ComboPooledDataSource">
6              <property name="driverClass" value="${jdbc.driverClass}"/>
7              <property name="jdbcUrl" value="${jdbc.jdbcUrl}"/>
8              <property name="user" value="${jdbc.user}"/>
9              <property name="password" value="${jdbc.password}"/>
10   </bean>
11   <!--注册 JdbcTemplate 类 -->
12   <bean name="jdbcTemplate"
13         class="org.springframework.jdbc.core.JdbcTemplate">
14         <property name="dataSource" ref="dataSource"/>
15   </bean>
```

在以上代码中，首先引入类路径下的 jdbc.properties 文件，如此一来，< bean >元素的子元素< property >中便可直接使用 jdbc.properties 文件中的内容，例如，第 6 行名称为 driverClass 的< property >元素，它的值 ${jdbc.driverClass} 就相当于 com.mysql.jdbc.Driver，因为 jdbc.properties 文件中已经声明了这一关系。在注册 JDBCTemplate 类时，由于 JDBCTemplate 依赖于 DataSource，因此将 DataSource 注入到 JDBCTemplate 类中。

（4）在 com.qfedu.test 包下创建测试类 TestJDBCTemplate02，具体代码如例 9-3 所示。

【例 9-3】 TestJDBCTemplate02.Java

```
package com.qfedu.test;
import org.springframework.context.ApplicationContext;
import org.springframework.context.support
    .ClassPathXmlApplicationContext;
import org.springframework.jdbc.core.JdbcTemplate;
public class TestJDBCTemplate02 {
    public static void main(String[] args) throws Exception {
        ApplicationContext context = new
            ClassPathXmlApplicationContext("applicationContext.xml");
        JdbcTemplate jdbcTemplate = context.getBean
            ("jdbcTemplate",JdbcTemplate.class);
        String sql = "insert into student (sname ,age,course) value (? , ?,?)";
        //调用 JDBCTemplate 类对象的 update()方法
        int result = jdbcTemplate.update(sql,"ZhaoLiu","21","Java");
        if (result>0) {
            System.out.println("成功添加" + result + "条数据");
        }else {
            System.out.println("数据添加失败");
        }
    }
}
```

在以上代码中，程序从 Spring 容器中获取 JDBCTemplate 类对象，然后通过调用 JDBCTemplate 类对象操作数据库。由于 JDBCTemplate 类由 Spring 容器管理，因此，和 TestJDBCTemplate01 类相比，此处不需要手动创建 JDBCTemplate 类对象，也无须手动为 JdbcTemplate 类对象设置数据源。

执行 TestJDBCTemplate02 类，执行结果如图 9.3 所示。

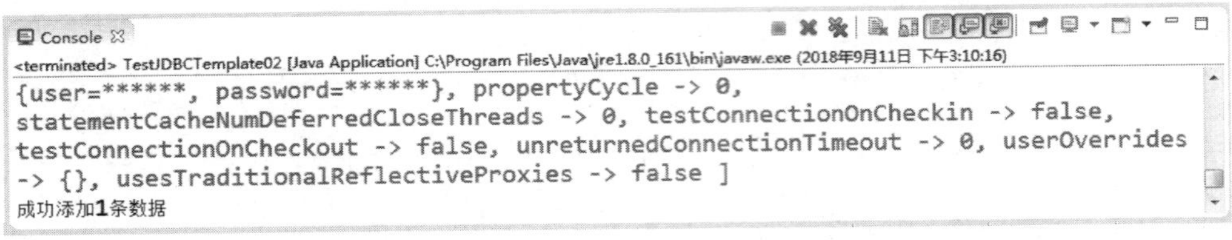

```
Console
&lt;terminated&gt; TestJDBCTemplate02 [Java Application] C:\Program Files\Java\jre1.8.0_161\bin\javaw.exe (2018年9月11日 下午3:10:16)
{user=******, password=******}, propertyCycle -> 0,
statementCacheNumDeferredCloseThreads -> 0, testConnectionOnCheckin -> false,
testConnectionOnCheckout -> false, unreturnedConnectionTimeout -> 0, userOverrides
-> {}, usesTraditionalReflectiveProxies -> false ]
成功添加1条数据
```

图 9.3　执行 TestJDBCTemplate02 类的结果

从以上运行结果可以看出，程序显示成功添加数据的提示信息。由此可见，Spring 自动完成了 JDBCTemplate 类对象的创建及数据源注入。由于将 JDBCTemplate 类交由 Spring 容器直接管理有诸多优势，因此实际应用中通常采用这种方式。

9.2　JDBCTemplate 操作数据库

9.2.1　JDBCTemplate 类实现 DDL 操作

9.1 节介绍了 JDBCTemplate 类的简单使用，本节将开始详细讲解通过 JDBCTemplate 类操作数据库。通常情况下，JDBCTemplate 类通过其 execute()方法完成 DDL 操作，接下来通过一个实例对 DDL 操作进行演示。

在 com.qfedu.test 包下创建测试类 TestJDBCTemplate03，具体代码如例 9-4 所示。

【例 9-4】　TestJDBCTemplate03.Java

```
package com.qfedu.test;
import org.springframework.context.ApplicationContext;
import org.springframework.context.support
    .ClassPathXmlApplicationContext;
import org.springframework.jdbc.core.JdbcTemplate;
public class TestJDBCTemplate03 {
    public static void main(String[] args) throws Exception {
        ApplicationContext context = new  ClassPathXmlApplicationContext
            ("applicationContext.xml");
        JdbcTemplate jdbcTemplate = context.getBean
            ("jdbcTemplate",JdbcTemplate.class);
        String sql = "CREATE TABLE stu(sid INT PRIMARY KEY AUTO_INCREMENT,"
                + "sname VARCHAR(20), age VARCHAR(20), "
                + "course VARCHAR(20) )";
        //调用 JDBCTemplate 类对象的 execute()方法
        jdbcTemplate.execute(sql);
    }
}
```

执行 TestJDBCTemplate03 类，由于 TestJDBCTemplate03 类的 execute()方法的返回值为 void，因此需要通过命令行窗口查看程序运行结果，运行结果如下所示。

```
mysql > DESC stu;
+--------+-------------+------+-----+---------+----------------+
| Field  | Type        | Null | Key | Default | Extra          |
+--------+-------------+------+-----+---------+----------------+
| sid    | int(11)     | NO   | PRI | NULL    | auto_increment |
| sname  | varchar(20) | YES  |     | NULL    |                |
| age    | varchar(20) | YES  |     | NULL    |                |
| course | varchar(20) | YES  |     | NULL    |                |
+--------+-------------+------+-----+---------+----------------+
4 rows in set (0.05 sec)
```

从以上运行结果可以看出，stu 表创建成功。由此可见，JDBCTemplate 类实现了 DDL 操作。

9.2.2 JDBCTemplate 类实现 DQL 操作

在 JDBCTemplate 类提供的方法中，query()、queryForObject()、queryForList()用于实现 DQL 操作。当需要查询单条记录时，通常使用 query()或 queryForObject()方法，当需要查询多条记录时，通常使用 query()或 queryForList()方法。当处理结果集时，如果要将返回结果映射为开发人员自定义的类，则需要使用 RowMapper < T >接口。

接下来通过一个实例演示单条记录的查询，具体步骤如下。

(1) 在 src 目录下创建一个 com. qfedu. test 包。

(2) 在 com. qfedu. test 包下创建测试类 TestJDBCTemplate04，具体代码如例 9-5 所示。

【例 9-5】 TestJDBCTemplate04. Java

```
package com.qfedu.test;
import org.springframework.context.ApplicationContext;
import org.springframework.context.support
    .ClassPathXmlApplicationContext;
import org.springframework.jdbc.core.JdbcTemplate;
public class TestJDBCTemplate04 {
    public static void main(String[] args) throws Exception {
        ApplicationContext context = new ClassPathXmlApplicationContext
            ("applicationContext.xml");
        JdbcTemplate jdbcTemplate = context.getBean
            ("jdbcTemplate",JdbcTemplate.class);
        String sql = "select count( * ) from student ";
        //调用 JDBCTemplate 类对象的 queryForObject()方法
        Integer count = jdbcTemplate.queryForObject(sql,Integer.class);
        System.out.println(count);
    }
}
```

在以上代码中，程序查询数据表 student 中所有记录的数量，由于查询结果是 int 类型，因此可使用 Integer 包装类映射查询结果。

执行 TestJDBCTemplate04 类，执行结果如图 9.4 所示。

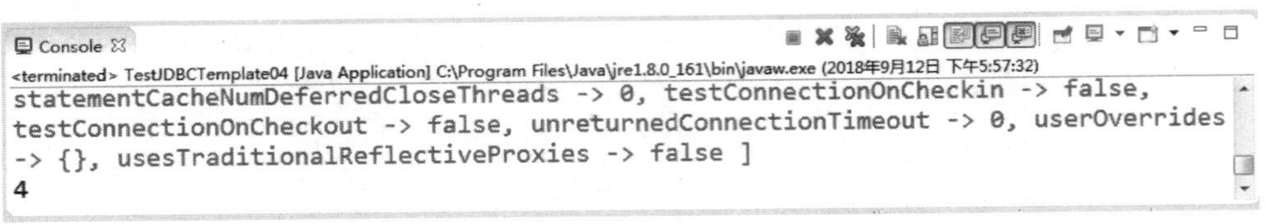

图 9.4　执行 TestJDBCTemplate04 类的结果

从以上运行结果可以看出，程序输出数据表 student 中的所有记录的数量。

在实际开发中，更多的需求是将结果集中的字段值映射为自定义类的属性，此时就要通过 RowMapper<T>接口。RowMapper<T>接口中定义了一个 mapRow()方法，该方法用于完成字段值和类的属性的映射。为了简化开发，Spring 提供了 RowMapper<T>接口的实现类 BeanPropertyRowMapper<T>，开发人员可直接使用该类完成以上功能。

接下来通过一个实例演示 BeanPropertyRowMapper<T>的使用，具体步骤如下。

(1) 在 src 目录下创建一个 com.qfedu.bean 包，在该包下创建一个 Student 类，具体代码参考第 1 章例 1.1，此处不再赘述。

(2) 在 com.qfedu.test 包下创建类 TestJDBCTemplate05，具体代码如例 9-6 所示。

【例 9-6】 TestJDBCTemplate05.Java

```
package com.qfedu.test;
import org.springframework.context.ApplicationContext;
import org.springframework.context.support
    .ClassPathXmlApplicationContext;
import org.springframework.jdbc.core.*;
import com.qfedu.bean.Student;
public class TestJDBCTemplate05 {
    public static void main(String[] args) throws Exception {
        ApplicationContext context = new  ClassPathXmlApplicationContext
            ("applicationContext.xml");
        JdbcTemplate jdbcTemplate = context.getBean
            ("jdbcTemplate",JdbcTemplate.class);
        String sql = "select  *  from student where sid = 1";
        //创建 BeanPropertyRowMapper<Student>类对象
        RowMapper<Student> rowMapper = new BeanPropertyRowMapper<Student>
            (Student.class);
        //调用 JDBCTemplate 类对象的 queryForObject()方法
        Student student = jdbcTemplate.queryForObject(sql,rowMapper);
        System.out.println(student);
    }
}
```

在以上代码中，程序首先创建一个 BeanPropertyRowMapper<Student>类对象，该对象用于完成数据表记录和 Student 类对象的映射。在通过 JDBCTemplate 类对象的

queryForObject()方法执行 SQL 语句时，需要将 BeanPropertyRowMapper<Student>类对象作为参数传入。

执行 TestJDBCTemplate05 类，执行结果如图 9.5 所示。

```
Console
<terminated> TestJDBCTemplate05 [Java Application] C:\Program Files\Java\jre1.8.0_161\bin\javaw.exe (2018年9月12日 下午6:49:12)
statementCacheNumDeferredCloseThreads -> 0, testConnectionOnCheckin -> false,
testConnectionOnCheckout -> false, unreturnedConnectionTimeout -> 0, userOverrides
-> {}, usesTraditionalReflectiveProxies -> false ]
Student [sid=1, sname=ZhangSan, age=20, course=Java]
```

图 9.5　执行 TestJDBCTemplate05 类的结果

从以上运行结果可以看出，程序输出 sid 为 1 的学生信息。由此可见，BeanPropertyRowMapper<Student>类对象实现了表记录和 Student 类对象的映射。

以上介绍了 JDBCTemplate 类对单条记录的查询，接下来开始演示对多条记录的查询。在 com.qfedu.test 包下新建 TestJDBCTemplate06 类，具体代码如例 9-7 所示。

【例 9-7】 TestJDBCTemplate06.Java

```
package com.qfedu.test;
import java.util.List;
import org.springframework.context.ApplicationContext;
importorg.springframework.context.support
    .ClassPathXmlApplicationContext;
import org.springframework.jdbc.core.*;
import com.qfedu.bean.Student;
public class TestJDBCTemplate06 {
    public static void main(String[] args) throws Exception {
        ApplicationContext context = new ClassPathXmlApplicationContext
            ("applicationContext.xml");
        JdbcTemplate jdbcTemplate = context.getBean
            ("jdbcTemplate",JdbcTemplate.class);
        String sql = "select * from student ";
        RowMapper<Student> rowMapper = new BeanPropertyRowMapper<Student>
            (Student.class);
        //调用 JDBCTemplate 类对象的 query()方法
        List<Student> list = jdbcTemplate.query(sql, rowMapper);
        for (Student student : list) {
            System.out.println(student);
        }
    }
}
```

在以上代码中，程序首先创建一个 BeanPropertyRowMapper<Student>类对象，然后调用 JDBCTemplate 类对象的 query()方法执行查询。在处理结果集时，BeanPropertyRowMapper<Student>类对象将结果集中的每一条记录映射为 Student 类的一个对象，最终这些对象被存入 List 集合。

执行 TestJDBCTemplate06 类，执行结果如图 9.6 所示。

从以上运行结果可以看出，程序输出数据表 student 中的所有学生信息，由此可见，

```
Console ⨯
<terminated> TestJDBCTemplate06 [Java Application] C:\Program Files\Java\jre1.8.0_161\bin\javaw.exe (2018年9月13日 上午9:54:31)
statementCacheNumDeferredCloseThreads -> 0, testConnectionOnCheckin -> false,
testConnectionOnCheckout -> false, unreturnedConnectionTimeout -> 0, userOverrides
-> {}, usesTraditionalReflectiveProxies -> false ]
Student [sid=1, sname=ZhangSan, age=20, course=Java]
Student [sid=2, sname=LiSi, age=21, course=Python]
Student [sid=3, sname=WangWu, age=21, course=Java]
Student [sid=4, sname=ZhaoLiu, age=21, course=Java]
```

图 9.6　执行 TestJDBCTemplate06 类的结果

JDBCTemplate 类实现了多条记录的查询。

9.2.3　JDBCTemplate 类实现 DML 操作

在 JDBCTemplate 类提供的方法中，update()方法用于实现 DML 操作。DML 操作包括对数据表中数据的添加、删除和更新。由于 9.1.3 节中已演示过使用 JDBCTemplate 类实现添加操作，此处不再赘述，接下来分别演示通过 JdbcTemplate 类实现更新和删除操作。

在 com.qfedu.test 包下创建 TestJDBCTemplate07 类，该类用于测试 JdbcTemplate 类更新数据表中数据的功能，具体代码如例 9-8 所示。

【例 9-8】　TestJDBCTemplate07.Java

```
package com.qfedu.test;
import org.springframework.context.ApplicationContext;
import org.springframework.context.support
    .ClassPathXmlApplicationContext;
import org.springframework.jdbc.core.*;
public class TestJDBCTemplate07 {
    public static void main(String[] args) throws Exception {
        ApplicationContext context = new  ClassPathXmlApplicationContext
            ("applicationContext.xml");
        JdbcTemplate jdbcTemplate = context.getBean
            ("jdbcTemplate",JdbcTemplate.class);
        String sql = "update student set sname = 'SunQi' where sid = 4 ";
        //调用 JDBCTemplate 类的 update()方法执行更新操作
        int result = jdbcTemplate.update(sql);
        if (result > 0) {
            System.out.println("成功更新" + result + "条数据");
        }else {
            System.out.println("数据更新失败");
        }
    }
}
```

执行 TestJDBCTemplate07 类，执行结果如图 9.7 所示。

从以上运行结果可以看出，程序输出成功更新了一条数据的提示信息，由此可见，JDBCTemplate 类实现了对数据表中数据的更新。

在 com.qfedu.test 包下创建 TestJDBCTemplate08 类，该类用于测试 JDBCTemplate

```
Console ⊠
<terminated> TestJDBCTemplate07 [Java Application] C:\Program Files\Java\jre1.8.0_161\bin\javaw.exe (2018年9月13日 下午6:21:07)
maxPoolSize -> 15, maxStatements -> 0, maxStatementsPerConnection -> 0,
minPoolSize -> 3, numHelperThreads -> 3, preferredTestQuery -> null, properties ->
{user=******, password=******}, propertyCycle -> 0,
statementCacheNumDeferredCloseThreads -> 0, testConnectionOnCheckin -> false,
testConnectionOnCheckout -> false, unreturnedConnectionTimeout -> 0, userOverrides
-> {}, usesTraditionalReflectiveProxies -> false ]
成功更新1条数据
```

图 9.7　执行 TestJDBCTemplate07 类的结果

类删除数据表中数据的功能，具体代码如例 9-9 所示。

【例 9-9】　TestJDBCTemplate08. Java

```
package com.qfedu.test;
import org.springframework.context.ApplicationContext;
import org.springframework.context.support
    .ClassPathXmlApplicationContext;
import org.springframework.jdbc.core.*;
public class TestJDBCTemplate08 {
    public static void main(String[] args) throws Exception {
        ApplicationContext context = new  ClassPathXmlApplicationContext
            ("applicationContext.xml");
        JdbcTemplate jdbcTemplate = context.getBean
            ("jdbcTemplate",JdbcTemplate.class);
        String sql = "delete from student where sid = 4 ";
        //调用 JDBCTemplate 类的 update()方法执行删除操作
        int result = jdbcTemplate.update(sql);
        if (result>0) {
            System.out.println("成功删除" + result + "条数据");
        }else {
            System.out.println("数据删除失败");
        }
    }
}
```

执行 TestJDBCTemplate08 类，执行结果如图 9.8 所示。

```
Console ⊠
<terminated> TestJDBCTemplate08 [Java Application] C:\Program Files\Java\jre1.8.0_161\bin\javaw.exe (2018年9月13日 下午6:22:43)
maxPoolSize -> 15, maxStatements -> 0, maxStatementsPerConnection -> 0,
minPoolSize -> 3, numHelperThreads -> 3, preferredTestQuery -> null, properties ->
{user=******, password=******}, propertyCycle -> 0,
statementCacheNumDeferredCloseThreads -> 0, testConnectionOnCheckin -> false,
testConnectionOnCheckout -> false, unreturnedConnectionTimeout -> 0, userOverrides
-> {}, usesTraditionalReflectiveProxies -> false ]
成功删除1条数据
```

图 9.8　执行 TestJDBCTemplate08 类的结果

从以上运行结果可以看出，程序输出成功删除一条数据的提示信息，由此可见，JDBCTemplate 类实现了对数据表中数据的删除。

9.3 使用 Spring JDBC 完成 Dao 封装

9.3.1 通过直接注入 JDBCTemplate 的方式

在实际开发中,程序与数据库的交互通常由 Dao 完成,如果想要在 Dao 中使用 JDBCTemplate 类对象,那么就需要将 JDBCTemplate 类对象注入到 Dao 类中,进而通过 JDBCTemplate 类对象完成对数据库的操作。

接下来通过一个实例对以上过程进行演示,具体步骤如下。

(1) 在 src 目录下创建 com.qfedu.dao 包,在 com.qfedu.dao 包下创建接口 UserDao,具体代码如例 9-10 所示。

【例 9-10】 UserDao.Java

```
package com.qfedu.dao;
import java.util.List;
import com.qfedu.bean.Student;
public interface UserDao {
    public void insert(Student student);          //添加
    public void delete(Integer sid);              //删除
    public void update(Student role);             //更新
    public Student selectOne(Integer sid);        //查询单个
    public List<Student> selectAll();             //查询所有
}
```

在以上代码中,UserDao 接口定义了对数据表中数据进行添加、删除、更新、查询的方法。

(2) 在 com.qfedu.dao 包下创建接口 UserDao 的实现类 UserDaoImpl01,具体代码如例 9-11 所示。

【例 9-11】 UserDaoImpl01.Java

```
package com.qfedu.dao;
import java.util.List;
import org.springframework.jdbc.core.*;
import com.qfedu.bean.Student;
public class UserDaoImpl01 implements UserDao {
    private JdbcTemplate jdbcTemplate;
    public void setJdbcTemplate(JdbcTemplate jdbcTemplate) {
        this.jdbcTemplate = jdbcTemplate;
    }
    @Override
    public void insert(Student student) {
        String sql = "insert into student(sname,age,course) value (?,?,?) ";
        jdbcTemplate.update(sql,
                student.getSname(),student.getAge(),student.getCourse());
    }
```

```java
    @Override
    public void delete(Integer sid) {
        String sql = "delete from student where sid = ?";
        jdbcTemplate.update(sql, sid);
    }
    @Override
    public void update(Student student) {
        String sql = "update student set sname = ? ,age = ? ,course = ?";
        jdbcTemplate.update(sql,
                student.getSname(),student.getAge(),student.getCourse());
    }
    @Override
    public Student selectOne(Integer sid) {
        String sql = "select * from student where sid = ?";
        BeanPropertyRowMapper<Student> rowMapper = new
                BeanPropertyRowMapper<Student>(Student.class);
        return jdbcTemplate.queryForObject(sql,rowMapper,sid);
    }
    @Override
    public List<Student> selectAll() {
        String sql = "select * from student";
        BeanPropertyRowMapper<Student> rowMapper = new
                BeanPropertyRowMapper<Student>(Student.class);
        return jdbcTemplate.query(sql,rowMapper);
    }
}
```

在以上代码中，UserDaoImpl01 类实现了对数据表中数据进行添加、删除、更新、查询的方法，由于需要使用 JDBCTemplate 类对象，因此 UserDaoImpl01 类将 JDBCTemplate 类对象作为属性并提供了 set()方法。

(3) 在 applicationContext.xml 文件的< bean >元素中添加配置信息，具体代码如下所示。

```xml
<bean name="student" class="com.qfedu.bean.Student"/>
<bean name="userDao01" class="com.qfedu.dao.UserDaoImpl01">
    <property name="jdbcTemplate" ref="jdbcTemplate"/>
</bean>
```

根据以上配置信息，Spring 将 JDBCTemplate 类对象注入到 UserDaoImpl01 类中。

(4) 在 com.qfedu.test 包下创建测试类 TestUserDao01，具体代码如例 9-12 所示。

【例 9-12】 TestUserDao01.Java

```java
package com.qfedu.test;
import org.springframework.context.ApplicationContext;
import org.springframework.context.support
        .ClassPathXmlApplicationContext;
import com.qfedu.bean.Student;
```

```
import com.qfedu.dao.UserDao;
public class TestUserDao01 {
    public static void main(String[] args) {
        ApplicationContext context = new
              ClassPathXmlApplicationContext("applicationContext.xml");
        UserDao userDao = context.getBean("userDao01",UserDao.class);
        Student student = context.getBean("student",Student.class);
        student.setSname("ZhouBa");
        student.setAge("20");
        student.setCourse("Java");
        userDao.insert(student);
    }
}
```

在以上代码中,程序首先从 Spring 容器中获取 UserDaoImpl01 类的对象,然后调用 insert()方法向数据表插入数据。

执行 TestUserDao01 类,由于 UserDaoImpl01 类的 insert()方法的返回值为 void,因此需要通过命令行窗口查看程序运行结果,运行结果如下所示。

```
mysql> select * FROM student;
+-----+----------+------+--------+
| sid | sname    | age  | course |
+-----+----------+------+--------+
|   1 | ZhangSan | 20   | Java   |
|   2 | LiSi     | 21   | Python |
|   3 | WangWu   | 21   | Java   |
|   5 | ZhouBa   | 20   | Java   |
+-----+----------+------+--------+
4 rows in set (0.00 sec)
```

从以上运行结果可以看出,student 表中增加了一条 sid 值为 5 的记录。由此可见,将 JDBCTemplate 类对象直接注入 UserDaoImpl01 类可以成功实现 Dao 的封装。

9.3.2 通过继承 JDBCDaoSupport 类的方式

在具体应用中,除了将 JDBCTemplate 类对象直接注入 Dao 类之外,还可以使用继承 JDBCDaoSupport 类的方式实现 Dao 封装。JDBCDaoSupport 类是 Spring JDBC 提供的类,它内部定义了 JDBCTemplate 类型的成员变量,接下来演示使用 JDBCDaoSupport 类实现 Dao 封装。

(1) 在 com.qfedu.dao 包下创建类 UserDaoImpl02,具体代码如例 9-13 所示。

【例 9-13】 UserDaoImpl02.Java

```
package com.qfedu.dao;
import java.util.List;
import org.springframework.jdbc.core.BeanPropertyRowMapper;
```

```
import org.springframework.jdbc.core.support.JdbcDaoSupport;
import com.qfedu.bean.Student;
public class UserDaoImpl02 extends JdbcDaoSupport implements UserDao {
    @Override
    public void insert(Student student) {
        String sql = "insert into student(sname,age,course) value (?,?,?)";
        getJdbcTemplate().update(sql,
        student.getSname(),student.getAge(),student.getCourse());
    }
    @Override
    public void delete(Integer sid) {
        String sql = "delete from student where sid = ?";
        getJdbcTemplate().update(sql, sid);
    }
    @Override
    public void update(Student student) {
        String sql = "update student set sname = ? ,age = ? ,course = ?";
        getJdbcTemplate().update(sql,
                student.getSname(),student.getAge(),student.getCourse());
    }
    @Override
    public Student selectOne(Integer sid) {
        String sql = "select * from student where sid = ?";
        BeanPropertyRowMapper<Student> rowMapper = new
                BeanPropertyRowMapper<Student>(Student.class);
        return getJdbcTemplate().queryForObject(sql,rowMapper,sid);
    }
    @Override
    public List<Student> selectAll() {
        String sql = "select * from student";
        BeanPropertyRowMapper<Student> rowMapper = new
                BeanPropertyRowMapper<Student>(Student.class);
        return getJdbcTemplate().query(sql,rowMapper);
    }
}
```

在以上代码中，因为 UserDaoImpl02 类继承了 JDBCDaoSupport 类，因此可以在 UserDaoImpl02 类中通过 getJDBCTemplate()方法获取 JDBCTemplate 类对象。

(2) 在 applicationContext.xml 文件的<bean>元素中添加配置信息，具体代码如下所示。

```
<bean name="userDao02" class="com.qfedu.dao.UserDaoImpl02">
        <property name="jdbcTemplate" ref="jdbcTemplate"/>
</bean>
```

根据以上配置信息，Spring 将 JDBCTemplate 类对象注入到 UserDaoImpl02 类中。

(3) 在 com.qfedu.test 包下创建测试类 TestUserDao02，具体代码如例 9-14 所示。

【例 9-14】 TestUserDao02.Java

```
package com.qfedu.test;
import org.springframework.context.ApplicationContext;
import org.springframework.context.support
      .ClassPathXmlApplicationContext;
import com.qfedu.bean.Student;
import com.qfedu.dao.UserDao;
public class TestUserDao02 {
    public static void main(String[] args) {
        ApplicationContext context = new ClassPathXmlApplicationContext
             ("applicationContext.xml");
        UserDao userDao = context.getBean("userDao02",UserDao.class);
        Student student = context.getBean("student",Student.class);
        student.setSname("WuJiu");
        student.setAge("19");
        student.setCourse("Java");
        userDao.insert(student);
    }
}
```

执行 TestUserDao02 类，由于 UserDaoImpl02 类的 insert()方法的返回值为 void，因此需要通过命令行窗口查看程序运行结果，运行结果如下所示。

```
mysql > SELECT * FROM student;
+-----+----------+------+--------+
| sid | sname    | age  | course |
+-----+----------+------+--------+
|   1 | ZhangSan | 20   | Java   |
|   2 | LiSi     | 21   | Python |
|   3 | WangWu   | 21   | Java   |
|   5 | ZhouBa   | 20   | Java   |
|   6 | WuJiu    | 19   | Java   |
+-----+----------+------+--------+
5 rows in set (0.00 sec)
```

从以上运行结果可以看出，student 表中增加了一条 sid 值为 6 的记录。由此可见，通过继承 JDBCDaoSupport 类可以成功实现 Dao 的封装。

当通过继承 JDBCDaoSupport 类实现 Dao 封装时，也可以采用直接注入数据源的形式，修改 applicationContext.xml 文件中 name 属性为 userDao02 的< bean >元素的配置信息，修改后的代码如下所示。

```
<bean name = "userDao02" class = "com.qfedu.dao.UserDaoImpl02">
    <property name = "dataSource" ref = "dataSource"/>
</bean>
```

再次执行 TestUserDao02 类，通过命令行查看运行结果，运行结果如下所示。

```
mysql > SELECT * FROM student;
+-----+----------+------+--------+
| sid | sname    | age  | course |
+-----+----------+------+--------+
|   1 | ZhangSan | 20   | Java   |
|   2 | LiSi     | 21   | Python |
|   3 | WangWu   | 21   | Java   |
|   5 | ZhouBa   | 20   | Java   |
|   6 | WuJiu    | 19   | Java   |
|   7 | WuJiu    | 19   | Java   |
+-----+----------+------+--------+
6 rows in set (0.00 sec)
```

从以上运行结果可以看出，student 表中增加了一条 sid 值为 7 的记录。

9.4 本章小结

本章首先介绍了 Spring JDBC 的基础知识，包括 Spring JDBC 的基本概况、JDBCTemplate 类的核心 API、JDBCTemplate 类的简单应用以及在 Spring 中管理 JDBCTemplate 类等；其次通过实例演示了 JDBCTemplate 类操作数据库的方法，包括 DDL 操作、DQL 操作、DML 操作等；最后讲解了使用 Spring JDBC 完成 Dao 封装的两种方法。通过本章知识的学习，大家应该能理解 Spring JDBC 的概念和优势，理解 JDBCTemplate 类的常用 API，掌握通过 JDBCTemplate 类操作数据库的方法，能够通过 Spring JDBC 完成 Dao 的封装。

9.5 习　　题

1. 填空题

(1) Spring JDBC 的核心类指的是________________。

(2) 当执行 DQL 操作时，通常调用 JDBCTemplate 类的________方法和________方法。

(3) 当执行 DML 操作时，通常调用 JDBCTemplate 类的________方法和________方法。

(4) 当执行 DDL 操作时，通常调用 JDBCTemplate 类的________方法。

(5) 使用 Spring JDBC 封装 Dao 时，可以通过继承________类的方式实现。

2. 选择题

(1) 关于 Spring JDBC，下列选项错误的是(　　)。

A. Spring JDBC 是对传统 JDBC 的改善和增强

B. Spring JDBC 和传统 JDBC 完全没有关联

C. Spring JDBC 的 core 包负责提供核心功能

D. Spring JDBC 的 support 包负责提供支持类

(2) 在下列选项中,不属于 C3P0 连接池必配属性的是(　　)。

A. driverClass　　B. jdbcUrl

C. user　　D. initialPoolSize

(3) 在 JDBCTemplate 类提供的方法中,用于查询单条记录的是(　　)。

A. batchUpdate()　　B. queryForList()

C. queryForObject()　　D. update()

(4) 关于 JDBCTemplate 类提供的方法,下列说法错误的是(　　)。

A. execute()常用于执行 DQL 操作

B. queryForList()常用于执行 DQL 操作

C. update()常用于执行 DML 操作

D. queryForObject 常用于执行 DQL 操作

(5) 关于 Spring JDBC 中的结果集处理,下列选项错误的是(　　)。

A. 如果要将表记录映射为自定义的类,则需要使用 RowMapper<T>接口

B. 如果查询返回的结果为 int 类型,则需要使用 RowMapper<T>接口完成映射

C. RowMapper<T>接口定义了 mapRow()方法并通过该方法提供功能

D. BeanPropertyRowMapper<T>是 RowMapper<T>接口的实现类

3. 思考题

(1) 简述 Spring JDBC 的概念。

(2) 简述使用 Spring JDBC 封装 Dao 的方法。

4. 编程题

请通过 Spring JDBC 编写程序并完成以下步骤。

(1) 在数据库 chapter09 中创建一张名称为 teacher 的数据表,具体表结构如表 9.3 所示。

表 9.3　teacher 数据表

字段名称	数据类型	备注说明
tid	INT	教师编号,主键,自增长
tname	VARCHAR(20)	教师姓名
age	VARCHAR(20)	教师年龄
course	VARCHAR(20)	课程

(2) 向数据表 teacher 中插入一条数据,具体字段值如表 9.4 所示。

表 9.4　插入数据

tname	age	course
ZhangSan	32	Java

(3) 查询数据表 teacher 中的记录条数。

第10章 Spring管理数据库事务

本章学习目标

- 理解 Spring 对数据库事务的支持
- 理解 Spring 管理数据库事务的核心 API
- 掌握以编程方式管理数据库事务
- 掌握以 XML 文件方式完成声明式事务管理
- 掌握以注解方式完成声明式事务管理

在项目开发中,数据库事务与程序的并发性能以及程序正确、安全读取数据息息相关。随着项目规模的扩大和业务逻辑的增多,事务管理成为开发人员在编码过程中的一项耗时且烦琐的工作。为了简化事务管理的过程,提升开发效率,Spring 提供了通用的事务管理的解决方案,这也是 Spring 作为一站式企业应用平台的优势所在。接下来,本章将对 Spring 管理数据库事务涉及的相关知识进行详细讲解。

10.1 Spring与事务管理

10.1.1 Spring对事务管理的支持

事务是指数据库中的一个操作序列,它由一系列的 SQL 指令组成。在 Java EE 程序开发中,事务管理是一个影响范围较广的领域,在程序与数据库交互时,保证事务的正确执行尤为重要。由于实际开发中事务管理存在的诸多弊端,Spring 框架针对事务管理提供了自己的解决方案。

对于事务管理,Spring 采用的方式是通过在高层次建立事务抽象,然后在此基础上提供一个统一的编程模型,这意味着,Spring 具有在多种环境中配置和使用事务的能力,无论是 Spring JDBC,还是以 MyBatis 为代表的 ORM 框架,Spring 都能够使用统一的编程模型对事务进行管理并为事务管理提供通用的支持。

基于 Spring IOC 和 Spring AOP,Spring 提供了声明式事务管理的方式,它允许开发人员直接在 Spring 容器中定义事务的边界和属性,除此之外,Spring 还实现了事务管理和数据访问的分离,在这种条件下,开发人员只需关注对当前事务的界定,其余工作将由 Spring 框架自动完成。

10.1.2 事务管理的核心接口

Spring 主要通过三个接口实现事务抽象,这三个接口分别是 TransactionDefinition、

TransactionStatus 和 PlatformTransactionManager，它们都位于 org. springframework. transaction 包中，其中，TransactionDefinition 用于定义事务的属性，TransactionStatus 用于界定事务的状态，PlatformTransactionManager 根据属性管理事务。

1. TransactionDefinition

TransactionDefinition 接口主要用于定义事务的属性，这些属性包括事务的隔离级别、事务的传播行为、事务的超时时间、是否为只读事务等。

1）事务的隔离级别

事务的隔离级别是指事务之间的隔离程度，TransactionDefinition 定义了五种隔离级别，具体如表 10.1 所示。

表 10.1 TransactionDefinition 定义的隔离级别

隔离级别	描述
ISOLATION_DEFAULT	采用当前数据库默认的隔离级别
ISOLATION_READ_UNCOMMITTED	允许读取尚未提交的数据变更，可能会导致脏读、幻读或不可重复读
ISOLATION_READ_COMMITTED	允许读取已经提交的数据变更，可以避免脏读，无法避免幻读或不可重复读
ISOLATION_REPEATABLE_READ	允许可重复读，可以避免脏读、不可重复读，资源消耗上升
ISOLATION_SERIALIZABLE	事务串行执行，资源消耗最大

表 10.1 列举了 TransactionDefinition 定义的五种隔离级别，除了 ISOLATION_DEFAULT 是 TransactionDefinition 特有的之外，其余四个分别与 java. sql. Connection 接口定义的隔离级别相对应。

2）事务的传播行为

事务的传播行为是指事务处理过程所跨越的对象将以什么样的方式参与事务，TransactionDefinition 定义了七种事务传播行为，具体如表 10.2 所示。

表 10.2 TransactionDefinition 定义的事务传播行为

传播行为	描述
PROPAGATION_REQUIRED	默认的事务传播行为。如果当前存在一个事务，则加入该事务；如果当前没有事务，则创建一个新的事务
PROPAGATION_SUPPORTS	如果当前存在一个事务，则加入该事务；如果当前没有事务，则直接执行
PROPAGATION_MANDATORY	当前必须存在一个事务，如果没有，就抛出异常
PROPAGATION_REQUIRES_NEW	创建一个新的事务，如果当前已存在一个事务，将已存在的事务挂起
PROPAGATION_NOT_SUPPORTED	不支持事务，在没有事务的情况下执行，如果当前已存在一个事务，则将已存在的事务挂起
PROPAGATION_NEVER	永远不支持当前事务，如果当前已存在一个事务，则抛出异常
PROPAGATION_NESTED	如果当前存在事务，则在当前事务的一个子事务中执行

表 10.2 列举了 TransactionDefinition 定义的事务传播行为，Spring 中声明式事务对传播行为依赖较大，开发人员可根据实际需要选择使用。

3）事务的超时时间和是否只读

事务的超时时间是指事务执行的时间界限，超过这个时间界限，事务将会回滚。TransactionDefinition 接口提供了 TIMEOUT_DEFAULT 的常量定义，用来指定事务的超时时间。

当事务的属性为只读时，该事务不修改任何数据，只读事务有助于提升性能，如果在只读事务中修改数据，可能会引发异常。

4）TransactionDefinition 接口的方法

TransactionDefinition 接口提供了一系列方法来获取事务的属性，具体如表 10.3 所示。

表 10.3　TransactionDefinition 接口的方法

方　　法	描　　述
IntgetPropagationBehavior()	返回事务的传播行为
int getIsolationLevel()	返回事务的隔离层次
int getTimeout()	返回事务的超时属性
boolean isReadOnly()	判断事务是否为只读
String getName()	返回定义的事务名称

表 10.3 列举了 TransactionDefinition 接口提供的方法，程序可通过 TransactionDefinition 接口的这些方法获取当前事务的属性。

2. TransactionStatus

TransactionStatus 接口主要用于界定事务的状态，通常情况下，编程式事务中使用该接口较多。

TransactionStatus 接口中提供了一系列返回事务状态信息的方法，具体如表 10.4 所示。

表 10.4　TransactionStatus 接口的方法

方　　法	描　　述
boolean isNewTransaction()	判断当前事务是否为新事务
boolean hasSavepoint()	判断当前事务是否创建了一个保存点
boolean isRollbackOnly()	判断当前事务是否被标记为 rollback-only
void setRollbackOnly()	将当前事务标记为 rollback-only
boolean isCompleted()	判断当前事务是否已经完成（提交或回滚）
void flush()	刷新底层的修改到数据库

表 10.4 列举了 TransactionStatus 接口提供的方法，事务管理器可以通过该接口提供的方法获取事务运行的状态信息，除此之外，事务管理器可以通过 setRollbackOnly() 方法间接回滚事务。

3. PlatformTransactionManager

PlatformTransactionManager 接口是 Spring 事务管理的中心接口，它真正执行了事务管理的职能，并针对不同的持久化技术封装了对应的实现类。

PlatformTransactionManager 接口提供了一系列方法用于管理事务，具体如表 10.5 所示。

表 10.5 PlatformTransactionManager 接口的方法

方　法	描　述
TransactionStatus getTransaction (TransactionDefinition definition)	根据事务定义获取一个已存在的事务或创建一个新的事务，并返回这个事务的状态
void commit(TransactionStatus status)	根据事务的状态提交事务
void rollback(TransactionStatus status)	根据事务的状态回滚事务

表 10.5 列举了 PlatformTransactionManager 接口提供的方法，在实际应用中，Spring 事务管理实际是由具体的持久化技术来完成的，而 PlatformTransactionManager 接口只提供统一的抽象方法。为了应对不同持久化技术的差异性，Spring 为它们提供了具体的实现类，例如，Spring 为 Spring JDBC 或 MyBatis 等依赖于 DataSource 的持久化技术提供了实现类 DataSourceTransactionManager，该类位于 org. springframework. jdbc. datasource 包中，如此一来，Spring JDBC 或 MyBatis 等持久化技术的事务管理由 DataSourceTransactionManager 来实现，而且 Spring 可以通过 PlatformTransactionManager 接口实现统一管理。

10.2 编程式事务管理

在 Spring 中，事务管理的方式主要有两种，它们分别是编程式事务管理和声明式事务管理。编程式事务管理通过开发人员手动编码实现事务管理，声明式事务管理是基于 Spring AOP 技术将事务管理的逻辑抽取，然后再织入到业务类中。在实际开发中，与编程式事务管理相比，声明式事务管理的应用范围更为广泛。

由于编程式事务管理的部分步骤是 Spring 自动完成的，为了让大家更加深入地理解 Spring 事务管理并能够完成一些特殊场景下的事务管理，本节首先讲解 Spring 的编程式事务管理。

Spring 为编程式事务管理提供了一个模板类 TransactionTemplate，该类位于 org. springframework. transaction. support 包中，TransactionTemplate 类封装了与 PlatformTransactionManager 相关的事务界定操作，开发人员可以通过对应的 callback()接口实现具体内容的界定。

TransactionTemplate 类提供了一系列方法，具体如表 10.6 所示。

表 10.6 TransactionTemplate 类的方法

方　法	描　述
void setTransactionManager (PlatformTransactionManager transactionManager)	设置事务管理器
PlatformTransactionManager getTransactionManager()	返回事务管理器
T execute(TransactionCallback<T> action)	在 TransactionCallback 接口中定义需要以事务方式组织的数据访问逻辑
void rollbackOnException(TransactionStatus status, Throwable ex)	处理事务遇到异常时调用的方法

表 10.6 列举出了 TransactionTemplate 类提供的方法，其中 execute()方法中需要传入 TransactionCallback<T>类型的参数，这也是开发者在编写程序时经常使用的。

日常生活中大家经常使用网银转账，当执行转账操作后，转出账户要减去相应的金额，转入账户要增加相应的金额。通常情况下，后台程序中的减去和增加这两次操作会构成一个事务，因为银行证券类系统对数据安全性、一致性要求较高，因此保证事务的正确执行显得尤为重要。接下来，本节将通过实例演示编程式事务管理。

(1) 在 MySQL 中创建数据库 chapter10 和数据表 account，SQL 语句如下所示。

```
DROP DATABASE IF EXISTS chapter10;
CREATE DATABASE chapter10;
USE chapter10;
CREATE TABLE account(
id INT PRIMARY KEY AUTO_INCREMENT, #ID
aname VARCHAR(20),      #姓名
money DOUBLE            #余额
);
```

(2) 向数据表 account 添加数据，SQL 语句如下所示。

```
INSERT INTO account(aname,money) VALUES('ZhangSan',5000);
INSERT INTO account(aname,money) VALUES('LiSi',20000);
```

(3) 通过 SQL 语句测试数据是否添加成功，运行结果如下所示。

```
mysql> select * from account;
+----+----------+-------+
| id | aname    | money |
+----+----------+-------+
| 1  | ZhangSan | 5000  |
| 2  | LiSi     | 20000 |
+----+----------+-------+
2 rows in set (0.00 sec)
```

从以上运行结果可以看出，数据添加成功。

(4) 在 Eclipse 中创建 Web 工程 chapter10，将事务管理所需的相关 jar 包添加到 lib 目录下，完成导包。本章所用 jar 包与第 9 章所用 jar 包相同，具体如图 9.1 所示，此处不再赘述。

(5) 在工程的 src 目录下创建包 com.qfedu.dao 包，在该包下创建接口 AccountDao，该接口定义了增加金额和减少金额的方法，具体代码如例 10-1 所示。

【例 10-1】 AccountDao.Java

```
package com.qfedu.dao;
public interface AccountDao {
    void increaseMoney(Integer id , Double money);
    void decreaseMoney(Integer id , Double money);
}
```

(6) 在 com. qfedu. dao 包下创建 AccountDao 接口的实现类 AccountDaoImpl，该接口实现了增加金额和减少金额的方法，具体代码如例 10-2 所示。

【例 10-2】 AccountDaoImpl. Java

```
package com.qfedu.dao;
import org.springframework.jdbc.core.JdbcTemplate;
public class AccountDaoImpl implements AccountDao {
    private JdbcTemplate jdbcTemplate;
    public void setJdbcTemplate(JdbcTemplate jdbcTemplate) {
        this.jdbcTemplate = jdbcTemplate;
    }
    public void increaseMoney(Integer id, Double money) {
        jdbcTemplate.update("update account set money = money + ? "
            + "where id = ? ;",money,id);
    }
    public void decreaseMoney(Integer id, Double money) {
        jdbcTemplate.update("update account set money = money - ? "
            + "where id = ? ;",money,id);
    }
}
```

在以上代码中，程序通过 JdbcTemplate 类对象的 update()方法操作数据库，进而实现账户金额的增加或减少。

(7) 在工程 chapter10 的 src 目录下创建 com. qfedu. service 包，在 com. qfedu. service 包下创建接口 AccountService，该接口定义了转账的方法，具体代码如例 10-3 所示。

【例 10-3】 AccountService. Java

```
package com.qfedu.service;
public interface AccountService {
    void transfer(Integer from , Integer to , Double money);
}
```

(8) 在 com. qfedu. service 包下创建 AccountService 接口的实现类 AccountServiceImpl01，该接口实现了转账的方法，具体代码如例 10-4 所示。

【例 10-4】 AccountServiceImpl01. Java

```
package com.qfedu.service;
import org.springframework.transaction.TransactionStatus;
import org.springframework.transaction.support.*;
import com.qfedu.dao.AccountDao;
public class AccountServiceImpl01 implements  AccountService{
    private AccountDao  accountDao;
    //事务管理类
    private TransactionTemplate transactionTemplate;
    public void setAccountDao(AccountDao accountDao) {
        this.accountDao = accountDao;
```

```
    }
    public void setTransactionTemplate(TransactionTemplate
        transactionTemplate) {
        this.transactionTemplate = transactionTemplate;
    }
    @Override
    public void transfer(Integer from, Integer to, Double money) {
            //调用 TransactionTemplate 类对象执行 execute()方法
            transactionTemplate.execute(new
            TransactionCallbackWithoutResult() {
            @Override
            //需要在事务环境中执行的代码
            protected void doInTransactionWithoutResult(TransactionStatus
                transactionStatus) {
                accountDao.decreaseMoney(from,money);
                accountDao.increaseMoney(to,money);
            }
        });
    }
}
```

在以上代码中，程序通过 TransactionTemplate 类实现事务管理。在调用 execute()方法时，传入的参数类型为 TransactionCallbackWithoutResult，TransactionCallbackWithoutResult 类是 TransactionCallback 接口的实现类，它提供有一个 doInTransactionWithoutResult()方法，该方法用于调用需要在事务环境中执行的代码。如此一来，当外部类调用 AccountServiceImpl01 的 transfer()方法时，增加账户金额和减少账户金额的操作将会在事务环境中执行。

(9) 在工程 chapter10 的 src 目录下创建 applicationContext_1.xml 文件，具体代码如下所示。

```
1  <?xml version="1.0" encoding="UTF-8"?>
2  <beans xmlns="http://www.springframework.org/schema/beans"
3      xmlns:xsi="http://www.w3.org/2001/XMLSchema-instance"
4      xmlns:context="http://www.springframework.org/schema/context"
5      xmlns:aop="http://www.springframework.org/schema/aop"
6      xmlns:tx="http://www.springframework.org/schema/tx"
7      xsi:schemaLocation="http://www.springframework.org/schema/beans
8          http://www.springframework.org/schema/beans/spring-beans.xsd
9          http://www.springframework.org/schema/context
10         http://www.springframework.org/schema/context/spring-context.xsd
11         http://www.springframework.org/schema/aop
12         http://www.springframework.org/schema/aop/spring-aop.xsd
13         http://www.springframework.org/schema/tx
14         http://www.springframework.org/schema/tx/spring-tx.xsd">
15 </beans>
```

在以上代码中，第 6 行、第 13 行、第 14 行用于引入事务管理的命名空间。

(10) 在 applicationContext_1.xml 文件中的< bean >元素中添加关于数据源和 JdbcTemplate 类的配置信息，具体代码如下所示。

```
<!-- 引入外部 properties 文件 -->
<context:property-placeholder location="classpath:jdbc.properties"/>
<!-- 注册数据源 -->
<bean name="dataSource" class="com.mchange.v2.c3p0.ComboPooledDataSource">
    <property name="driverClass" value="${jdbc.driverClass}"/>
    <property name="jdbcUrl" value="${jdbc.jdbcUrl}"/>
    <property name="user" value="${jdbc.user}"/>
    <property name="password" value="${jdbc.password}"/>
</bean>
  <!-- 注册 JdbcTemplate 类 -->
  <bean name="jdbcTemplate"
class="org.springframework.jdbc.core.JdbcTemplate">
      <property name="dataSource" ref="dataSource"/>
</bean>
```

(11) 在 applicationContext_1.xml 文件中的< bean >元素中添加关于事务管理器和 TransactionTemplate 类的配置信息，具体代码如下所示。

```
 <!-- 注册事务管理器 -->
 <bean id="transactionManager"
class="org.springframework.jdbc.datasource.DataSourceTransactionManager">
      <property name="dataSource" ref="dataSource"/>
 </bean>
   <!-- 注册 TransactionTemplate 类 -->
   <bean id="transactionTemplate" class="org.springframework.
        transaction.support.TransactionTemplate">
      <property name="transactionManager" ref="transactionManager" />
   </bean>
```

在以上代码中，DataSourceTransactionManager 类是 Spring 为 Spring JDBC 或 MyBatis 提供的事务管理器，DataSourceTransactionManager 类要依赖数据源才能实现功能，因此需要为该类注入数据源。与此类似，当使用 Spring JDBC 时，TransactionTemplate 类要依赖 DataSourceTransactionManager 类实现功能，因此需要在配置文件中完成相应的注入。完成以上配置后，程序即可通过 TransactionTemplate 完成事务管理。

(12) 在 applicationContext_1.xml 文件中的< bean >元素中添加关于 AccountDaoImpl 类和 AccountServiceImpl01 类的配置信息，具体代码如下所示。

```
<bean name="accountDao"  class="com.qfedu.dao.AccountDaoImpl">
      <property name="jdbcTemplate" ref="jdbcTemplate"></property>
</bean>
<bean name="accountService"
class="com.qfedu.service.AccountServiceImpl01">
    <property name="accountDao" ref="accountDao"></property>
```

```
        <property name="transactionTemplate"
ref="transactionTemplate"></property>
</bean>
```

在以上代码中，AccountDaoImpl 类需要依赖 JdbcTemplate 类实现功能，AccountServiceImpl01 类要依赖 AccountDaoImpl 类和 TransactionTemplate 类实现功能，因此需要在配置文件中完成相应的配置。

(13) 在工程 chapter10 的 src 目录下新建文件 jdbc. properties，具体代码如例 10-5 所示。

【例 10-5】 jdbc. properties

```
jdbc.driverClass=com.mysql.jdbc.Driver
jdbc.jdbcUrl=jdbc:mysql://localhost:3306/chapter10
jdbc.user=root
jdbc.password=root
```

(14) 在工程 chapter10 的 src 目录下新建包 com. qfedu. test，在该包下新建类 TestAccountService01，具体代码如例 10-6 所示。

【例 10-6】 TestAccountService01. java

```
package com.qfedu.test;
import org.springframework.context.ApplicationContext;
import org.springframework.context.support
    .ClassPathXmlApplicationContext;
import com.qfedu.service.AccountService;
public class TestAccountService01 {
    public static void main(String[] args) {
        ApplicationContext context = new
            ClassPathXmlApplicationContext("applicationContext_1.xml");
        AccountService accountService = context.getBean("accountService",
             AccountService.class);
        accountService.transfer(1, 2, 1000.0);
    }
}
```

在以上代码中，程序首先通过 Spring 容器获取 AccountService 实例，然后调用其 transfer()方法执行转账操作。

执行 TestAccountService01 类，由于 AccountService 的 transfer()方法的返回值为 void，因此需要通过命令行窗口查看程序运行结果，运行结果如下所示。

```
mysql> select * from account;
+----+----------+-------+
| id | aname    | money |
+----+----------+-------+
| 1  | ZhangSan | 4000  |
| 2  | LiSi     | 21000 |
```

```
+----+----------+-------+
2 rows in set (0.03 sec)
```

从以上运行结果可以看出，数据表中的相关数据已经被更新，由此可见，事务得到正确执行。

修改 AccountServiceImpl01 类，在第 25 行后添加代码，如下所示。

```
accountDao.decreaseMoney(from,money);
int num = 1/0;
accountDao.increaseMoney(to,money);
```

在以上代码中，1/0 将会导致程序异常，当程序出现异常时，这段代码所在的事务将会回滚，事务中对数据表的操作不会被提交。

再次执行 TestAccountService01 类，此时程序出现异常，通过命令行窗口查看程序运行结果，运行结果如下所示。

```
mysql> select * from account;
+----+----------+-------+
| id | aname    | money |
+----+----------+-------+
| 1  | ZhangSan |  4000 |
| 2  | LiSi     | 21000 |
+----+----------+-------+
2 rows in set (0.03 sec)
```

从以上运行结果可以看出，数据表中的相关数据没有发生变化，由此可见，事务回滚，事务中对数据表的操作没有被提交。

10.3 声明式事务管理

10.3.1 使用 XML 配置声明式事务

在编程式事务管理中，事务管理代码和逻辑代码混合在一起，这降低了程序的可维护性，除此之外，反复调用 TransactionTemplate 类会导致大量重复代码的出现，而声明式事务管理可以避免这些问题。声明式事务管理建立在 Spring AOP 之上，可以通过 XML 或注解两种方式实现，本节首先讲解以 XML 形式实现声明式事务管理。

在使用 XML 文件完成声明式事务管理时，首先要引入 tx 命名空间。在引入 tx 命名空间之后，可以使用<tx:advice>元素来配置事务管理的通知，进而通过 Spring AOP 实现事务管理。

<tx:advice>元素包含两个属性，它们分别是 id 属性和 transaction-manager 属性，其中 id 属性用于配置<tx:advice>元素的 id 值，transaction-manager 属性用于配置 TransactionManager。

除此之外，<tx:advice>元素还包含有子元素<tx:attributes>，<tx:attributes>元素中

可以配置多个<tx:method>子元素，这些<tx:method>子元素主要用于配置事务的属性。

<tx:method>元素提供了一系列属性用于事务的定义，具体如表 10.7 所示。

表 10.7　<tx:advice>元素的属性

属　　性	描　　述
name	用于指定方法名的匹配模式
propagation	用于指定事务的传播行为
isolation	用于指定事务的隔离级别
timeout	用于指定事务的超时时间
read-only	用于指定事务是否为只读
no-rollback-for	用于指定不会导致事务回滚的异常
rollback-for	用于指定会导致事务回滚的异常

表 10.7 列举了<tx:advice>元素的属性，其中 name、propagation 和 isolation 这三个属性的应用范围较广，开发人员可根据实际需要选择使用。

接下来用一个实例演示以 XML 形式实现声明式事务管理，具体步骤如下所示。

(1) 在 com.qfedu.service 包下创建 AccountService 接口的实现类 AccountServiceImpl02，具体代码如例 10-7 所示。

【例 10-7】　AccountServiceImpl02.java

```
package com.qfedu.service;
import com.qfedu.dao.AccountDao;
public class AccountServiceImpl02 implements AccountService{
    private AccountDao accountDao;
    public void setAccountDao(AccountDao accountDao) {
        this.accountDao = accountDao;
    }
    @Override
    public void transfer(Integer from, Integer to, Double money) {
        accountDao.decreaseMoney(from,money);
        accountDao.increaseMoney(to,money);
    }
}
```

在以上代码中，transfer()方法中封装了增加金额和减少金额的操作，如果想要让 transfer()方法在事务环境中执行，需要在 Spring 的配置文件中完成相应的配置。

(2) 在工程 chapter10 的 src 目录下创建 applicationContext_2.xml 文件，具体代码如下所示。

```
<?xml version="1.0" encoding="UTF-8"?>
<beans xmlns="http://www.springframework.org/schema/beans"
    xmlns:xsi="http://www.w3.org/2001/XMLSchema-instance"
    xmlns:context="http://www.springframework.org/schema/context"
    xmlns:aop="http://www.springframework.org/schema/aop"
    xmlns:tx="http://www.springframework.org/schema/tx"
```

```
    xsi:schemaLocation = "http://www.springframework.org/schema/beans
        http://www.springframework.org/schema/beans/spring-beans.xsd
        http://www.springframework.org/schema/context
        http://www.springframework.org/schema/context/spring-context.xsd
        http://www.springframework.org/schema/aop
        http://www.springframework.org/schema/aop/spring-aop.xsd
        http://www.springframework.org/schema/tx
        http://www.springframework.org/schema/tx/spring-tx.xsd">
    <!-- 引入外部 properties 文件 -->
    <context:property-placeholder location = "classpath:jdbc.properties"/>
    <!-- 注册数据源 -->
    <bean name = "dataSource"
            class = "com.mchange.v2.c3p0.ComboPooledDataSource">
            <property name = "driverClass" value = "${jdbc.driverClass}"/>
            <property name = "jdbcUrl" value = "${jdbc.jdbcUrl}"/>
            <property name = "user" value = "${jdbc.user}"/>
            <property name = "password" value = "${jdbc.password}"/>
    </bean>
    <!-- 注册 JdbcTemplate 类 -->
    <bean name = "jdbcTemplate"
            class = "org.springframework.jdbc.core.JdbcTemplate">
        <property name = "dataSource" ref = "dataSource"/>
    </bean>
    <!-- 注册事务管理器 -->
    <bean id = "transactionManager" class =
            "org.springframework.jdbc.datasource.DataSourceTransactionManager">
        <property name = "dataSource" ref = "dataSource"/>
    </bean>
</beans>
```

以上代码用于实现 tx 命名空间的导入、数据源的注册、JdbcTemplate 类的注册以及事务管理器的注册。

(3) 在 applicationContext_2.xml 文件的< bean >元素中添加关于 AccountDaoImpl 类和 AccountServiceImpl02 类的配置信息，具体代码如下所示。

```
<bean name = "accountDao"     class = "com.qfedu.dao.AccountDaoImpl">
    <property name = "jdbcTemplate" ref = "jdbcTemplate"/>
</bean>
<bean name = "accountService"
class = "com.qfedu.service.AccountServiceImpl02">
    <property name = "accountDao" ref = "accountDao"/>
</bean>
```

(4) 在 applicationContext_2.xml 文件的< bean >元素中添加关于事务管理的配置信息，具体代码如下所示。

```
<!-- 事务通知 -->
    <tx:advice id = "txAdvice" transaction-manager = "transactionManager">
        <tx:attributes>
            <tx:method name = "save*" propagation = "REQUIRED" />
            <tx:method name = "insert*" propagation = "REQUIRED" />
            <tx:method name = "add*" propagation = "REQUIRED" />
            <tx:method name = "create*" propagation = "REQUIRED" />
            <tx:method name = "delete*" propagation = "REQUIRED" />
            <tx:method name = "update*" propagation = "REQUIRED" />
            <tx:method name = "transfer*" propagation = "REQUIRED" />
        </tx:attributes>
    </tx:advice>
    <!-- AOP配置 -->
<aop:config>
        <aop:pointcut id = "txPointCut"
            expression = "execution(* com..service..*.*(..))" />
        <aop:advisor advice-ref = "txAdvice" pointcut-ref = "txPointCut"/>
    </aop:config>
```

在以上代码中,<tx:advice>元素不但配置了 id 属性为 txAdvice 的事务通知,而且还引入了事务管理器 DataSourceTransactionManager,此处需要说明的是,<tx:advice>元素的 transaction-manager 属性值与 DataSourceTransactionManager 类的 id 值是相同的。除此之外,<tx:attributes>元素的多个<tx:method>子元素执行了执行事务的方法,这些方法的名称分别以 save、insert、add、create、delete、update、transfer 开头,它们适用事务的传播级别为 REQUIRED。因此,如果当前存在事务,这些方法直接在事务中执行;如果当前不存在事务,则创建一个新的事务。使用<tx:advice>元素配置事务通知后,通过<aop:config>元素完成 AOP 声明。

(5) 在 com.qfedu.test 包下创建 TestAccountService02,具体代码如例 10-8 所示。

【例 10-8】 TestAccountService02.java

```
package com.qfedu.test;
import org.springframework.context.ApplicationContext;
import org.springframework.context.support
        .ClassPathXmlApplicationContext;
import com.qfedu.service.AccountService;
public class TestAccountService02 {
    public static void main(String[] args) {
        ApplicationContext context = new
            ClassPathXmlApplicationContext("applicationContext_2.xml");
        AccountService accountService = context.getBean("accountService",
            AccountService.class);
        accountService.transfer(1, 2, 1000.0);
    }
}
```

在以上代码中，程序首先通过 Spring 容器获取 AccountService 实例，然后调用其 transfer()方法执行转账操作。

执行 TestAccountService02 类，由于 AccountService 的 transfer()方法的返回值为 void，因此需要通过命令行窗口查看程序运行结果，运行结果如下所示。

```
mysql > select * from account;
+----+----------+-------+
| id | aname    | money |
+----+----------+-------+
| 1  | ZhangSan | 3000  |
| 2  | LiSi     | 22000 |
+----+----------+-------+
2 rows in set (0.00 sec)
```

从以上运行结果可以看出，数据表中的相关数据已经被更新，由此可见，事务得到正确执行。

修改 AccountServiceImpl02 类，在第 10 行后添加代码，如下所示。

```
accountDao.decreaseMoney(from,money);
int num = 1/0;
accountDao.increaseMoney(to,money);
```

和 10.2 节类似，以上代码用于测试程序出现异常时事务的执行情况，如果程序出现异常，那么事务将会回滚。

再次执行 TestAccountService02 类，此时程序出现异常，通过命令行窗口查看程序运行结果，运行结果如下所示。

```
mysql > select * from account;
+----+----------+-------+
| id | aname    | money |
+----+----------+-------+
| 1  | ZhangSan | 3000  |
| 2  | LiSi     | 22000 |
+----+----------+-------+
2 rows in set (0.00 sec)
```

从以上运行结果可以看出，数据表中的相关数据没有发生变化，由此可见，程序实现了声明式事务管理。

10.3.2 使用注解配置声明式事务

在 Spring 中，除了基于 XML 配置声明式事务，还可以通过注解配置声明式事务。@Transactional 是 Spring 提供的配置声明式事务的主要注解，它可以实现和 XML 文件中 <tx:advice>元素相同的功能。

@Transactional 注解提供了一系列属性用于配置事务，具体如表 10.8 所示。

表 10.8 @Transactional 注解的属性

属　　性	描　　述
value	用于指定使用的事务管理器
propagation	用于指定事务的传播行为
isolation	用于指定事务的隔离级别
timeout	用于指定事务的超时时间
readonly	用于指定事务是否为只读
rollbackFor	用于指定导致事务回滚的异常类数组
rollbackForClassName	用于指定导致事务回滚的异常类名称数组
noRollbackFor	用于指定不会导致事务回滚的异常类数组
noRollbackForClassName	用于指定不会导致事务回滚的异常类名称数组

表 10.8 列举了@Transactional 注解的属性，@Transactional 注解可以用于接口、接口方法、类或类方法上，当用于类时，该类的所有 public 方法都将具有同样类型的事务属性，当用于类中的方法时，如果该类的定义处也有@Transactional 注解，那么类中方法的注解将会覆盖类定义处的注解。在实际应用中，@Transactional 注解通常应用在业务实现类上，除此之外，value、propagation 和 isolation 这三个属性的应用范围较广，开发人员可根据实际需要选择使用。

当使用@Transactional 注解时，还需在 Spring 的 XML 文件中通过< tx:annotation-driven>元素配置事务注解驱动，< tx:annotation-driven >元素中有一个常用属性 transaction-manager，该属性用于指定事务管理器。

接下来以一个实例演示使用注解配置声明式事务，具体步骤如下。

（1）在 com.qfedu.service 包下创建 AccountService 接口的实现类 AccountServiceImpl03，具体代码如例 10-9 所示。

【例 10-9】 AccountServiceImpl03.java

```
package com.qfedu.service;
import org.springframework.transaction.annotation.Transactional;
import com.qfedu.dao.AccountDao;
@Transactional
public class AccountServiceImpl03 implements AccountService{
    private AccountDao accountDao;
    public void setAccountDao(AccountDao accountDao) {
        this.accountDao = accountDao;
    }
    @Override
    public void transfer(Integer from, Integer to, Double money) {
        accountDao.decreaseMoney(from,money);
        accountDao.increaseMoney(to,money);
    }
}
```

在以上代码中，AccountServiceImpl03 类的定义处标记了一个@Transactiona 注解，因此，该类中的 transfer()方法将在事务环境中执行。

(2) 在工程 chapter10 的 src 目录下创建 applicationContext_3. xml 文件,具体代码如下所示。

```
<?xml version = "1.0" encoding = "UTF-8"?>
<beans xmlns = "http://www.springframework.org/schema/beans"
    xmlns:xsi = "http://www.w3.org/2001/XMLSchema-instance"
    xmlns:context = "http://www.springframework.org/schema/context"
    xmlns:aop = "http://www.springframework.org/schema/aop"
    xmlns:tx = "http://www.springframework.org/schema/tx"
    xsi:schemaLocation = "http://www.springframework.org/schema/beans
        http://www.springframework.org/schema/beans/spring-beans.xsd
        http://www.springframework.org/schema/context
        http://www.springframework.org/schema/context/spring-context.xsd
        http://www.springframework.org/schema/aop
        http://www.springframework.org/schema/aop/spring-aop.xsd
        http://www.springframework.org/schema/tx
        http://www.springframework.org/schema/tx/spring-tx.xsd">
    <!-- 引入外部 properties 文件 -->
    <context:property-placeholder location = "classpath:jdbc.properties"/>
    <!-- 注册数据源 -->
    <bean name = "dataSource"
            class = "com.mchange.v2.c3p0.ComboPooledDataSource">
            <property name = "driverClass" value = "${jdbc.driverClass}"/>
            <property name = "jdbcUrl" value = "${jdbc.jdbcUrl}"/>
            <property name = "user" value = "${jdbc.user}"/>
            <property name = "password" value = "${jdbc.password}"/>
    </bean>
    <!-- 注册 JdbcTemplate 类 -->
    <bean name = "jdbcTemplate"
            class = "org.springframework.jdbc.core.JdbcTemplate">
            <property name = "dataSource" ref = "dataSource"/>
    </bean>
    <!-- 注册事务管理器 -->
    <bean id = "transactionManager" class =
            "org.springframework.jdbc.datasource.DataSourceTransactionManager">
           <property name = "dataSource" ref = "dataSource"/>
    </bean>
</beans>
```

(3) 在 applicationContext_3. xml 文件的< bean >元素中添加关于 AccountDaoImpl 类和 AccountServiceImpl03 类的配置信息,具体代码如下所示。

```
<bean name = "accountDao"      class = "com.qfedu.dao.AccountDaoImpl">
      <property name = "jdbcTemplate" ref = "jdbcTemplate"/>
</bean>
<bean name = "accountService"
class = "com.qfedu.service.AccountServiceImpl03">
      <property name = "accountDao" ref = "accountDao"/>
</bean>
```

(4) 在 applicationContext_3. xml 文件的< bean >元素中添加关于事务注解驱动的配置信息，具体代码如下所示。

```
<tx:annotation-driven transaction-manager="transactionManager"/>
```

(5) 在 com. qfedu. test 包下创建测试类 TestAccountService03，具体代码如例 10-10 所示。

【例 10-10】 TestAccountService03. java

```
package com.qfedu.test;
import org.springframework.context.ApplicationContext;
import org.springframework.context.support
    .ClassPathXmlApplicationContext;
import com.qfedu.service.AccountService;
public class TestAccountService03 {
    public static void main(String[] args) {
        ApplicationContext context = new
            ClassPathXmlApplicationContext("applicationContext_3.xml");
        AccountService accountService = context.getBean("accountService",
            AccountService.class);
        accountService.transfer(1, 2, 1000.0);
    }
}
```

执行 TestAccountService03 类，通过命令行窗口查看程序运行结果，运行结果如下所示。

```
mysql>  select * from account;
+----+----------+-------+
| id | aname    | money |
+----+----------+-------+
| 1  | ZhangSan | 2000  |
| 2  | LiSi     | 23000 |
+----+----------+-------+
2 rows in set (0.00 sec)
```

从以上运行结果可以看出，数据表中的相关数据已经被更新，由此可见，事务得到正确执行。

修改 AccountServiceImpl03 类，在第 12 行后添加代码如下所示。

```
accountDao.decreaseMoney(from,money);
int num = 1/0;
accountDao.increaseMoney(to,money);
```

和 10.2 节类似，以上代码用于测试程序出现异常时事务的执行情况，如果程序出现异常，那么事务将会回滚。

再次执行 TestAccountService03 类，此时程序出现异常，通过命令行窗口查看程序运行结果,运行结果如下所示。

```
mysql>  select * from account;
+----+----------+-------+
| id | aname    | money |
+----+----------+-------+
|  1 | ZhangSan |  2000 |
|  2 | LiSi     | 23000 |
+----+----------+-------+
2 rows in set (0.00 sec)
```

从以上运行结果可以看出,数据表中的相关数据没有发生变化,由此可见,程序实现了声明式事务管理。

10.4 本章小结

本章首先介绍了 Spring 与事务管理的基础知识,包括 Spring 对事务管理的支持、Spring 为事务管理提供的核心接口;其次通过实例演示了 Spring 的编程式事务管理;最后详细讲解了 Spring 的声明式事务管理,其中,Spring 可以通过 XML 文件或注解这两种形式实现声明式事务管理。通过本章知识的学习,大家应该能理解 Spring 完成事务管理的理念、Spring 完成事务管理的核心接口、通过 Spring 进行编程式事务管理的方法,掌握通过 Spring 进行声明式事务管理的方法。

10.5 习　　题

1. 填空题

(1) 事务的四大特性分别是________、________、________、________。

(2) 在 Spring 提供的接口中,________________用于定义事务的属性。

(3) 在 Spring 提供的接口中,________________用于界定事务的状态。

(4) 在 Spring 提供的接口中,________________用于根据属性管理事务。

(5) 为实现编程式事务管理,Spring 提供的模板类是________________。

2. 选择题

(1) 关于 Spring 管理数据库事务,下列选项错误的是(　　)。

A. 为了便于事务管理,Spring 提供了自己的解决方案

B. Spring 的编程式事务管理完全避免了事务管理和数据访问的耦合

C. Spring 的声明式事务管理需要基于 Spring AOP 实现

D. Spring 的声明式事务管理降低了事务管理和数据访问的耦合

(2) 在下列隔离级别中,允许可重复读并且可以避免脏读、不可重复读的是(　　)。

A. ISOLATION_READ_UNCOMMITTED

B. ISOLATION_READ_COMMITTED

C. ISOLATION_REPEATABLE_READ

D. ISOLATION_SERIALIZABLE

(3) 在 TransactionDefinition 接口提供的方法中，用于返回事务的传播行为的是(　　)。

A. getPropagationBehavior()　　B. getIsolationLevel()

C. getTimeout()　　D. isReadOnly()

(4) 在 TransactionStatus 接口提供的方法中，用于返回当前事务是否已完成的是(　　)。

A. isNewTransaction()　　B. hasSavepoint()

C. isCompleted()　　D. flush()

(5) 在 TransactionTemplate 类提供的方法中，用于定义需要在事务环境中执行的数据访问逻辑的是(　　)。

A. execute()　　B. getTransactionManager()

C. rollbackOnException()　　D. commit()

3. 思考题

(1) 简述 Spring 提供的用于事务管理的核心接口。

(2) 简述使用注解配置声明式事务的步骤。

4. 编程题

使用 Spring 框架编写程序，请完成以下步骤。

(1) 在数据库 chapter10 中创建一张名称为 stu 的数据表，具体表结构如表 10.9 所示。

表 10.9　stu 的表结构

字段名称	数据类型	备注说明
sid	INT	学生编号，主键，自增长
sname	VARCHAR(20)	学生姓名
age	VARCHAR(20)	学生年龄
course	VARCHAR(20)	学科

(2) 向数据表 stu 中插入一条数据，具体字段值如表 10.10 所示。

表 10.10　要插入的数据

sname	age	course
ZhangSan	21	Java

(3) 删除数据表 stu 中 sname 为 ZhangSan 的数据，再次插入一条数据，具体字段值如表 10.11 所示，这两项操作需在同一个事务中执行。

表 10.11　再次要插入的数据

sname	age	course
LiSi	20	Java

第11章 Spring MVC 基础

本章学习目标

- 理解 Spring MVC 的功能组件
- 理解 Spring MVC 的工作流程
- 掌握 Spring MVC 的重要 API
- 掌握 Spring MVC 的简单应用
- 掌握 Spring MVC 的常用注解

在 Java EE 企业级开发中，MVC 是构建 Web 程序的最流行的架构方法。MVC 中的 M 表示 Model(模型)，V 表示 View(视图)，C 表示 Controller(控制器)。在 Java EE 倡导的 Model 2 模式中，C 通常由 Servlet 充当，V 通常由 JSP 充当，M 通常由 JavaBean 充当，然而随着技术的演变，传统开发中使用的原生代码逐渐满足不了企业项目在性能、扩展性等方面的要求。因此，以 Spring MVC 为代表的 MVC 框架开始崛起并最终获得了广泛应用。接下来，本章开始讲解关于 Spring MVC 的基础知识。

11.1 Spring MVC 概述

11.1.1 Spring MVC 简介

在第三方 MVC 框架获得广泛运用之前，开发中通常直接基于 Model 2 模式编写原生代码。为了避免大量使用原生代码带来的弊端，企业级开发中开始转向使用 MVC 框架，常用的 MVC 框架包括 Struts、Spring MVC 等。

Spring MVC 是 Spring 的一个模块，同时也是一个可用于构建 Web 程序的 MVC 框架。Spring MVC 改善了传统的 Model 2 模式，它提供了一个前端控制器来分发请求，同时，它还支持包括 JSP、FreeMarker、Velocity 等在内的多项视图技术。除此之外，Spring MVC 还提供有多样化配置、表单校验、自动绑定用户输入等功能。

Spring MVC 灵活、高效，配置方便，与其他的 MVC 框架相比，它可以与 Spring 无缝集成并使用 Spring 的功能，表现出更好的复用性和扩展性。

11.1.2 Spring MVC 的功能组件

Spring 通过一系列组件实现功能，这些组件包括 DispatcherServlet(前端控制器)、HandlerMapping(处理器映射器)、Handler(处理器)、HandlerAdapter(处理器适配器)、ViewResolver(视图解析器)等。

1. DispatcherServlet(前端控制器)

前端控制器负责拦截客户端请求并分发给其他组件,它是整个流程控制的中心,负责调度其他组件的执行,降低各组件之间的耦合,提升整体效率。前端控制器由框架提供,在程序运行过程中自动实现功能。

2. HandlerMapping(处理器映射器)

处理器映射器负责根据客户端请求的 URL 寻找处理器,Spring MVC 中提供了配置文件、注解等映射方式,这些映射方式将由相应的处理器映射器负责处理。处理器映射器由框架提供,在程序运行过程中自动实现功能。

3. Handler(处理器)

处理器负责对客户端的请求进行处理,由于处理请求涉及具体的业务逻辑,因此,开发者需要在处理器中编写处理业务逻辑的代码。

4. HandlerAdapter(处理器适配器)

处理器适配器负责根据特定的规则对处理器进行执行,它可以执行多种类型的处理器,是设计模式中适配器模式的具体应用。处理器适配器由框架提供,在程序运行过程中自动实现功能。

5. ViewResolver(视图解析器)

视图解析器负责视图解析,它可以将处理结果生成 View(视图)并展示给用户。视图解析器由框架提供,在程序运行过程中自动实现功能,但是 View 需要由开发者根据具体需求编写。

11.1.3 Spring MVC 的工作流程

Spring MVC 的工作流程主要围绕 DispatcherServlet(前端控制器)展开,前端控制器负责拦截客户端发送的请求并将它分发给对应的处理器处理。因此,前端控制器是 Spring MVC 整个流程控制的中心。

Spring MVC 的工作流程,如图 11.1 所示。

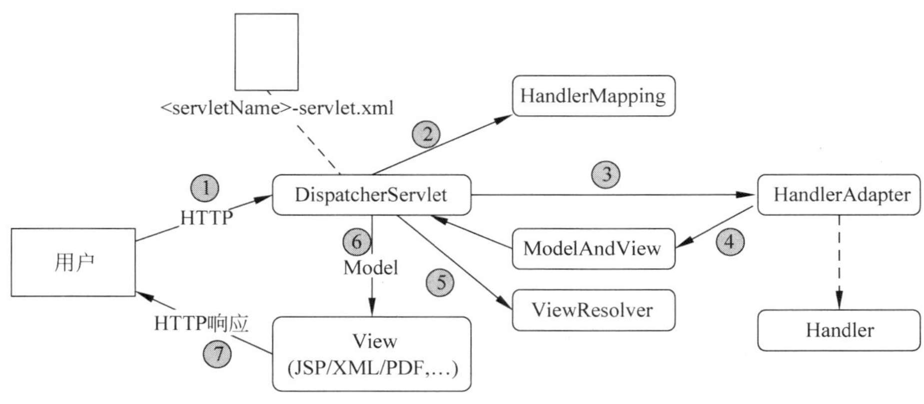

图 11.1 Spring MVC 的工作流程

图 11.1 展示了 Spring MVC 从接收请求到响应的工作流程。在运行过程中,Spring MVC 框架的众多组件协同工作,各司其职,共同支撑了 Spring MVC 框架的功能。通常情

况下，Spring MVC 在处理客户端请求时要完成以下步骤。

(1) 客户端发出一个 HTTP 请求，Web 应用服务器接收这个请求，如果 Web 应用的配置文件中指定有 DispatcherServlet 的映射路径，那么服务器将该请求交给 DispatchserServlet 处理。

(2) DispatchserServlet 接收到请求后，将根据包括 URL、方法、报文头和参数在内的请求信息以及 HandlerMapping 的配置解析出目标 Handler。

(3) 在解析出目标 Handler 后，DispatchserServlet 将通过相应的 HandlerAdapter 来调用 Handler 并完成业务逻辑的处理。

(4) 在完成业务逻辑处理后，代表处理结果的 ModelAndView 会被返回到 DispatchserServlet，ModelAndView 中包含逻辑视图名和模型数据信息。

(5) DispatcherServlet 通过 ViewResolver 完成逻辑视图名到真实 View 对象的解析。

(6) 获得真实的 View 对象后，DispatcherServlet 将模型数据传给 View 对象并通过 View 对象对模型数据进行视图渲染。

(7) DispatcherServlet 将最终的 View 对象响应给客户端并展示给用户。

以上 7 个步骤是 Spring MVC 在处理客户端请求时的基本流程，这些步骤中的大部分是 Spring MVC 自动完成的。由于 Handler 需要处理具体业务逻辑，View 需要向用户展示页面，因此，开发者只需编写与 Handler 和 View 相关的内容。

11.2 Spring MVC 的重要 API

11.2.1 DispatcherServlet 类

11.1.3 节介绍过，DispatcherServlet 是前端控制器，是整个流程控制的中心。实际上，作为 Spring MVC 程序的入口，DispatcherServlet 首先是一个普通的 Servlet 类，因此，要想让 DispatcherServlet 类实现功能，首先要在 web.xml 文件中配置 DispatcherServlet。

DispatcherServlet 类的配置和普通 Servlet 类相同，具体如下所示。

```
<servlet>
    <servlet-name>springMVC</servlet-name>
    <servlet-class>org.springframework.web.servlet.DispatcherServlet
</servlet-class>
<init-param>
        <param-name>contextConfigLocation</param-name>
        <param-value>/WEB-INF/springMVC-config.xml</param-value>
    </init-param>
    <load-on-startup>1</load-on-startup>
</servlet>
 <!-- 访问 DispatcherServlet 对应的路径 -->
 <servlet-mapping>
      <servlet-name>springMVC</servlet-name>
      <url-pattern>/</url-pattern>
</servlet-mapping>
```

在以上代码中，< init-param >元素配置了 DispatcherServlet 的初始化参数，contextConfigLocation 表示引入 springMVC 配置文件所在位置，< load-on-startup >元素表示 DispatcherServlet 在 Web 程序启动时初始化的优先级，< url-pattern >元素的值为"/"，这意味着将所有的请求都映射到 DispatcherServlet。

除了 contextConfigLocation，还可以通过< param-name >的其他三个值来配置 DispatcherServlet，它们分别是 namespace、publishContext、publishEvents，其中，namespace 用于修改 DispatcherServlet 对应的命名空间；publishContext 用于指定是否将 WebApplicationContext 发布到 ServletContext 的属性列表中，如果允许发布，那么开发者可以通过 ServletContext 获取 WebApplicationContext 实例；publishEvents 用于指定当 DispatcherServlet 处理完一个请求后，是否需要向容器发布一个 ServletRequestHandledEvent 事件。

11.2.2 DispatcherServlet 类的辅助 API

为了帮助 DispatcherServlet 类实现功能，Spring MVC 提供了一些辅助 API，这些辅助 API 是和 Spring MVC 中的组件相对应的，具体如表 11.1 所示。

表 11.1 DispatcherServlet 类的辅助 API

名称	描述
Controller	接口，封装了处理客户端请求的方法，开发者可以在实现该接口的类中编写处理具体业务逻辑的代码
HandlerMapping	用于实现请求到处理器的映射，当映射成功时返回 HandlerExecutionChain 对象，该对象中封装了一个 Handler 对象、多个 HandlerInterceptor 对象
HandlerAdapter	适配器设计模式的具体应用，支持多种类型的处理器，调用处理器的 handleRequest()方法进行功能处理
HandlerExceptionResolver	用于处理器异常解析，可将异常提醒转至统一的错误界面
ViewResolver	用于将逻辑视图名解析为具体的 View 对象
LocaleResolver & LocaleContextResolver	用于解析客户端使用的地区、时区信息，然后据此提供视图国家化的支持
MultipartResolver	用于处理 multi-part 请求，例如，文件上传等
ModelAndView	用于封装逻辑视图名和模型数据信息

表 11.1 中列出了 DispatcherServlet 类的辅助 API，在一些应用场景中，这些 API 需要被配置到 Spring 的 IOC 容器中，开发者可根据具体情况完成配置。

11.2.3 Controller 接口

Controller 接口位于 org. springframework. web. servlet. mvc 包中，它提供了一个 handleRequest()方法，开发者在编写 Controller 的实现类时，可以在 handleRequest()方法中添加处理业务逻辑的代码。在 Spring MVC 的工作流程中，Controller 的实现类充当 Handler 的功能，在程序运行过程中，DispatcherServlet 类会间接调用 Controller 实现类中的 handleRequest()方法并完成业务处理，最终 handleRequest()方法将以 ModelAndView

的形式返回处理结果。

11.2.4 ModelAndView 类

ModelAndView 类用于封装 Controller 的处理结果，在 ModelAndView 类提供的方法中，有一些是开发中经常使用的，具体如表 11.2 所示。

表 11.2 ModelAndView 类的常用方法

方　法	描　述
ModelAndView()	默认的构造方法
ModelAndView(String viewName)	需要传入 View 名称的构造方法
ModelAndView addObject(Object attributeValue)	向模型添加属性
ModelAndView addObject(String attributeName, Object attributeValue)	通过 name-value 的形式向模型添加属性
voidsetView(View view)	设置 View 对象
voidsetViewName(String viewName)	设置 View 的名称
boolean hasView()	返回 ModelAndView 是否具有 View 名称或 View 对象

表 11.2 中列出了 ModelAndView 类的常用方法，开发者可根据实际情况选择使用。

11.3 Spring MVC 的简单应用

本章前面讲解了 Spring MVC 的基础知识，接下来，本节通过实例演示使用 Spring MVC 编写程序，具体步骤如下。

(1) 在 Eclipse 中新建 Web 工程 chapter11，将工程所需 jar 包添加到 lib 目录下，完成包的导入。本次要导入的所有 jar 包如图 11.2 所示。

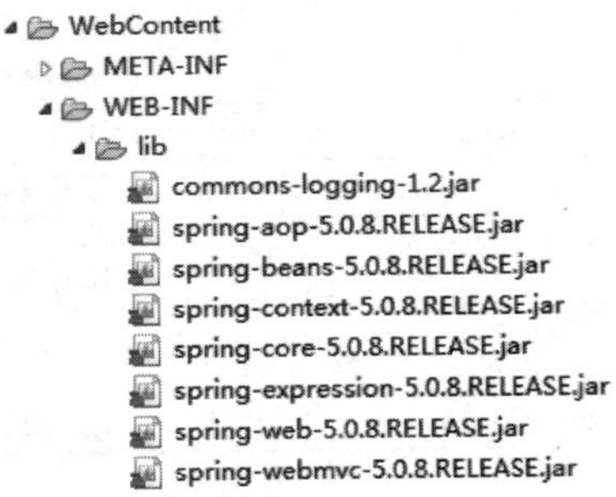

```
WebContent
  META-INF
  WEB-INF
    lib
      commons-logging-1.2.jar
      spring-aop-5.0.8.RELEASE.jar
      spring-beans-5.0.8.RELEASE.jar
      spring-context-5.0.8.RELEASE.jar
      spring-core-5.0.8.RELEASE.jar
      spring-expression-5.0.8.RELEASE.jar
      spring-web-5.0.8.RELEASE.jar
      spring-webmvc-5.0.8.RELEASE.jar
```

图 11.2 要导入的 jar 包

(2) 在工程 chapter11 的 web.xml 文件中添加 DispatcherServlet 类的配置信息，具体代码如 11.2.1 节中所示，此处不再赘述。

(3) 在工程 chapter11 的 WebContent/WEB-INF 目录下新建 Spring MVC 的配置文件 springMVC-config.xml，具体代码如下所示。

```
<?xml version = '1.0' encoding = 'UTF - 8'?>
<beans xmlns = "http://www.springframework.org/schema/beans"
xmlns:xsi = "http://www.w3.org/2001/XMLSchema - instance"
xmlns:context = "http://www.springframework.org/schema/context"
        xsi:schemaLocation = "http://www.springframework.org/schema/beans
        http://www.springframework.org/schema/beans/spring - beans.xsd
        http://www.springframework.org/schema/context
        http://www.springframework.org/schema/context/spring - context.xsd">
```

```
    <!-- 配置 Controller 类 -->
        <bean name="/controller01"
class="com.qfedu.controller.MyController01"/>
<!-- 配置处理器映射器 -->
<bean class="org.springframework.web.servlet.handler.BeanNameUrlHandlerMapping"/>
<!-- 配置处理器适配器 -->
<bean class="org.springframework.web.servlet.mvc.SimpleControllerHandlerAdapter"/>
    <!-- 配置视图解析器 -->
    <bean class="org.springframework.web.servlet.view.InternalResourceViewResolver"/>
</beans>
```

以上配置信息为程序配置了 Controller 实现类、处理器映射器、处理器适配器、视图解析器。

(4) 在工程 chapter11 的 src 目录下新建 com.qfedu.controller 包，在该包下新建类 MyController01，具体代码如例 11-1 所示。

【例 11-1】 MyController01.java

```
package com.qfedu.controller;
import javax.servlet.http.*;
import org.springframework.web.servlet.ModelAndView;
import org.springframework.web.servlet.mvc.Controller;
public class MyController01 implements Controller {
    @Override
    public ModelAndView handleRequest(HttpServletRequest request,
        HttpServletResponse response) throws Exception {
        ModelAndView mv = new ModelAndView();
        mv.addObject("msg","Hello Word");
        mv.setViewName("/WEB-INF/page/page01.jsp");
        return mv;
    }
}
```

在以上代码中，MyController01 类的 handleRequest()方法用于处理客户端请求，在 handleRequest()方法中，首先新建一个 ModelAndView 对象，然后为 ModelAndView 对象设置模型数据和 View 名称，最后将该 ModelAndView 对象返回。

(5) 在工程 chapter11 的 WebContent/WEB-INF 目录下新建 page 子目录，在 page 子目录下新建 page01.jsp，具体代码如例 11-2 所示。

【例 11-2】 page01.jsp

```
<%@ page language="java" contentType="text/html; charset=UTF-8"
    pageEncoding="UTF-8"%>
<!DOCTYPE html PUBLIC "-//W3C//DTD HTML 4.01 Transitional//EN"
    "http://www.w3.org/TR/html4/loose.dtd">
<html>
<head>
<meta http-equiv="Content-Type" content="text/html; charset=UTF-8">
```

```
<title>page01</title>
</head>
<body>
    ${requestScope.msg}
</body>
</html>
```

(6) 将工程 chapter11 添加到 Tomcat，启动 Tomcat，使用浏览器访问 http://localhost:8080/chapter11/controller01，浏览器显示的页面如图 11.3 所示。

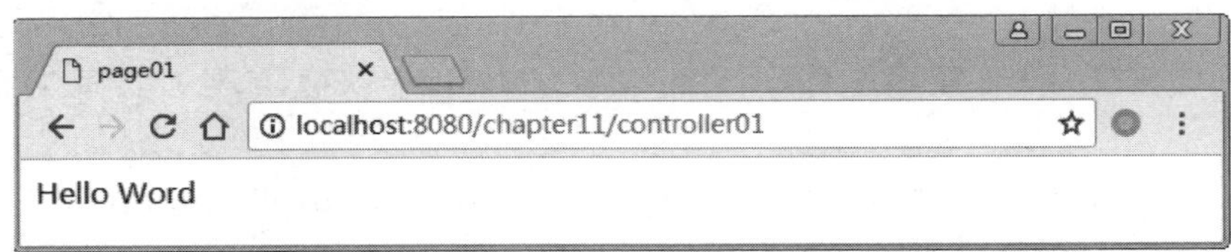

图 11.3　浏览器显示的页面(1)

从图 11.3 中可以看出，程序显示出模型信息中封装的内容，由此可见，程序通过 Spring MVC 完成了对客户端请求的处理及响应。

以上介绍了编写一个 Spring MVC 程序的基本步骤，包括配置 DispatcherServlet、编写处理业务逻辑的 Controller 类，将处理器映射器、处理器适配器、视图解析器配置到 Spring 容器中，编写向用户显示信息的 View 等。

除此之外，Spring MVC 还支持通过注解的形式编写程序。当使用注解编写 Spring MVC 程序时，开发人员无须手动实现 Controller 接口，只需通过@Controller 和@RequestMapping 注解完成相应的标注即可。

@Controller 注解表示一个 Controller 类，Spring 会把@Controller 注解标注的类作为处理器，@RequestMapping 指定了对应请求的路径映射。在程序运行过程中，Spring MVC 会使用这些信息去寻找对应的 Controller。接下来通过实例演示基于注解编写 Spring MVC 程序。

(1) 在 com.qfedu.controller 包下新建类 MyController02，具体代码如例 11-3 所示。

【例 11-3】 MyController02.java

```
package com.qfedu.controller;
import org.springframework.stereotype.Controller;
import org.springframework.web.bind.annotation.RequestMapping;
import org.springframework.web.servlet.ModelAndView;
@Controller
public class MyController02  {
    @RequestMapping("/execute")
    public ModelAndView execute() {
        ModelAndView mv = new ModelAndView();
        mv.addObject("msg","Hello Word");
        mv.setViewName("page01");
        return mv;
```

```
    }
}
```

在以上代码中,@Controller 注解将 MyController02 类作为一个 Controller 类并配置到 Spring 容器,@RequestMapping 注解指定了 execute()方法的映射路径。

(2) 在工程 chapter11 的 WebContent/WEB-INF 目录下新建 Spring MVC 的配置文件 springMVC-config02. xml,具体代码如下所示。

```
<?xml version = '1.0' encoding = 'UTF - 8'?>
< beans xmlns = "http://www.springframework.org/schema/beans"
xmlns:xsi = "http://www.w3.org/2001/XMLSchema - instance"
        xmlns:context = "http://www.springframework.org/schema/context"
        xsi:schemaLocation = "http://www.springframework.org/schema/beans
        http://www.springframework.org/schema/beans/spring - beans.xsd
        http://www.springframework.org/schema/context
        http://www.springframework.org/schema/context/spring - context.xsd">
    <! -- 扫描 Controller -- >
    < context:component - scan base - package = "com.qfedu.controller"/>
    <! -- 配置视图解析器 -- >
    < bean class = "org.springframework.web.servlet.
            view.InternalResourceViewResolver">
        < property name = "prefix" value = "/WEB - INF/page/"/>
        < property name = "suffix" value = ".jsp"/>
    </bean >
</beans >
```

在以上代码中,< context:component-scan >元素的 base-package 属性用于定义要扫描的包,此处为 com. qfedu. controller,在程序启动以后,com. qfedu. controller 包内的 Controller 和其他的一些组件将会被加载。在 InternalResourceViewResolver 类的属性中,prefix 指定 View 名称的前缀,suffix 指定 View 名称的后缀,这样视图解析器就能匹配到对应的 JSP 文件并响应到客户端。

(3) 修改工程 chapter11 的 web. xml 文件,将 DispatcherServlet 配置信息中的< init-param >元素修改为如下所示。

```
< init - param >
    < param - name > contextConfigLocation </param - name >
    < param - value >/WEB - INF/springMVC - config02.xml </param - value >
</init - param >
```

在以上配置信息中,contextConfigLocation 的值被修改为/WEB-INF/springMVC-config02. xml,这将意味着 Spring MVC 将从 springMVC-config02. xml 文件中获取配置信息。

(4) 重启 Tomcat,使用浏览器访问 http://localhost:8080/chapter11/execute,浏览器显示的页面如图 11. 4 所示。

从图 11. 4 中可以看出,程序显示出模型信息中封装的内容,由此可见,程序通过 Spring

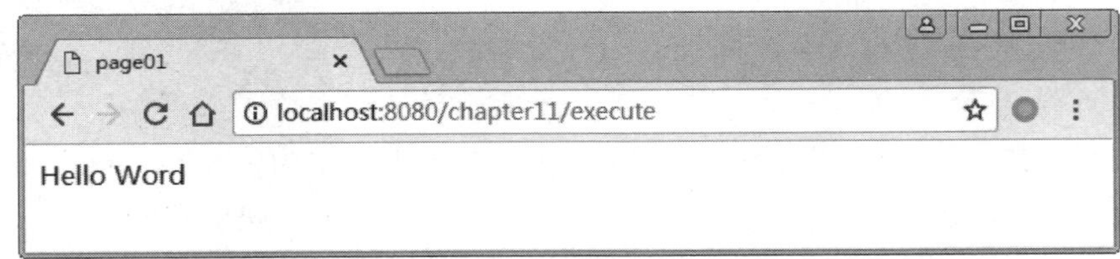

图 11.4　浏览器显示的页面(2)

MVC 完成了对客户端请求的处理及响应。

11.4　Spring MVC 的常用注解

11.4.1　@RequestMapping 注解

为了简化开发，Spring MVC 提供有一系列注解供开发人员使用。除了@Controller、@RequestMapping 之外，Spring MVC 还提供有@RequestParam、@PathVariable、@CookieValue、@RequestHeader 等注解。由于前面的小节已演示过@Controller 注解的使用方法，此处不再赘述，接下来开始详细讲解关于@RequestMapping 的知识。

@RequestMapping 用于处理请求地址映射，当@RequestMapping 用于一个 Controller 类时，表示类中的所有响应请求的方法都是以该注解指定的地址作为父路径，当@RequestMapping 用于 Controller 类中的一个方法时，该方法将成为处理请求的方法。

@RequestMapping 注解提供了一系列属性，具体如表 11.3 所示。

表 11.3　@RequestMapping 注解提供的属性

属　　性	描　　述
value	指定请求的实际地址
method	指定该方法可以处理的 HTTP 请求方式，如果没有指定 method 属性值，则默认映射所有 HTTP 请求方式
consumes	指定处理请求的提交内容类型
produces	指定返回的内容类型，返回的内容类型必须是请求头中所包含的类型
params	指定请求中必须包含某些参数值，才让该方法处理
headers	指定请求中必须包含某些特性的 header 值，才能让该方法处理
name	为映射的地址指定别名

表 11.3 列举了@RequestMapping 注解支持的属性，其中最为常用的是 value 和 method，在实际开发中，开发人员可根据具体需求选择使用。

接下来以一个实例演示@RequestMapping 注解的使用，具体步骤如下。

(1) 在 com. qfedu. controller 包下新建一个类 MyController03，具体代码如例 11-4 所示。

【例 11-4】　MyController03. java

```
package com.qfedu.controller;
import org.springframework.stereotype.Controller;
```

```java
import org.springframework.web.bind.annotation.*;
import org.springframework.web.servlet.ModelAndView;
@Controller
public class MyController03  {
    //访问 page02.jsp
    @RequestMapping(value = "/toWelcome")
    public ModelAndView toWelcome() {
        ModelAndView mv = new ModelAndView();
        mv.setViewName("page02");
        return mv;
    }
    //访问 page01.jsp
    @RequestMapping(value = "/welcome",method = RequestMethod.POST)
    public ModelAndView welcome() {
        ModelAndView mv = new ModelAndView();
        mv.addObject("msg","Hello Word");
        mv.setViewName("page01");
        return mv;
    }
}
```

在以上代码中，第 1 个@RequestMapping 注解的 value 值为"/toWelcome"，这意味着访问路径"/toWelcome"的请求将由 toWelcome ()方法处理，处理完毕后，请求被转发到 page02.jsp；第 2 个@RequestMapping 注解的 value 值为"/welcome"，这意味着访问路径"/welcome"的请求将由 welcome()方法处理，method 属性值为 RequestMethod.POST，这意味着 welcome()方法仅处理 POST 方式的请求，如果接收的为 POST 请求，处理完毕后，请求被转发到 page01.jsp，如果接收的为 GET 请求，则返回相应的提示信息。

(2) 在 page 目录下新建 page02.jsp，具体代码如例 11-5 所示。

【例 11-5】 page02.jsp

```jsp
<%@ page language = "java" contentType = "text/html; charset = UTF-8"
    pageEncoding = "UTF-8" %>
<!DOCTYPE html PUBLIC "-//W3C//DTD HTML 4.01 Transitional//EN"
    "http://www.w3.org/TR/html4/loose.dtd">
<html>
<head>
<meta http-equiv = "Content-Type" content = "text/html; charset = UTF-8">
<title>page02</title>
</head>
<body>
<form action = "${pageContext.request.contextPath}/welcome" method = "post">
    <input type = "submit" value = "POST 方式访问欢迎页面">
</form>
<hr>
<a href = "${pageContext.request.contextPath}/welcome">GET 方式访问欢迎页面</a>
</body>
</html>
```

(3) 重启 Tomcat,在浏览器中访问 http://localhost:8080/chapter11/toWelcome,浏览器显示的页面如图 11.5 所示。

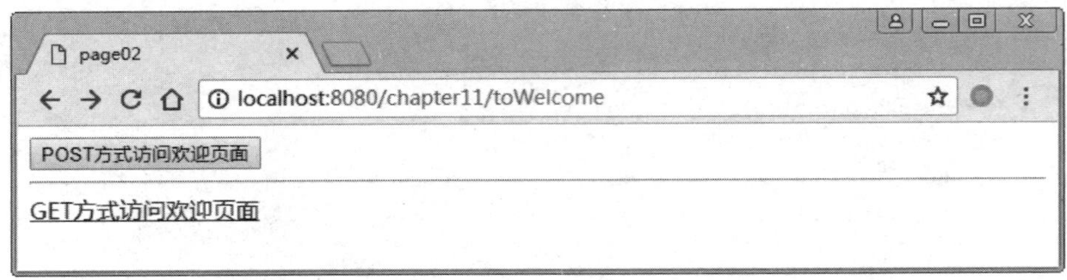

图 11.5 浏览器显示的页面(3)

(4) 单击"POST 方式访问欢迎页面"按钮,浏览器显示的页面如图 11.6 所示。

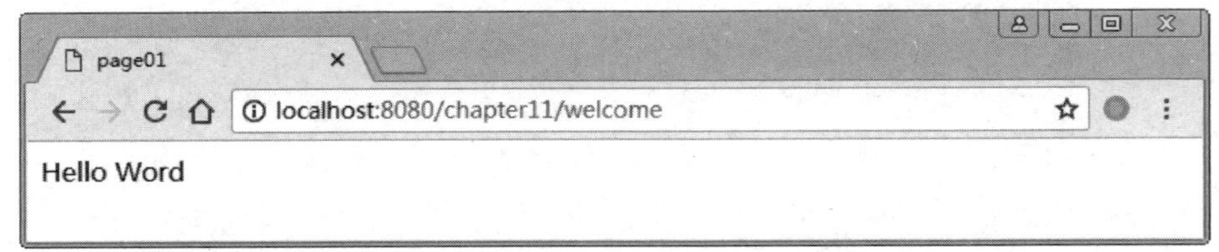

图 11.6 浏览器显示的页面(4)

从图 11.6 中可以看出,浏览器显示出 page01.jsp 页面,这就说明,以 POST 方式访问成功。

(5) 单击"返回"按钮,浏览器显示如图 11.5 所示页面,单击"GET 方式访问欢迎页面"超链接,浏览器显示的页面如图 11.7 所示。

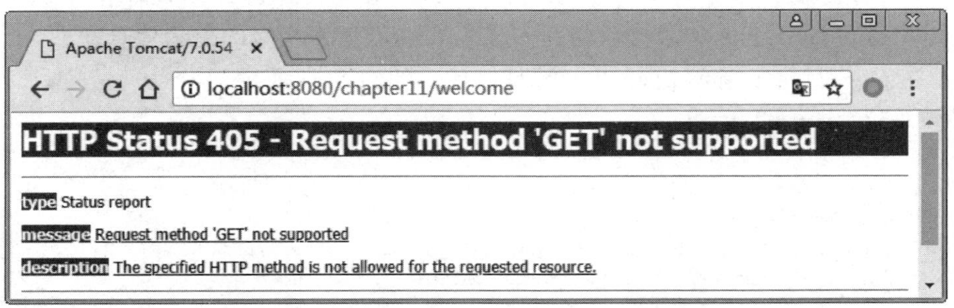

图 11.7 浏览器显示的页面(5)

从图 11.7 中可以看出,页面显示出 405 提示信息,这就说明,以 GET 方式访问 page01.jsp 失败。

11.4.2 @RequestParam 注解

@RequestParam 注解用于获取请求参数的值,它可以将请求参数赋值给方法中的形参,进而完成对请求参数的处理。

@RequestParam 注解提供了若干属性,具体如表 11.4 所示。

表 11.4　@RequestParam 注解提供的属性

属　　性	描　　述
value	指定请求参数的名称
required	指定参数是否必须绑定
defaultValue	指定参数的默认值

表 11.4 列举了@RequestParam 注解支持的属性，其中最为常用的是 value 和 defaultValue，在实际开发中，开发人员可根据具体需求选择使用。

接下来以一个实例演示@RequestParam 注解的使用，具体步骤如下。

（1）在 com.qfedu.controller 包下新建一个类 MyController04，具体代码如例 11-6 所示。

【例 11-6】　MyController04.Java

```
package com.qfedu.controller;
import org.springframework.stereotype.Controller;
import org.springframework.web.bind.annotation.*;
import org.springframework.web.servlet.ModelAndView;
@Controller
public class MyController04  {
    //访问 page03.jsp
    @RequestMapping(value = "/toLogin")
    public ModelAndView toLogin() {
        ModelAndView mv  =  new ModelAndView();
        mv.setViewName("page03");
        return mv;
    }
    //处理登录请求
    @RequestMapping(value = "/login")
    public ModelAndView login(@RequestParam(value = "username",
        defaultValue = "xiaofeng" )String username,
        @RequestParam(value = "password",defaultValue = "123abc")
        String password) {
        System.out.println("用户名: " + username);
        System.out.println("密码: " + password);
        return null;
    }
}
```

在以上代码中，第 1 个@RequestParam 注解的 value 属性值为 username，defaultValue 属性值为 xiaofeng，这意味着请求参数 username 将赋值给 login()方法中的形参 username，当请求参数 username 没有提交值时，login()方法中的形参 username 将被赋给默认值 xiaofeng；第 2 个@RequestParam 注解的 value 属性值为 password，defaultValue 属性值为 123abc，这意味着请求参数 password 将赋值给 login()方法中的形参 password，当请求参数 password 没有提交值时，login()方法中的形参 password 将被赋给默认值 123abc。

（2）在 page 目录下新建 page03.jsp，具体代码如例 11-7 所示。

【例 11-7】 page03.jsp

```
<%@ page language="java" contentType="text/html; charset=UTF-8"
    pageEncoding="UTF-8"%>
<!DOCTYPE html PUBLIC "-//W3C//DTD HTML 4.01 Transitional//EN"
"http://www.w3.org/TR/html4/loose.dtd">
<html>
<head>
<meta http-equiv="Content-Type" content="text/html; charset=UTF-8">
<title>page03</title>
</head>
<body>
    <form action="${pageContext.request.contextPath}/login" method="post">
        用户名<input type="text" name="username"><br>
        密    码<input type="password" name="password"><br>
        <input type="submit" value="登录">
    </form>
</body>
</html>
```

(3) 重启 Tomcat，在浏览器中访问 http://localhost:8080/chapter11/toLogin，浏览器显示的页面如图 11.8 所示。

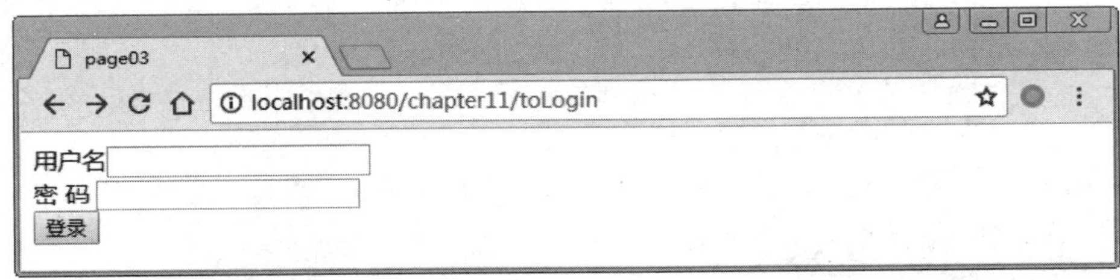

图 11.8 浏览器显示的页面(6)

(4) 在用户名文本框中输入 xiaoqian，在密码文本框中输入 123456，单击“登录”按钮，此时控制台窗口显示出页面提交的用户名和密码信息，具体如图 11.9 所示。

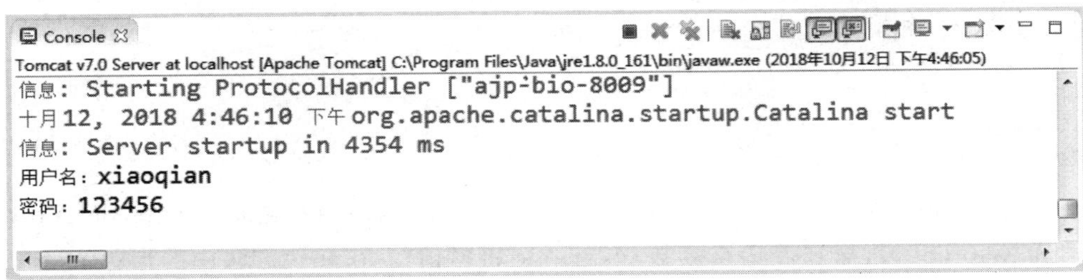

图 11.9 控制台窗口(1)

从图 11.9 中可以看出，控制台显示的用户名为 xiaoqian，密码为 123456，这就说明 MyController04 类中的@RequestParam 注解成功获取了请求参数的值。

(5) 在浏览器中重新访问 http://localhost:8080/chapter11/toLogin，浏览器显示如图

11.8 所示页面，不填写用户名和密码信息，直接单击登录按钮，此时控制台窗口显示出用户名和密码信息，具体如图 11.10 所示。

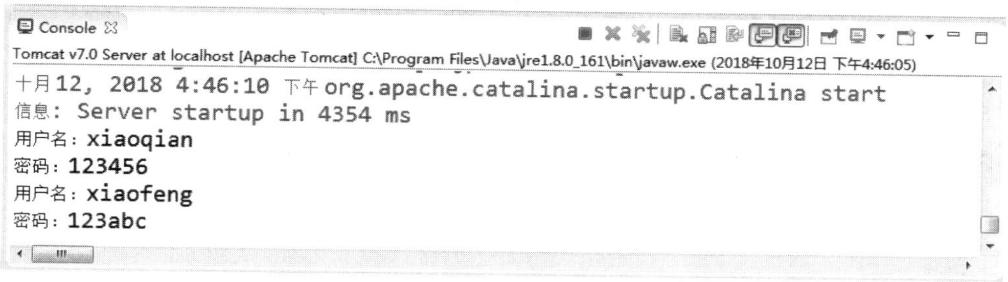

```
Console
Tomcat v7.0 Server at localhost [Apache Tomcat] C:\Program Files\Java\jre1.8.0_161\bin\javaw.exe (2018年10月12日 下午4:46:05)
十月 12, 2018 4:46:10 下午 org.apache.catalina.startup.Catalina start
信息: Server startup in 4354 ms
用户名: xiaoqian
密码: 123456
用户名: xiaofeng
密码: 123abc
```

图 11.10　控制台窗口(2)

从图 11.10 中可以看出，控制台显示的用户名为 xiaofeng，密码为 123abc，这就说明，当请求参数 username、password 没有提交值时，login()方法中的形参 username、password 将会被赋给@RequestParam 注解指定的默认值。

11.4.3　@PathVariable 注解

@PathVariable 注解用于获取 URL 中的动态参数，它支持动态 URL 访问并可以将请求 URL 中的动态参数映射到功能处理方法的形参上。

@PathVariable 注解提供有 value、required 属性，其中，value 属性是最常使用的，它用于指定将要映射的参数名称；required 属性用于指定参数是否为必须绑定的参数。

接下来以一个实例演示@PathVariable 注解的使用，具体步骤如下。

（1）在 com.qfedu.controller 包下新建一个类 MyController05，具体代码如例 11-8 所示。

【例 11-8】　MyController05.Java

```
package com.qfedu.controller;
import org.springframework.stereotype.Controller;
import org.springframework.web.bind.annotation.*;
@Controller
@RequestMapping(value = "/claList/{cid}")
public class MyController05  {
    @RequestMapping(value = "/stuList/{sid}")
    public String findStudnt(@PathVariable(value = "cid") Integer cid,
        @PathVariable(value = "sid") Integer sid) {
        System.out.println("班级 ID 为: " + cid);
        System.out.println("学生 ID 为: " + sid);
        return null;
    }
}
```

在以上代码中，第 1 个@RequestMapping 注解的 value 属性中使用了动态参数 cid，第 2 个@RequestMapping 注解的 value 属性中使用了动态参数 sid，第 1 个@PathVariable 注解将动态参数 cid 映射为 findStudnt()方法中的形参 cid，第 2 个@PathVariable 注解将动态参数 sid 映射为 findStudnt()方法中的形参 sid。

(2) 重启 Tomcat，在浏览器中访问 http://localhost:8080/chapter11/claList/1/stuList/2，此时控制台窗口显示 URL 中的动态参数的信息，具体如图 11.11 所示。

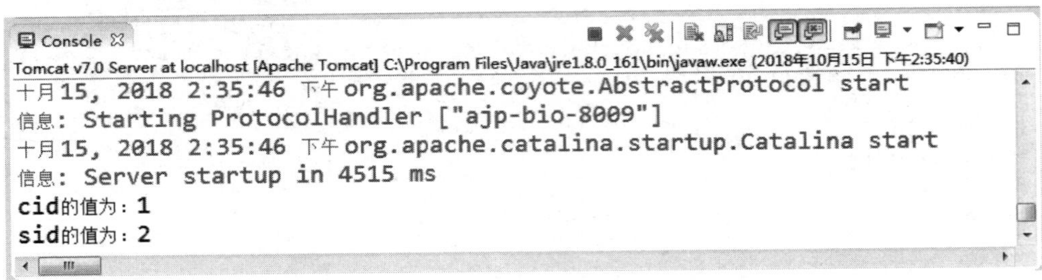

图 11.11 控制台窗口(3)

从图 11.11 中可以看出，cid 的值是 1，sid 的值是 2，这与 URL 中动态参数 cid、动态参数 sid 的值是对应的。

(3) 在浏览器中访问 http://localhost:8080/chapter11/claList/3/stuList/4，此时控制台窗口显示 URL 中的动态参数的信息，具体如图 11.12 所示。

```
Console
Tomcat v7.0 Server at localhost [Apache Tomcat] C:\Program Files\Java\jre1.8.0_161\bin\javaw.exe (2018年10月15日 下午2:35:40)
十月15, 2018 2:35:46 下午 org.apache.catalina.startup.Catalina start
信息: Server startup in 4515 ms
cid的值为：1
sid的值为：2
cid的值为：3
sid的值为：4
```

图 11.12 控制台窗口(4)

从图 11.12 中可以看出，cid 的值是 3，sid 的值是 4，这与 URL 中动态参数 cid、动态参数 sid 的值是对应的。

11.4.4 @CookieValue 注解

@CookieValue 注解用于获取 Cookie 数据，它可以将 Cookie 数据映射到功能处理方法的形参上。

@CookieValue 注解提供了若干属性，具体如表 11.5 所示。

表 11.5 @CookieValue 注解提供的属性

属性	描述
value	指定 Cookie 的名称
required	指定参数是否必须绑定
defaultValue	指定参数的默认值

表 11.5 列举了@CookieValue 注解支持的属性，在实际开发中，开发人员可根据具体需求选择使用。

接下来以一个实例演示@CookieValue 注解的使用，具体步骤如下。

（1）在 com. qfedu. controller 包下新建一个类 MyController06，具体代码如例 11-9 所示。

【例 11-9】 MyController06. Java

```
package com.qfedu.controller;
import org.springframework.stereotype.Controller;
import org.springframework.web.bind.annotation.*;
@Controller
public class MyController06  {
    @RequestMapping(value = "/testCookie")
    public String testCookie(@CookieValue("value = "JSESSIONID")String
        cookie){
        System.out.println("JSESSIONID:" + cookie);
        return null;
    }
}
```

在以上代码中，@CookieValue 注解将 Cookie 信息中 JSESSIONID 的值映射为 testCookie()方法中形参 cookie 的值。

（2）重启 Tomcat，在浏览器中访问 http://localhost:8080/chapter11/testCookie，此时控制台窗口显示 Cookie 信息中 JSESSIONID 的值，具体如图 11.13 所示。

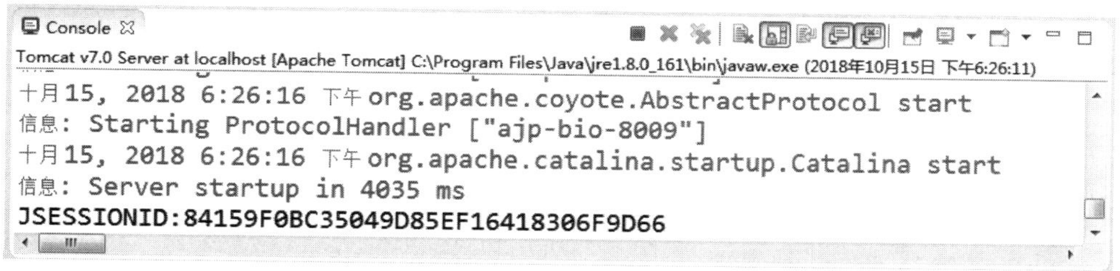

图 11.13 控制台窗口(5)

11.4.5 @RequestHeader 注解

@RequestHeader 注解用于获取请求头中的数据，它可以将请求头中的数据映射到功能处理方法的形参上。

@RequestHeader 注解提供的属性和@CookieValue 相同，此处不再具体列举其功能，接下来以一个实例演示@RequestHeader 注解的使用，具体步骤如下。

（1）在 com. qfedu. controller 包下新建一个类 MyController07，具体代码如例 11-10 所示。

【例 11-10】 MyController07. Java

```
package com.qfedu.controller;
import org.springframework.stereotype.Controller;
import org.springframework.web.bind.annotation.*;
```

```
@Controller
public class MyController07  {
    @RequestMapping(value = "/testRequestHeader")
    public String testRequestHeader(@RequestHeader(value = "Host")String
            host,@RequestHeader(value = "Connection")String connection){
        System.out.println("Host:" + host);
        System.out.println("Connection:" + connection);
        return null;
    }
}
```

在以上代码中，第 1 个 @RequestHeader 注解将请求头 Host 的值映射为 testRequestHeader()方法中形参 host 的值，第 2 个 @RequestHeader 注解将请求头 Connection 的值映射为 testRequestHeader()方法中形参 connection 的值。

（2）重启 Tomcat，在浏览器中访问 http://localhost:8080/chapter11/testRequestHeader，此时控制台窗口显示本次请求的请求头 Host 和 Connection 的值，具体如图 11.14 所示。

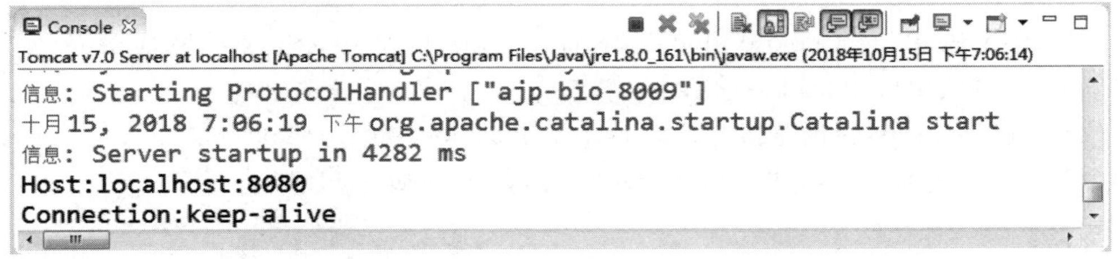

图 11.14　控制台窗口(6)

从图 11.14 可以看出，控制台窗口显示出 Host 和 Connection 的值分别为 localhost:8080 和 keep-alive，由此可见，程序可通过@RequestHeader 注解获取请求头的值。

11.5　本章小结

本章首先介绍了 Spring MVC 的理论基础，包括 Spring MVC 简介、Spring MVC 的功能组件、Spring MVC 的工作流程；然后介绍了 Spring MVC 的重要 API，其中重点讲解了 DispatcherServlet 类；接下来通过一个实例演示了 Spring MVC 的简单应用；最后详细讲解了 Spring MVC 的常用注解。通过本章知识的学习，大家应该能理解 Spring MVC 的特性和优势、Spring MVC 各功能组件的功能和工作流程，掌握 Spring MVC 常用注解的使用，能够使用 Spring MVC 编写简单的应用程序。

11.6　习　题

1. 填空题

（1）Spring MVC 提供的________注解用于处理请求地址映射。

（2）Spring MVC 提供的________注解可以将请求参数赋值给方法中的形参。

(3) Spring MVC 提供的________注解用于获取 URL 中的动态参数。
(4) Spring MVC 提供的________注解用于获取 Cookie 数据。
(5) Spring MVC 提供的________注解用于获取请求头中的数据。

2. 选择题

(1) 关于 Spring MVC 的功能组件,下列选项错误的是(　　)。

A. 前端控制器负责拦截客户端请求并分发给其他组件
B. 处理器适配器负责根据客户端请求的 URL 寻找处理器
C. 处理器负责对客户端的请求进行处理
D. 视图解析器负责视图解析,它可以将处理结果生成 View(视图)

(2) 关于 Spring MVC 的工作流程,下列选项错误的是(　　)。

A. 客户端发出 HTTP 请求,请求将首先由 DispatchserServlet 处理
B. DispatchserServlet 接收到请求后,通过相应 HandlerAdapter 解析出目标 Handler
C. DispatcherServlet 通过 ViewResolver 完成逻辑视图名到真实 View 对象的解析
D. DispatcherServlet 将最终的 View 对象响应给客户端并展示给用户

(3) 在 ModelAndView 类提供的 API 中,用于设置视图名称的是(　　)。

A. setView()　　　　B. setViewName()
C. hasView()　　　　D. addObject()

(4) 下列类中,充当处理器映射器的是(　　)。

A. HandlerMapping　　　　B. Handler
C. HandlerAdapter　　　　D. ViewResolver

(5) 在@RequestMapping 注解提供的属性中,用于指定该方法可以处理的 HTTP 请求方式的是(　　)。

A. value　　　　B. method
C. consumes　　　　D. params

3. 思考题

(1) 简述 Spring MVC 的功能组件。
(2) 简述 Spring MVC 的工作流程。

4. 编程题

分别编写一个登录页面和一个欢迎页面,当用户登录成功后,页面跳转到欢迎页面,欢迎页面如图 11.15 所示,要求使用 Spring MVC 框架完成后台业务处理。

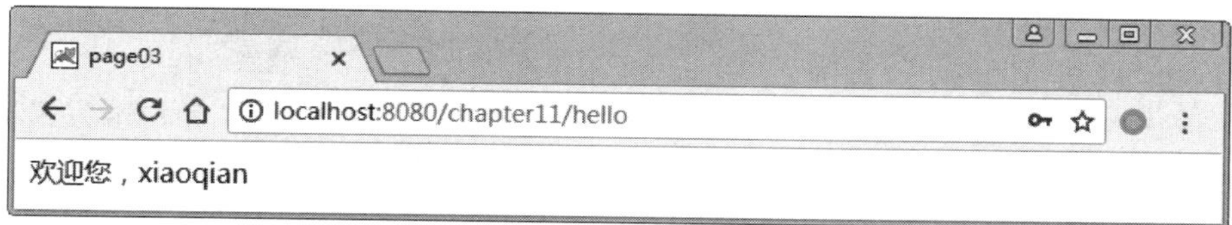

图 11.15　欢迎页面

第 12 章 Spring MVC 的参数绑定

本章学习目标

- 理解数据绑定
- 掌握简单数据绑定
- 掌握复杂数据绑定

在 Spring MVC 使用的过程中，经常需要编写控制器来接收和响应客户端的请求，客户端发送的请求一般都是包含数据信息的，那么这些数据是如何到达控制器的呢？不同类型的数据又该如何绑定呢？本章将通过学习 Spring MVC 的参数绑定来解决此类问题。

12.1 Spring MVC 数据绑定

在 Spring MVC 中，提交请求的数据是通过方法形参来接收的。从客户端请求的 key/value 数据，经过参数绑定，将 key/value 数据绑定到 Controller 的形参上，然后在 Controller 就可以直接使用该形参。Spring MVC 内置了很多参数转换器，只有在极少数情况下需要自定义参数转换器。

12.2 简单数据绑定

12.2.1 绑定默认数据类型

Spring MVC 有支持的默认参数类型，直接在形参上给出这些默认类型的声明，就能直接使用。支持的默认参数如下：

1）HttpServletRequest 对象

HttpServletRequest 对象代表客户端的请求，当客户端通过 HTTP 访问服务器时，HTTP 请求头中的所有信息都封装在这个对象中，通过这个对象提供的方法，可以获得客户端请求的所有信息。

2）HttpServletResponse 对象

HttpServletResponse 对象是服务器的响应。这个对象中封装了向客户端发送数据、发送响应头和发送响应状态码的方法。

3）HttpSession 对象

HttpSession 是一个用户第一次访问某个网站时，通过在 HttpServletRequest 中调用 getSession 方法创建的，可以用来记录用户信息。

4）Model/ModelMap 对象

ModelMap 对象主要用来传递控制器方法中的数据信息到结果页面，该对象的用法类似 request 对象的 setAttribute 方法，而 Model/ModelMap 则是通过 addAttribute 方法向页面传递参数的。

在控制方法的形参上直接声明 HttpServletRequest 类型，实现数据的绑定。控制器中具体实现代码如例 12-1 所示。

【例 12-1】 MyController01.java

```java
package com.qfedu.controller;
import java.io.IOException;
import javax.servlet.http.HttpServletRequest;
import javax.servlet.http.HttpServletResponse;
import org.springframework.stereotype.Controller;
import org.springframework.web.bind.annotation.RequestMapping;
import org.springframework.web.bind.annotation.RequestMethod;
@Controller
public class MyController01 {
    @RequestMapping(value = "/defaultparam",method = RequestMethod.POST)
    public void param1(HttpServletRequest request,HttpServletResponse
response) throws IOException {
        String msg = request.getParameter("msg");
        System.out.println("HttpServletRequest: " + msg);
        response.setContentType("text/html;charset = UTF - 8");
        response.getWriter().print("HttpServletResponse: 响应内容");
    }
}
```

通过 JSP 页面的 form 表单输入文本内容，在浏览器中完成数据赋值，发送 post 请求，跳转至例 12-1 中的代码块，完成数据绑定。具体 JSP 页面代码如例 12-2 所示。

【例 12-2】 page01.jsp

```jsp
<%@ page language = "java" contentType = "text/html; charset = UTF - 8"
    pageEncoding = "UTF - 8" %>
<!DOCTYPE html PUBLIC " - //W3C//DTD HTML 4.01 Transitional//EN" "http://www.w3.org/TR/
    html4/loose.dtd">
<html>
<head>
<meta http - equiv = "Content - Type" content = "text/html; charset = UTF - 8">
<title> page02 </title>
</head>
<body>
    <form action = "defaultparam" method = "post">
        <input name = "msg" placeholder = "请输入消息……"><br /><input
            type = "submit" value = "请求">
    </form>
</body>
</html>
```

通过浏览器访问 page01.jsp 页面，可以自定义输入 form 表单中 name 属性的值，如图 12.1 所示。

图 12.1　表单请求页面

图 12.1 中所赋的值“千锋 java”在控制面板打印的结果如图 12.2 所示。

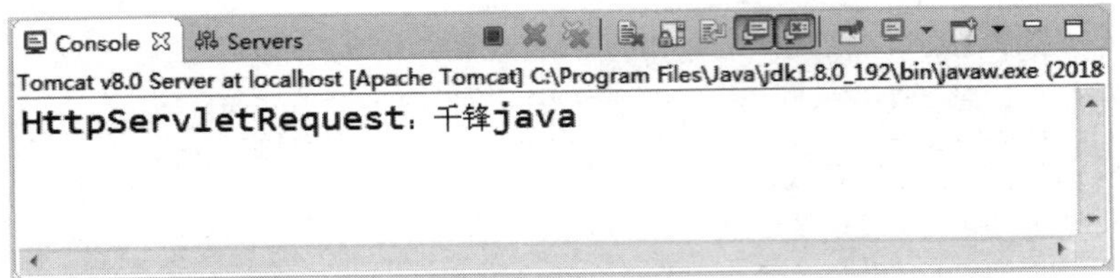

图 12.2　请求响应结果

在控制方法的形参上直接声明 HttpSession 和 Model 类型，实现数据的绑定。控制器中具体实现代码如例 12-3 所示。

【例 12-3】　MyController02.java

```
package com.qfedu.controller;
import java.io.IOException;
import java.util.Map;
import javax.servlet.http.HttpSession;
import org.springframework.stereotype.Controller;
import org.springframework.ui.Model;
import org.springframework.web.bind.annotation.RequestMapping;
import org.springframework.web.bind.annotation.RequestMethod;
@Controller
public class MyController02 {
    @RequestMapping(value = "/defaultparam2",method = RequestMethod.GET)
    public String param2(HttpSession session,Model model,Map<String,Object> map) throws
            IOException {
        session.setAttribute("session1", "JAVA 核心框架");
        model.addAttribute("model1", "SpringMVC");
        map.put("map1", "参数绑定");
        return "page03";
    }
}
```

通过浏览器访问 page02.jsp 页面，单击页面中的超链接，程序跳转至例 12-3 中的代码块，完成数据绑定，具体 JSP 页面代码如例 12-4 所示。

【例 12-4】 page02.jsp

```
<%@ page language="java" contentType="text/html; charset=UTF-8"
    pageEncoding="UTF-8"%>
<!DOCTYPE html PUBLIC "-//W3C//DTD HTML 4.01 Transitional//EN" "http://www.w3.org/TR/
    html4/loose.dtd">
<html>
<head>
<meta http-equiv="Content-Type" content="text/html; charset=UTF-8">
<title>page02</title>
</head>
<body>
<%
   String path = request.getContextPath();
   String basePath = request.getScheme() + "://" + request.getServerName() + ":
      " + request.getServerPort() + path + "/";
    %>
<a href="<%=basePath%>defaultparam2">默认类型绑定</a>
</body>
</html>
```

通过 EL 表达式获取例 12-3 中绑定的值实现在浏览器中打印，具体 JSP 页面代码如例 12-5 所示。

【例 12-5】 page03.jsp

```
<%@ page language="java" contentType="text/html; charset=UTF-8"
    pageEncoding="UTF-8"%>
<!DOCTYPE html PUBLIC "-//W3C//DTD HTML 4.01 Transitional//EN" "http://www.w3.org/TR/
    html4/loose.dtd">
<html>
<head>
<meta http-equiv="Content-Type" content="text/html; charset=UTF-8">
<title>page03</title>
</head>
<body>
<h3>Model\Map\Session</h3>
<h4>
HttpSession:<label>${session1}</label><br/>
Model:<label>${model1}</label><br/>
Map:<label>${map1}</label><br/>
</h4>
</body>
</html>
```

通过浏览器访问 page02.jsp 页面，单击超链接完成请求跳转，如图 12.3 所示。

通过 HttpSession 对象和 Model/ModelMap 对象，实现数据的绑定，在控制面板打印的结果如图 12.4 所示。

图 12.3　超链接请求页面

图 12.4　数据绑定结果展示

12.2.2　绑定简单数据类型

Java 的数据类型分为 2 种，基本类型和引用类型，而 Spring MVC 是支持基本类型自动转换的。

简单数据类型的绑定在控制器中具体实现的代码如例 12-6 所示。

【例 12-6】　MyController03.java

```
package com.qfedu.controller;
import java.io.IOException;
import javax.servlet.http.HttpServletResponse;
import org.springframework.stereotype.Controller;
import org.springframework.web.bind.annotation.RequestMapping;
import org.springframework.web.bind.annotation.RequestMethod;
@Controller
public class MyController03 {
    @RequestMapping(value = "/simpleparam",method = RequestMethod.GET)
    public void param1(int param1,float param2,double param3,char param4,boolean param5,
          long param6,HttpServletResponse response) throws IOException {
        System.out.println("int: " + param1);
        System.out.println("float: " + param2);
        System.out.println("double: " + param3);
        System.out.println("char: " + param4);
        System.out.println("boolean: " + param5);
        System.out.println("long: " + param6);
        response.getWriter().print("OK");
    }
}
```

实现简单数据类型的绑定，可以通过 JSP 页面的 form 表单对属性动态赋值进行验证，具体页面代码如例 12-7 所示。

【例 12-7】 page04.jsp

```
<%@ page language = "java" contentType = "text/html; charset = UTF - 8"
    pageEncoding = "UTF - 8" %>
<!DOCTYPE html PUBLIC " - //W3C//DTD HTML 4.01 Transitional//EN" "http://www.w3.org/TR/html4/loose.dtd">
<html>
<head>
<meta http - equiv = "Content - Type" content = "text/html; charset = UTF - 8">
<title>page04</title>
</head>
<body>
<h3>简单类型演示</h3>
<form action = "simpleparam">
    <label>int 类型：</label><input name = "param1"><br/>
    <label>float 类型：</label><input name = "param2"><br/>
    <label>double 类型：</label><input name = "param3"><br/>
    <label>char 类型：</label><input name = "param4"><br/>
    <label>boolean 类型：</label><input name = "param5"><br/>
    <label>long 类型：</label><input name = "param6"><br/>
    <input type = "submit" value = "确认">
</form>
</body>
</html>
```

通过浏览器访问 page04.jsp 页面，对简单数据类型赋值，如图 12.5 所示。

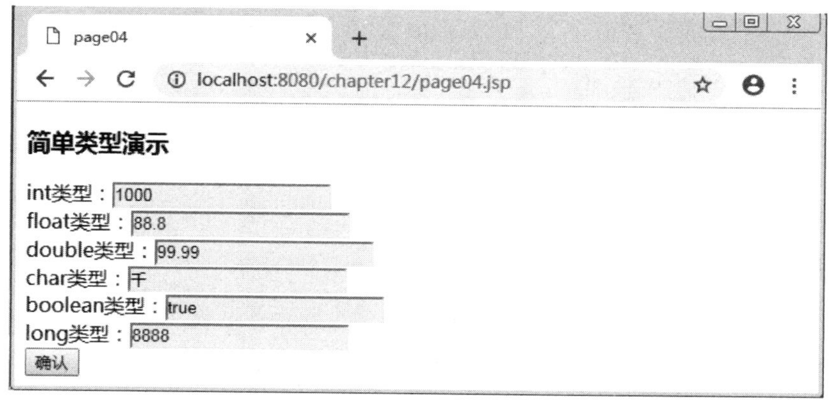

图 12.5 简单数据类型绑定

图 12.5 中所赋的值确认提交后，在控制面板打印的结果如图 12.6 所示。

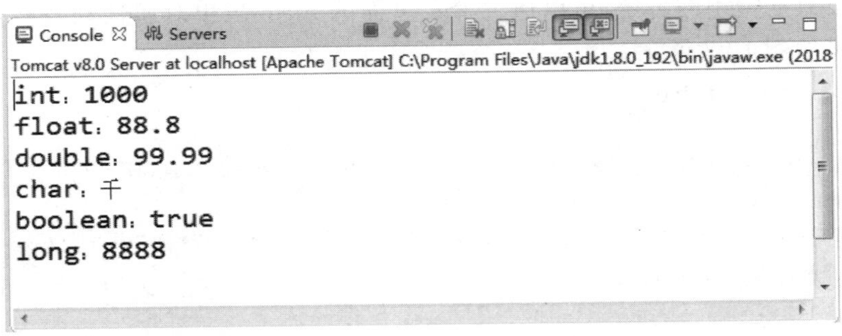

```
Console ✕  Servers
Tomcat v8.0 Server at localhost [Apache Tomcat] C:\Program Files\Java\jdk1.8.0_192\bin\javaw.exe (2018
int: 1000
float: 88.8
double: 99.99
char: 千
boolean: true
long: 8888
```

图 12.6　简单数据类型绑定结果展示

12.2.3　绑定 POJO 类型

Spring MVC 的控制器支持自定义类 POJO(Plain Ordinary Java Object)的自动转换，使用的时候 POJO 类中的属性名称就是请求参数的名称，如果名称不一致可以使用注解@RequestParam 注解标注。

自定义简单对象 Student 具体代码，如例 12-8 所示。

【例 12-8】　Student.java

```
package com.qfedu.pojo;
public class Student {
    private String no;
    private String name;
    private String classNo;
    public String getNo() {
        return no;
    }
    public void setNo(String no) {
        this.no = no;
    }
    public String getName() {
        return name;
    }
    public void setName(String name) {
        this.name = name;
    }
    public String getClassNo() {
        return classNo;
    }
    public void setClassNo(String classNo) {
        this.classNo = classNo;
    }
    @Override
    public String toString() {
        return "学生信息: \r\n no = " + no + " \r\n name = "
     + name + " \r\n classNo = " + classNo + "\r\n";
```

```
    }
}
```

对于 POJO 类型的数据绑定，在控制器中具体实现代码如例 12-9 所示。

【例 12-9】 MyController04.java

```java
package com.qfedu.controller;
import java.io.IOException;
import javax.servlet.http.HttpServletResponse;
import org.springframework.stereotype.Controller;
import org.springframework.web.bind.annotation.RequestMapping;
import org.springframework.web.bind.annotation.RequestMethod;
import com.qfedu.pojo.Student;
@Controller
public class MyController04 {
    @RequestMapping(value = "/pojoparam", method = RequestMethod.POST)
    public void param1(Student stu, HttpServletResponse response) throws
IOException {
        System.out.println("POJO: " + stu);
        response.getWriter().print("OK");
    }
}
```

通过 JSP 页面的 form 表单对 Student 对象的属性赋值，具体代码如例 12-10 所示。

【例 12-10】 page05.jsp

```jsp
<%@ page language = "java" contentType = "text/html; charset = UTF - 8"
    pageEncoding = "UTF - 8" %>
<!DOCTYPE html PUBLIC " - //W3C//DTD HTML 4.01 Transitional//EN" "http://www.w3.org/TR/
        html4/loose.dtd">
<html>
<head>
<meta http - equiv = "Content - Type" content = "text/html; charset = UTF - 8">
<title>page05</title>
</head>
<body>
<h3>POJO 类型演示</h3>
<form action = "pojoparam" method = "post">
    <label>学号：</label><input name = "no"><br/>
    <label>姓名：</label><input name = "name"><br/>
    <label>班级：</label><input name = "classNo"><br/>
    <input type = "submit" value = "确认">
</form>
</body>
</html>
```

浏览器访问 page05.jsp 页面，对学号、姓名、班级赋值，如图 12.7 所示。

图 12.7 中所赋的值确认提交后，在控制面板打印的结果如图 12.8 所示。

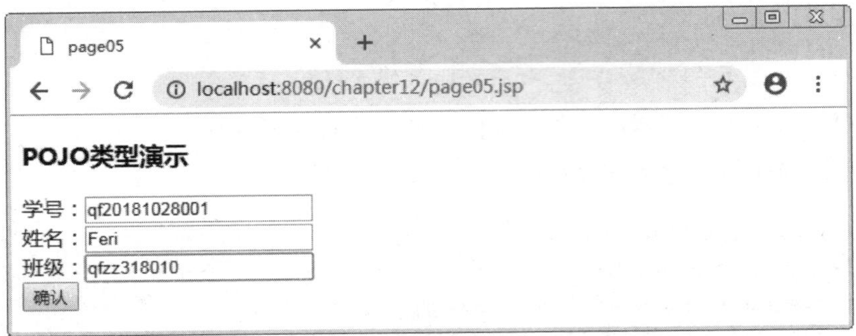

图 12.7 对象类型绑定

```
Console  Servers
Tomcat v8.0 Server at localhost [Apache Tomcat] C:\Program Files\Java\jdk1.8.0_192\bin\javaw.exe (2018
POJO：学生信息：
 no=qf20181028001
 name=Feri
 classNo=qfzz318010
```

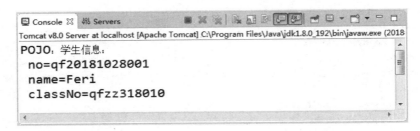

图 12.8 对象类型绑定结果展示

12.2.4 绑定包装 POJO

包装类型如 Integer、Long、Byte、Double、Float、Short 分别对应八种基本类型，Spring MVC 控制器支持基本类型自动转换，也支持对应的包装类的自动转换。

对于包装类型的数据绑定，控制器中的具体实现代码如例 12-11 所示。

【例 12-11】 MyController05.java

```java
package com.qfedu.controller;
import java.io.IOException;
import javax.servlet.http.HttpServletResponse;
import org.springframework.stereotype.Controller;
import org.springframework.web.bind.annotation.RequestMapping;
import org.springframework.web.bind.annotation.RequestMethod;
@Controller
public class MyController05 {
    @RequestMapping(value = "/packageparam",method = RequestMethod.GET)
    public void param1(Integer param1,Long param2,Double param3,Character
            param4,HttpServletResponse response) throws IOException {
        System.out.println("Integer: " + param1);
        System.out.println("Long: " + param2);
        System.out.println("Double: " + param3);
        System.out.println("Character" + param4);
        response.getWriter().print("OK");
    }
}
```

通过 jsp 页面的 form 表单对包装类型赋值，具体代码如例 12-12 所示。

【例 12-12】 page06.jsp

```
<%@ page language = "java" contentType = "text/html; charset = UTF - 8"
     pageEncoding = "UTF - 8" %>
<!DOCTYPE html PUBLIC " - //W3C//DTD HTML 4.01 Transitional//EN" "http://www.w3.org/TR/
        html4/loose.dtd">
<html>
<head>
<meta http - equiv = "Content - Type" content = "text/html; charset = UTF - 8">
<title>page06</title>
</head>
<body>
<h3>包装类型演示</h3>
<form action = "packageparam" method = "get">
     <label>Integer: </label><input name = "param1"><br/>
     <label>Long: </label><input name = "param2"><br/>
     <label>Double: </label><input name = "param3"><br/>
     <label>Character: </label><input name = "param4"><br/>
     <input type = "submit" value = "确认">
</form>
</body>
</html>
```

浏览器访问 page06.jsp 页面，对 Integer、Long、Double、Character 类型赋值，如图 12.9 所示。

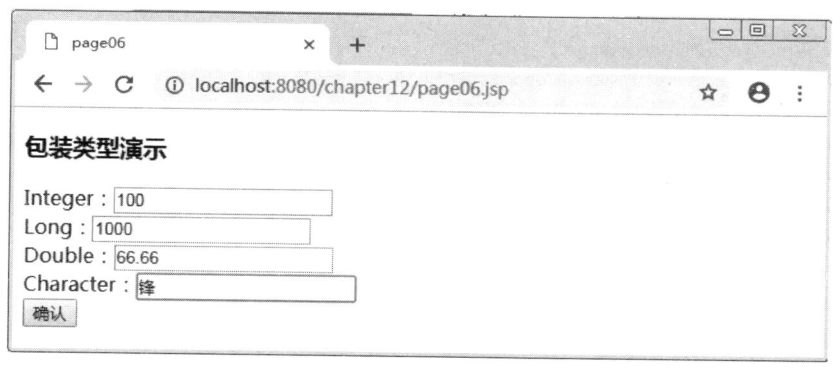

图 12.9 包装类型绑定

图 12.9 中所赋的值确认提交后，在控制面板打印的结果如图 12.10 所示。

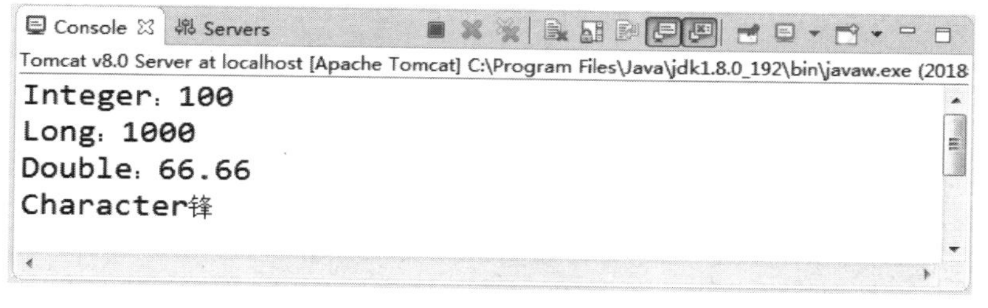

图 12.10 包装类型绑定结果展示

需要注意的是参数对应的类型，在传递参数的时候需要为对应的类型传递对应的值，比如为 Integer 的时候，只能为整型，当传递的值为数字时不可以是浮点型或者是非数字类型，否则直接引发 400 异常(参数错误)。

12.2.5 自定义数据绑定

在实际开发中，有时会遇到无法解析的数据类型，比如日期、加密的参数、部分参数等各种复杂类型，那么在这种情况下，就需要用到自定义参数解析接口 HandlerMethodArgumentResolver。

HandlerMethodArgumentResolver 接口包含两个接口方法：

```
public interface HandlerMethodArgumentResolver {
    boolean supportsParameter(MethodParameter var1);
    @Nullable
    Object resolveArgument(MethodParameter var1, @Nullable
ModelAndViewContainer var2, NativeWebRequest var3, @Nullable
WebDataBinderFactory var4) throws Exception;
```

接下来通过代码完成自定义参数解析器实现的数据绑定，该方式主要用于解析 request 请求参数，并绑定数据到 Controller 的入参上，实现接口 HandlerMethodArgumentResolver，并且重写 supportsParameter 和 resolveArgument 方法，最后在配置文件中加入 resolver 配置。

新建 com.qfedu.annotation 包，并创建 MyParam 类，具体代码如例 12-13 所示。

【例 12-13】 MyParam.java

```
package com.qfedu.annotation;
import java.lang.annotation.Documented;
import java.lang.annotation.ElementType;
import java.lang.annotation.Retention;
import java.lang.annotation.RetentionPolicy;
import java.lang.annotation.Target;
@Target(value = {ElementType.PARAMETER})
@Retention(value = RetentionPolicy.RUNTIME)
@Documented
public @interface MyParam {
    String name();
}
```

上述代码中通过实现自定义注解来标记需要进行自定义的参数解析器。

在 com.qfedu.resolver 包中，新建 MyResolver 类，具体代码如例 12-14 所示。

【例 12-14】 MyResolver.java

```
package com.qfedu.resolver;
import javax.servlet.http.HttpServletRequest;
import org.springframework.core.MethodParameter;
import org.springframework.web.bind.support.WebDataBinderFactory;
import org.springframework.web.context.request.NativeWebRequest;
```

```java
import org.springframework.web.method.support.HandlerMethodArgumentResolver;
import org.springframework.web.method.support.ModelAndViewContainer;
import com.qfedu.annotation.MyParam;
public class MyResolver implements HandlerMethodArgumentResolver {
    /**
     * 将 request 中的请求参数解析到当前 Controller 参数上
     * @param parameter
     * 需要被解析的 Controller 参数,此参数必须首先传给 supportsParameter
     * 并返回 true
     * @param mavContainer
     *              当前 request 的 ModelAndViewContainer
     * @param webRequest
     *              当前 request
     * @param binderFactory
     *              生成实例的工厂
     * @return 解析后的 Controller 参数
     */
    public Object resolveArgument(MethodParameter parameter,
        ModelAndViewContainer mavContainer,
        NativeWebRequest webRequest, WebDataBinderFactory
        binderFactory)
throws Exception {
        System.out.println("MethodParameter:" + parameter.toString());
        MyParam myParam = parameter.getParameterAnnotation(MyParam.class);
        String pn = myParam.name();
        if (pn == null) {
            pn = parameter.getParameterName();
        }
        System.out.println(pn);
        // 简单的案例:如果客户端未传值,就设置默认值
        Object res = webRequest.getNativeRequest
            (HttpServletRequest.class).getParameter(pn);
        return res == null ? "默认值哟" : res;
    }
    /**
     * 解析器是否支持当前参数 判断 Controller 层中的参数,是否满足条件,满足条件则执
     * 行 resolveArgument 方法,不满足则跳过
     */
    public boolean supportsParameter(MethodParameter parameter) {
         //如果方法包含自定义注解,那就走 resolveArgument()方法
        return parameter.hasParameterAnnotation(MyParam.class);
    }
}
```

在上述代码中,实现了自定义参数解析器,新建 MyResolver 类实现了 HandlerMethodArgumentResolver 接口,并且重写了 supportsParameter 方法(校验是否需要进行参数解析,通过携带自定义注解@MyParam),重写 resolveArgument 方法(实现参数解析的核心),主要是实现了参数是否传递数据,如果传递就返回原参数数据,否则就返回一个默认值。

在项目的 WEB-INF 目录下，创建一个 SpringMVC 的配置文件，命名为 springMVC-config.xml，内容如例 12-15 所示。

【例 12-15】 springMVC-config.xml

```
<?xml version='1.0' encoding='UTF-8'?>
<beans xmlns="http://www.springframework.org/schema/beans"
    xmlns:xsi="http://www.w3.org/2001/XMLSchema-instance"
    xmlns:context="http://www.springframework.org/schema/context"
    xmlns:mvc="http://www.springframework.org/schema/mvc"
    xsi:schemaLocation="http://www.springframework.org/schema/beans
        http://www.springframework.org/schema/beans/spring-beans.xsd
        http://www.springframework.org/schema/context
        http://www.springframework.org/schema/context/spring-context.xsd
        http://www.springframework.org/schema/mvc
        http://www.springframework.org/schema/mvc/spring-mvc.xsd">
    <!-- 扫描 Controller -->
    <context:component-scan
        base-package="com.qfedu.controller" />
    <mvc:annotation-driven>
        <!-- 自定义参数解析器 -->
        <mvc:argument-resolvers>
            <bean class="com.qfedu.resolver.MyResolver"/>
        </mvc:argument-resolvers>
    </mvc:annotation-driven>
    <!-- 配置视图解析器 -->
    <bean class="org.springframework.web.servlet.view.
        InternalResourceViewResolver">

        <property name="suffix" value=".jsp" />
    </bean>
</beans>
```

注意：一定要在<mvc:annotation-driven>标签内部注册自定义的参数解析器。

在 com.qfedu.controller 包中新建 MyController06 类，代码如例 12-16 所示。

【例 12-16】 MyController06.java

```
package com.qfedu.controller;
import java.util.Map;
import org.springframework.stereotype.Controller;
import org.springframework.web.bind.annotation.RequestMapping;
import org.springframework.web.bind.annotation.ResponseBody;
import com.qfedu.annotation.MyParam;
@Controller
public class MyController06 {
    @ResponseBody
    @RequestMapping("/data")
    public Map<String, Object> data(@MyParam(name = "a") String a) {
        System.out.println("参数信息: " + a);
        return null;
    }
}
```

上述代码中，演示了自定义参数处理器的作用，需要带上参数，而且使用了自定义注解@MyParam 进行标记。

创建一个分别带有参数和不带参数请求的页面，如例 12-17 所示。

【例 12-17】 page07.jsp

```
<%@ page language="java" contentType="text/html; charset=UTF-8"
    pageEncoding="UTF-8"%>
<!DOCTYPE html PUBLIC "-//W3C//DTD HTML 4.01 Transitional//EN"
"http://www.w3.org/TR/html4/loose.dtd">
<html>
<head>
<meta http-equiv="Content-Type" content="text/html; charset=UTF-8">
<title>page07</title>
</head>
<body>
<h3>自定义数据类型绑定演示</h3>
<h4><a href="data">没有参数</a></h4>
<h4><a href="data?a=qianfeng">携带参数</a></h4>
</body>
</html>
```

代码编写结束，接下来在浏览器中访问 page07.jsp 页面，结果如图 12.11 所示。探前台中打印的结果如图 12.12 所示。

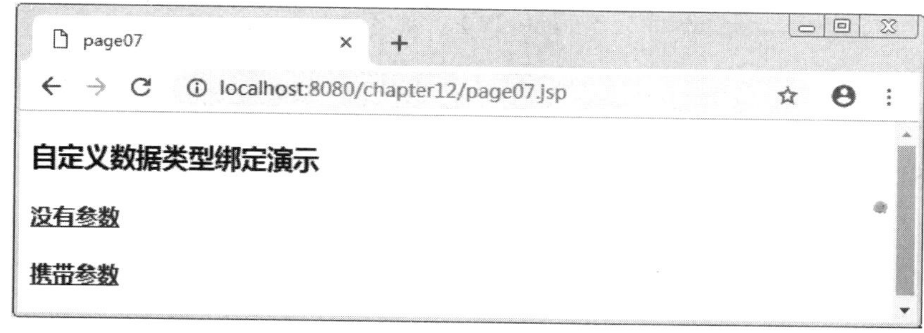

图 12.11 自定义数据类型绑定

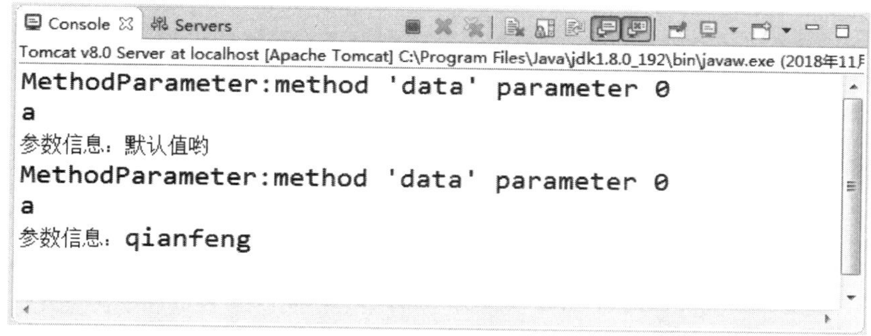

图 12.12 自定义数据类型绑定结果展示

12.3 复杂数据绑定

12.3.1 绑定数组

在进行数据传递的时候，可能会需要使用数组，那么 Spring MVC 的控制器是否支持数组的字段转换呢？我们通过代码演示一下。

对于数组类型的数据绑定，控制器中具体实现代码如例 12-18 所示。

【例 12-18】 MyController07.java

```
package com.qfedu.controller;
import java.io.IOException;
import java.util.Arrays;
import javax.servlet.http.HttpServletResponse;
import org.springframework.stereotype.Controller;
import org.springframework.web.bind.annotation.RequestMapping;
import org.springframework.web.bind.annotation.RequestMethod;
@Controller
public class MyController06 {
    @RequestMapping(value = "/arrayparam",method = RequestMethod.GET)
    public void param1(String[ ] hobby ,HttpServletResponse response) throws
IOException {
        System.out.println("您的爱好: " + Arrays.toString(hobby));
        response.getWriter().print("OK");
    }
}
```

通过 JSP 页面的 form 表单对数组类型赋值，具体代码如例 12-19 所示。

【例 12-19】 page08.jsp

```
<%@ page language = "java" contentType = "text/html; charset = UTF - 8"
    pageEncoding = "UTF - 8" %>
<!DOCTYPE html PUBLIC " - //W3C//DTD HTML 4.01 Transitional//EN"
"http://www.w3.org/TR/html4/loose.dtd">
<html>
<head>
<meta http - equiv = "Content - Type" content = "text/html; charset = UTF - 8">
<title>page07</title>
</head>
<body>
<h3>数组类型演示</h3>
<form action = "arrayparam" method = "get">
    <label>爱好: </label>
    <input type = "checkbox" name = "hobby" value = "JAVA">JAVA
    <input type = "checkbox" name = "hobby" value = "Html5">Html5
    <input type = "checkbox" name = "hobby" value = "Python">Python
    <input type = "checkbox" name = "hobby" value = "Go">Go
```

```
        <input type="checkbox" name="hobby" value="Ruby">Ruby
        <input type="checkbox" name="hobby" value="C++">C++<br/>
        <input type="submit" value="确认">
</form>
</body>
</html>
```

浏览器访问 page08.jsp 页面，在复选框中选中喜爱的课程名称，如图 12.13 所示。

图 12.13　数组类型绑定

图 12.13 中选中的课程名称在控制面板打印的结果如图 12.14 所示。

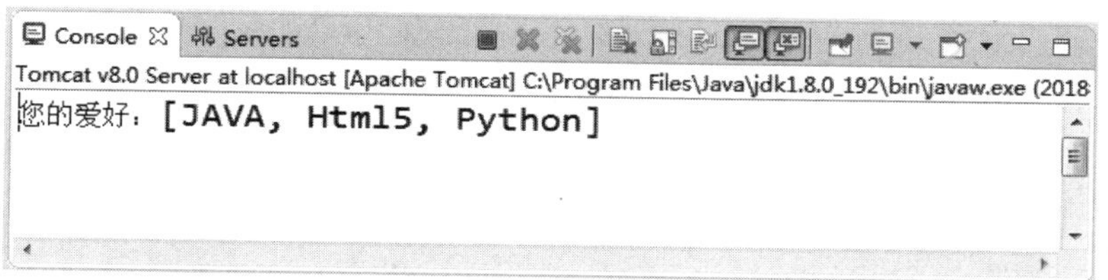

图 12.14　数组类型绑定结果展示

12.3.2　绑定集合

Spring MVC 的控制器对集合的支持也是可以的，对于传递的类型进行自动化转换。但是需要通过注解标记，否则转换异常。

对于集合类型的数据绑定，控制器中具体实现代码如例 12-20 所示。

【例 12-20】　MyController09.java

```
package com.qfedu.controller;
import java.io.IOException;
import java.util.Arrays;
import java.util.List;
import javax.servlet.http.HttpServletResponse;
import org.springframework.stereotype.Controller;
import org.springframework.web.bind.annotation.RequestMapping;
import org.springframework.web.bind.annotation.RequestMethod;
import com.qfedu.pojo.Student;
@Controller
public class MyController07 {
```

```
    @RequestMapping(value = "/gatherparam",method = RequestMethod.GET)
    public void param1(@RequestParam("foods") List<String>
foods ,HttpServletResponse response) throws IOException {
        System.out.println("集合: " + foods);
        response.getWriter().print("OK");
    }
}
```

通过 JSP 页面的 form 表单对集合类型赋值，具体代码如例 12-21 所示。

【例 12-21】 page010.jsp

```
<%@ page language = "java" contentType = "text/html; charset = UTF - 8"
    pageEncoding = "UTF - 8" %>
<!DOCTYPE html PUBLIC " - //W3C//DTD HTML 4.01 Transitional//EN"
"http://www.w3.org/TR/html4/loose.dtd">
<html>
<head>
<meta http - equiv = "Content - Type" content = "text/html; charset = UTF - 8">
<title>page05</title>
</head>
<body>
<h3>集合类型演示</h3>
<form action = "gatherparam" method = "get">
    <label>爱好: </label>
    <input type = "checkbox" name = "foods" value = "馒头">馒头
    <input type = "checkbox" name = "foods" value = "大米">大米
    <input type = "checkbox" name = "foods" value = "红烧肉">红烧肉
    <input type = "checkbox" name = "foods" value = "梅菜扣肉">梅菜扣肉<br/>
    <input type = "submit" value = "确认">
</form>
</body>
</html>
```

浏览器访问 page10.jsp 页面，在复选框中选中喜爱的食物名称，如图 12.15 所示。

图 12.15 集合类型绑定

图 12.15 中选中的食物名称在控制面板打印结果如图 12.16 所示。

```
Console ⊠  Servers
Tomcat v8.0 Server at localhost [Apache Tomcat] C:\Program Files\Java\jdk1.8.0_192\bin\javaw.exe (2018
集合：[馒头, 大米, 梅菜扣肉]
```

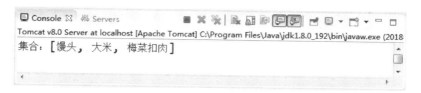

图 12.16　集合类型绑定结果展示

12.4 本章小结

本章首先介绍了 Spring 的参数绑定，包括默认类型、基本类型、包装类型和复杂类型的绑定；其次对各种类型的参数绑定都做了相关实例演示。通过本章节的学习可以了解控制器的映射方法接收参数时的多种形式以及编码过程中需要注意的地方。对于特殊的参数处理，还可以通过自定义参数解析器完成参数的绑定。

12.5 习　题

1. 填空题

(1) 在 Spring MVC 的控制器中，用户提交请求的数据是通过________来接收的。

(2) Spring MVC 默认支持的绑定参数类型有________、________、________、________。

(3) 在绑定 POJO 类型时，如果 POJO 类中的属性名称与请求参数的名称不一致可以使用________注解标注。

(4) Spring MVC 支持________种基本类型的自动转换。

(5) 在实际开发的过程中，当遇到无法解析的数据类型时可以自定义参数解析接口________。

2. 选择题

(1) 关于 Spring MVC 的参数绑定，下列选项正确的是(　　)。

A. Spring MVC 中默认支持的参数绑定类型，可以不用声明直接使用

B. 在参数传递时，如果参数名称不同，但是参数的类型、个数、顺序都相同，即可完成参数的绑定

C. 在参数传递时，如果参数名称不同，可以使用@RequestParam 注解标注

D. Spring MVC 在控制器类中只能通过 getParamter 方法接收参数

(2) HandlerMethodArgumentResolver 接口中包含(　　)两个方法。

A. supportsParameter、resolveArgument

B. addObject、handler

C. preHandle、postHandle

D. addAccount、updateAccount

(3) 关于 Spring MVC 支持的参数绑定类型，下列选项正确的是(　　)。

A. Spring MVC 不支持绑定自定义的数据类型

B. Spring MVC 不支持绑定集合类型

C. Spring MVC 不支持绑定数组类型

D. 以上说法均错误

(4) 自定义参数解析器编写完成后需要在 SpringMVC. xml(　　)标签中注册。

A. < mvc:annotation-driven >

B. < mvc:default-servlet-handler >

C. < bean ></bean >

D. < context:component-scan />

(5) 以下(　　)选项中都是 Spring MVC 控制器参数绑定所支持的自动转换类型。

A. Int、Long、Short、Char

B. Integer、String、Long、Byte

C. Double、Float

D. 以上都是

3. 思考题

(1) 什么是 Spring MVC 中的参数绑定?

(2) Spring MVC 中参数绑定是怎么实现的?

4. 编程题

自定义一个 person 对象,通过在浏览器中输入编号、姓名、电话、地址的信息,实现在控制器中绑定,并完成在控制台中打印输入结果。

第 13 章 异常处理和拦截器

本章学习目标

- 理解全局异常处理器
- 理解拦截器的概念
- 掌握拦截器的应用

企业级开发中，会时常遇到代码异常的问题，在代码异常的处理上，Spring MVC 提供了异常处理器进行全局的异常处理。之前在有关 Java Web 的书中学习过 Filter(过滤器)，本章将通过学习与之类似的技术，由 Spring MVC 框架提供的拦截器技术来解决此类问题。

13.1 全局异常处理器

项目开发的过程中，进行异常处理往往是项目开发流程中不可或缺的一部分。因为在程序发生异常时，不能把一些只有程序员才能看懂的错误代码抛给用户看，所以在这个时候进行统一的异常处理，展现一个较为友好的错误页面就显得十分有必要了。跟其他 MVC 框架一样，Spring MVC 也有自己的异常处理机制。Spring MVC 提供的异常处理主要有两种方式：一种是直接实现自己的 HandlerExceptionResolver 接口，当然这也包括使用 Spring 已经提供好的 SimpleMappingExceptionResolver 和 DefaultHandlerExceptionResolver；另一种是使用注解的方式实现一个专门用于处理异常的 Controller 即@ExceptionHandler 注解。除此之外还有第三种处理方式，使用@ControllerAdvice 注解。下面将通过代码来展示异常处理器的应用。

13.1.1 HandlerExceptionResolver

HandlerExceptionResolver 作为异常处理接口，它的内部有一个方法 resolveException 方法，提供了对异常进行解析的操作，该方法的参数说明如表 13.1 所示。

表 13.1 方法参数

参　　数	说　　明
HttpServletRequest	请求对象
HttpServletResponse	响应对象
Object	参数信息
Exception	异常信息对象

resolveException 方法中的参数及具体用法的代码实现如例 13-1 所示。

【例 13-1】 SpringExceptionResolver. java

```
package com.qfedu.resolver;
import javax.servlet.http.HttpServletRequest;
import javax.servlet.http.HttpServletResponse;
import org.slf4j.Logger;
import org.slf4j.LoggerFactory;
import org.springframework.web.servlet.HandlerExceptionResolver;
import org.springframework.web.servlet.ModelAndView;
public class SpringExceptionResolver implements HandlerExceptionResolver{
    private Logger
log = LoggerFactory.getLogger(SpringExceptionResolver.class);
    @Override
    public ModelAndView resolveException(HttpServletRequest request,
HttpServletResponse response, Object param,
            Exception exception) {
        ModelAndView mv = new ModelAndView();
        mv.setViewName("error");
        mv.addObject("errorMsg", "异常处理: " + exception.getMessage());
        return mv;
    }
}
```

注意:该类需要在 springmvc 的配置文件中进行配置:

```
<bean id = "springExceptionResolver"
```

class="com. qfedu. resolver. SpringExceptionResolver"></bean >,否则出现异常的时候找不到对应的类。

接下来实现控制器的代码编写,完成异常处理,案例将采用简单的数学运算进行演示,把 0 作为被除数,因此在映射方法中当程序执行到 1/0 时会主动抛出异常。控制器中的具体代码如例 13-2 所示。

【例 13-2】 MyController01. java

```
package com.qfedu.controller;
import java.io.IOException;
import javax.servlet.http.HttpServletResponse;
import org.springframework.stereotype.Controller;
import org.springframework.web.bind.annotation.RequestMapping;
@Controller
public class MyController01 {
    @RequestMapping("exceptionapp")
    public void handler(HttpServletResponse response) throws IOException{
        System.out.println(1/0);
        response.getWriter().print("OK");
    }
}
```

完成上述操作后再编写一个用来统一展示错误的 error.jsp 页面，完成全局异常在浏览器中的展现，页面的代码如例 13-3 所示。

【例 13-3】 error.jsp

```
<%@ page language="java" contentType="text/html; charset=UTF-8"
    pageEncoding="UTF-8" %>
<!DOCTYPE html PUBLIC "-//W3C//DTD HTML 4.01 Transitional//EN"
"http://www.w3.org/TR/html4/loose.dtd">
<html>
<head>
<meta http-equiv="Content-Type" content="text/html; charset=UTF-8">
<title>服务器暂时无法访问</title>
</head>
<body>
<h2>异常统一处理</h2>
<h4 style="color:red">${errorMsg}</h4>
</body>
</html>
```

在上述代码中，通过 EL 表达式获取在控制器中存储到 request 中的错误信息，并且设置错误信息内容的字体颜色为红色，访问浏览器可得到的结果如图 13.1 所示。

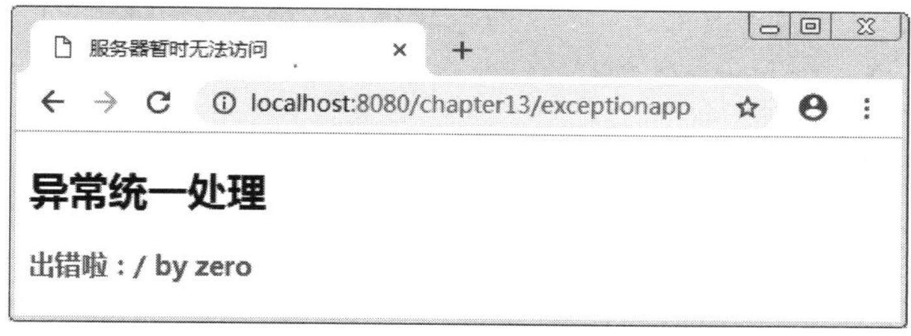

图 13.1　统一异常处理

其实只要是 HandlerExceptionResolver 接口的实现类即可实现全局异常的处理，无论是自定义还是 SimpleMappingExceptionResolver、DefaultHandlerExceptionResolver 都可以实现全局异常处理。

13.1.2　@ExceptionHandler

@ExceptionHandler 注解用在控制器内，进行异常处理的方法必须与出错的方法在同一个 Controller 里面。

创建一个 MyController02 类，具体代码如例 13-4 所示。

【例 13-4】 MyController02.java

```
package com.qfedu.controller;
import java.io.IOException;
import javax.servlet.http.HttpServletResponse;
import org.springframework.stereotype.Controller;
```

```
import org.springframework.ui.Model;
import org.springframework.web.bind.annotation.ExceptionHandler;
import org.springframework.web.bind.annotation.RequestMapping;
@Controller
public class MyController02 {
    @ExceptionHandler({Exception.class})
    public String exception(Exception e,Model model) {
        System.out.println(e.getMessage());
        model.addAttribute("errorMsg", e.getMessage());
        return "error";
    }
    @RequestMapping("exceptionapp2")
    public void handler(HttpServletResponse response) throws IOException {
        throw new RuntimeException("测试异常处理");
    }
}
```

在上述代码中，直接在映射方法中，抛出一个运行时异常，来触发异常处理器。在浏览器地址栏输入 http://localhost:8080/chapter13/exceptionapp2，即可得到如图 13.2 所示结果。

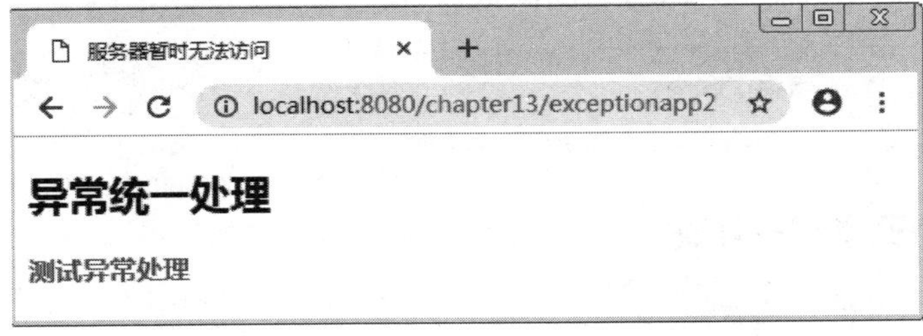

图 13.2 异常处理结果

13.1.3 @ControllerAdvice

@ExceptionHandler 注解，需要进行异常处理的方法必须与出错的方法在同一个 Controller 里面。如果在代码加入了 @ControllerAdvice，则不需要必须在同一个 Controller 中了，这也是 Spring 3.2 带来的新特性。从名字上可以看出注解的大体意思是控制器增强。也就是说，@ControllerAdvice + @ ExceptionHandler 也可以实现全局的异常捕捉。

创建一个 MyController03 类，具体代码如例 13-5 所示。

【例 13-5】 MyController03.java

```
package com.qfedu.controller;
import java.io.IOException;
import javax.servlet.http.HttpServletResponse;
import org.springframework.stereotype.Controller;
import org.springframework.web.bind.annotation.RequestMapping;
```

```java
@Controller
public class MyController03 {
    @RequestMapping("exceptionapp3")
    public void handler(HttpServletResponse response) throws IOException{
        throw new NullPointerException("测试注解异常处理");
    }
}
```

在上述代码中，方法内部主动抛出异常，触发异常处理器。

通过@ControllerAdvice + @ ExceptionHandler 注解实现全局的异常捕捉的代码。创建一个 GlobalExceptionHandler 类，具体代码如例 13-6 所示。

【例 13-6】 GlobalExceptionHandler. java

```java
package com.qfedu.resolver;
import org.slf4j.Logger;
import org.slf4j.LoggerFactory;
import org.springframework.web.bind.annotation.ControllerAdvice;
import org.springframework.web.bind.annotation.ExceptionHandler;
import org.springframework.web.servlet.ModelAndView;
@ControllerAdvice
public class GlobalExceptionHandler {
    private Logger
log = LoggerFactory.getLogger(GlobalExceptionHandler.class);
    @ExceptionHandler(Exception.class)
    public ModelAndView handleEx(Exception ex) {
        ModelAndView mv = new ModelAndView();
        mv.setViewName("error");
        mv.addObject("errorMsg", "出错啦: " + ex.getMessage());
        return mv;
    }
}
```

注意：GlobalExceptionHandler 类需要通过 IOC 创建对象，在 springMVC-config. xml 配置文件中进行装配，配置文件 springMVC-config. xml 中的内容如例 13-7 所示。

【例 13-7】 springMVC-config. xml

```xml
<?xml version="1.0" encoding="UTF-8"?>
<beans xmlns="http://www.springframework.org/schema/beans"
    xmlns:xsi="http://www.w3.org/2001/XMLSchema-instance"
    xmlns:context="http://www.springframework.org/schema/context"
    xmlns:mvc="http://www.springframework.org/schema/mvc"
    xsi:schemaLocation="
        http://www.springframework.org/schema/beans
        http://www.springframework.org/schema/beans/spring-beans.xsd
        http://www.springframework.org/schema/context
        http://www.springframework.org/schema/context/spring-context.xsd
        http://www.springframework.org/schema/mvc
```

```
            http://www.springframework.org/schema/mvc/spring-mvc.xsd">
    <!-- 启动自动扫描 -->
    <context:component-scan base-package="com.qfedu.*" />
    <!-- 注册 MVC 注解驱动 -->
    <mvc:annotation-driven />
    <!-- 静态资源可访问的设置方式 -->
    <mvc:default-servlet-handler />
    <!-- 配置视图解析器,如果不设置会依据 SpringMVC 的默认设置 -->
    <bean id="viewResolver"class="org.springframework.
            web.servlet.view.InternalResourceViewResolver">

        <property name="prefix" value="/" />
        <property name="suffix" value=".jsp" />
    </bean>
    <bean id="springExceptionResolver"
class="com.qfedu.resolver.GlobalExceptionHandler"></bean>
</beans>
```

在浏览器地址栏中输入 http://localhost:8080/chapter13/exceptionapp3，即可得到图 13.3 所示结果。

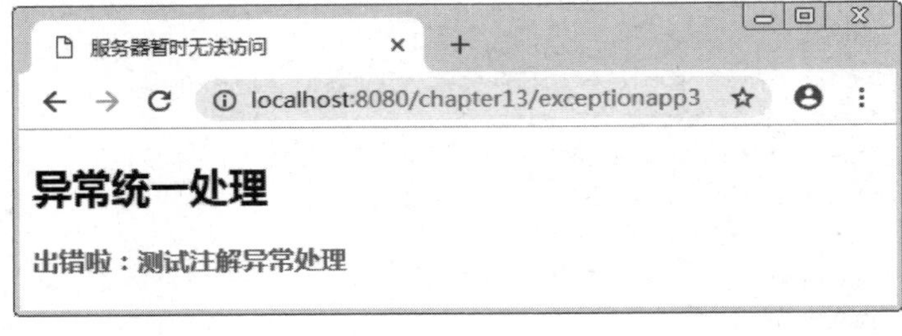

图 13.3　测试注解异常处理

通过以上页面中显示的结果，可以发现自定义的异常处理器起作用了，完成了异常的捕获。

13.2　拦截器定义与配置

拦截器 Interceptor 的拦截功能是基于 Java 的动态代理来实现的，Interceptor 拦截器用于拦截 Controller 层接口，表现形式有点像 Spring 的 AOP，但是 AOP 是针对单一的方法。Interceptor 是针对 Controller 接口以及可以处理的 request 和 response 对象。

在 Spring MVC 中定义一个 Interceptor 是比较简单的，主要有两种方式：第一种是实现 HandlerInterceptor 接口，或者是继承实现了 HandlerInterceptor 接口的类；第二种是实现 Spring 的 WebRequestInterceptor 接口，或者是继承实现了 WebRequestInterceptor 的类。

13.2.1 HandlerInterceptor 接口

HandlerInterceptor 接口中的三个方法，对应的作用说明如表 13.2 所示。

表 13.2 接口方法

方 法 名	返回值	说 明
preHandle (HttpServletRequest request, HttpServletResponse response, Object handle)	Boolean	该方法将在请求处理之前进行调用，只有该方法返回 true，才会继续执行后续的 Interceptor 和 Controller；当返回值为 true 时就会继续调用下一个 Interceptor 的 preHandle 方法；如果已经是最后一个 Interceptor 的时候就会调用当前请求的 Controller 方法
postHandle (HttpServletRequest request, HttpServletResponse response, Object handle, ModelAndView modelAndView)	Void	该方法将在请求处理之后，DispatcherServlet 进行视图返回渲染之前进行调用，可以在这个方法中对 Controller 处理之后的 ModelAndView 对象进行操作
afterCompletion (HttpServletRequest request, HttpServletResponse response, Object handle, Exception ex)	Void	该方法也是需要当前对应的 Interceptor 的 preHandle 方法的返回值为 true 时才会执行。该方法将在整个请求结束之后，也就是在 DispatcherServlet 渲染了对应的视图之后执行，用于进行资源清理

Spring MVC 还提供了 HandlerInterceptorAdapter 抽象类，该类也是 HandlerInterceptor 的子类，在实现了 HandlerInterceptor 的三个函数后还增加了一个函数。

(1) preHandle：在执行 controller 处理之前执行，返回值为 boolean ，返回值为 true 时接着执行 postHandle 和 afterCompletion，如果返回值为 false 则中断执行。

(2) postHandle：在执行 controller 处理之后 ModelAndView 处理之前执行。

(3) afterCompletion ：在 DispatchServlet 执行完 ModelAndView 之后执行。

(4) afterConcurrentHandlingStarted：这个方法会在 Controller 方法异步执行时开始执行，而 Interceptor 的 postHandle 方法则是需要等到 Controller 的异步执行完才能执行，只要继承这个类并实现其方法就可以了。接下来将通过案例演示实现拦截器的接口，并且重写对应的三个方法，通过在方法的内部分别输出一句话，标记该方法的执行。

通过 HandlerInterceptor 接口实现拦截器功能，具体代码如例 13-8 所示。

【例 13-8】 HelloInterceptor.java

```
package com.qfedu.interceptor;
import javax.servlet.http.HttpServletRequest;
import javax.servlet.http.HttpServletResponse;
import org.springframework.web.servlet.HandlerInterceptor;
import org.springframework.web.servlet.ModelAndView;
public class HelloInterceptor implements HandlerInterceptor {
    @Override
    public boolean preHandle(HttpServletRequest request,
        HttpServletResponse response, Object handler)
            throws Exception {
```

```
            System.out.println("preHandle 预处理……");
            return true;
        }
        @Override
        public void postHandle(HttpServletRequest request, HttpServletResponse
                response, Object handler,
                ModelAndView modelAndView) throws Exception {
            System.out.println("postHandle 后处理……");
        }
        @Override
        public void afterCompletion(HttpServletRequest request,
            HttpServletResponse response, Object handler, Exception ex)
                throws Exception {
            System.out.println("afterCompletion 请求结束……");
        }
}
```

注意：preHandle 方法的返回值表示是否继续触发后面的请求，即是否拦截，如果返回值为 true 则放行，否则拦截将不再继续运行。

在 WEB-INF/springMVC-config.xml 配置文件中，需要配置拦截器并且标注出拦截规则和对应的自定义拦截器，如例 13-9 所示。

【例 13-9】 springMVC-config.xml

```
<?xml version='1.0' encoding='UTF-8'?>
<beans xmlns="http://www.springframework.org/schema/beans"
    xmlns:xsi="http://www.w3.org/2001/XMLSchema-instance"
    xmlns:context="http://www.springframework.org/schema/context"
    xmlns:mvc="http://www.springframework.org/schema/mvc"
    xsi:schemaLocation="http://www.springframework.org/schema/beans
        http://www.springframework.org/schema/beans/spring-beans.xsd
        http://www.springframework.org/schema/context
        http://www.springframework.org/schema/context/spring-context.xsd
        http://www.springframework.org/schema/mvc
        http://www.springframework.org/schema/mvc/spring-mvc.xsd
        ">
    <!-- 扫描 Controller -->
    <context:component-scan base-package="com.qfedu" />
    <!-- 配置视图解析器 -->
    <bean class="org.springframework.
        web.servlet.view.InternalResourceViewResolver">

        <property name="suffix" value=".jsp" />
    </bean>
    <!-- 配置拦截器 -->
    <mvc:interceptors>
    <mvc:interceptor>
            <!-- 进行拦截：/** 表示拦截所有 controller -->
            <mvc:mapping path="/**" />
```

```
            <!-- 不进行拦截 -->
            <mvc:exclude-mapping path="/*.jsp"/>
            <bean class="com.qfedu.interceptor.HelloInterceptor"/>
        </mvc:interceptor>
    </mvc:interceptors>
</beans>
```

在控制器中打印一句自定义文本内容，测试代码执行的顺序，如例 13-10 所示。

【例 13-10】 MyController04.java

```
package com.qfedu.controller;
import java.io.IOException;
import javax.servlet.http.HttpServletResponse;
import org.springframework.stereotype.Controller;
import org.springframework.web.bind.annotation.RequestMapping;
@Controller
public class MyController04 {
    @RequestMapping("interceptorapp1")
    public void handler(HttpServletResponse response) throws IOException {
        System.out.println("通过拦截器进来了吗?");
    }
}
```

在浏览器地址栏中输入：http://localhost:8080/chapter13/interceptorapp1 进行测试，可得到如图 13.4 所示结果。

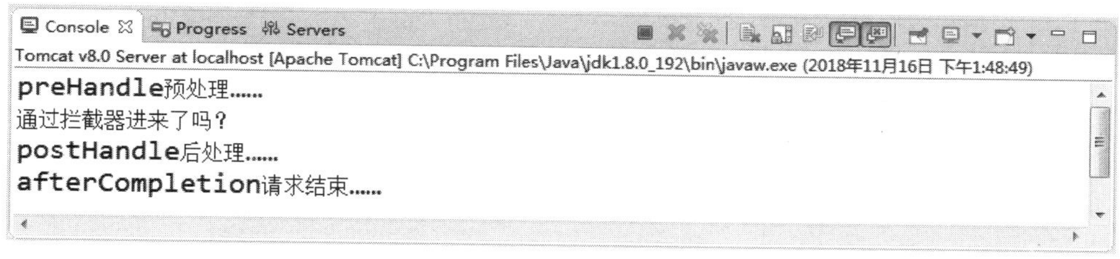

图 13.4 Handler Interceptor 实现拦截器

通过上图可以发现拦截器中的内容被打印出来，而且控制器内容的打印在 postHandler 方法的执行之前。

13.2.2 WebRequestInterceptor 接口

WebRequestInterceptor 接口中也定义了三个方法，同 HandlerInterceptor 接口的用法相似，它也是通过复写这三个方法来对用户的请求进行拦截处理的，不同的是 WebRequestInterceptor 接口中的 preHandle 方法没有返回值，而且 WebRequestInterceptor 的三个方法的参数都是 WebRequest。这个 WebRequest 到底是什么呢？其实 WebRequest 是 Spring 中定义的一个接口，它里面的方法定义跟 HttpServletRequest 类似，在 WebRequestInterceptor 中对 WebRequest 进行的所有操作都将同步到 HttpServletRequest 中，然后在当前请求中依次传递。

在 Spring 框架之中，还提供了一个和 WebRequestInterceptor 接口很像的抽象类，那就是 WebRequestInterceptorAdapter，其实现了 AsyncHandlerInterceptor 接口，并在内部调用了 WebRequestInterceptor 接口，如表 13.3 所示。

表 13.3　接口方法说明

方　　法	返回值	说　　明
preHandle(WebRequest request)	Void	该方法在请求处理之前，也就是在 Controller 中的方法调用之前被调用
postHandle (WebRequest request, ModelMap model)	Void	该方法在请求处理之后，也就是在 Controller 中的方法调用之后被调用，但是会在视图返回被渲染之前被调用
afterCompletion(WebRequest request, Exception ex)	Void	该方法会在整个请求处理完成，也就是在视图返回并被渲染之后执行

新建一个 MyWebRequestInterceptor 类实现 WebRequestInterceptor 接口来实现拦截器的功能，具体代码如例 13-11 所示。

【例 13-11】 MyWebRequestInterceptor.java

```
package com.qfedu.interceptor;
import org.springframework.ui.ModelMap;
import org.springframework.web.context.request.WebRequest;
import org.springframework.web.context.request.WebRequestInterceptor;
public class MyWebRequestInterceptor implements WebRequestInterceptor{
    @Override
    public void afterCompletion(WebRequest request, Exception ex) throws
Exception {
        // TODO Auto - generated method stub
        System.out.println("afterCompletion:执行了");
    }
    @Override
    public void postHandle(WebRequest request, ModelMap model) throws
Exception {
        // TODO Auto - generated method stub
        System.out.println("postHandle:?执行了");
    }
    @Override
    public void preHandle(WebRequest request) throws Exception {
        // TODO Auto - generated method stub
        System.out.println("preHandle:执行了");
    }
}
```

在上述代码中，定义类实现了 WebRequestInterceptor 接口，重写了对应的方法，并实现在控制台输出信息。

接下来在配置文件中更改拦截器，将之前的拦截器注释掉，配置上刚刚创建的拦截器，配置文件信息如例 13-12 所示。

【例 13-12】 **springMVC-config.xml**

```
<?xml version = '1.0' encoding = 'UTF - 8'?>
<beans xmlns = "http://www.springframework.org/schema/beans"
    xmlns:xsi = "http://www.w3.org/2001/XMLSchema - instance"
    xmlns:context = "http://www.springframework.org/schema/context"
    xmlns:mvc = "http://www.springframework.org/schema/mvc"
    xsi:schemaLocation = "http://www.springframework.org/schema/beans
        http://www.springframework.org/schema/beans/spring - beans.xsd
        http://www.springframework.org/schema/context
        http://www.springframework.org/schema/context/spring - context.xsd
        http://www.springframework.org/schema/mvc
        http://www.springframework.org/schema/mvc/spring - mvc.xsd">
    <!-- 扫描 Controller -->
    <context:component - scan base - package = "com.qfedu" />
    <!-- 配置视图解析器 -->
    <bean class = "org.springframework.web.
        servlet.view.InternalResourceViewResolver">

        <property name = "suffix" value = ".jsp" />
    </bean>
    <mvc:annotation - driven/>
    <!-- 配置拦截器 -->
    <mvc:interceptors>
    <mvc:interceptor>
            <mvc:mapping path = "/ *" />
            <bean class = "com.qfedu.interceptor.MyWebRequestInterceptor"/>
          <!-- <bean class = "com.qfedu.interceptor.HelloInterceptor"/>   -->
    </mvc:interceptor>
    </mvc:interceptors>
</beans>
```

在浏览器中访问该项目具体页面或者访问任意控制器时，都可以看到拦截器被触发时在控制台中打印的结果，可以明确代码执行的顺序，如图 13.5 所示。

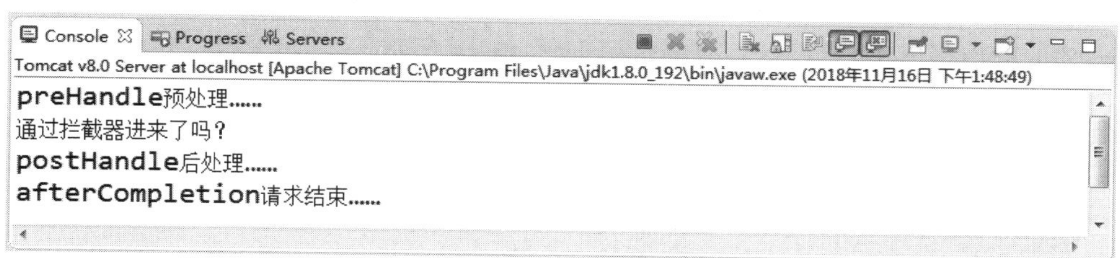

图 13.5 WebRequest Interceptor 实现拦截器

13.2.3 拦截器链

在日常开发中可能会遇到多个拦截器在一起发生作用，比如未登录拦截、编码格式转换等，这时 Spring MVC 就需要配置多个拦截器，而这些拦截器就组成了拦截器链。拦截器链中拦截器起作用的顺序即为在配置文件中声明的先后顺序，一定要注意顺序关系，否则可能会引起拦截器无效。

接下来将写两个拦截器，来演示多个拦截器一起作用时的效果。首先创建第一个拦截器，实现接口，重写拦截方法，具体代码如例 13-13 所示。

【例 13-13】 MyInterceptor1. java

```
package com.qfedu.interceptor;
import javax.servlet.http.HttpServletRequest;
import javax.servlet.http.HttpServletResponse;
import org.springframework.web.servlet.HandlerInterceptor;
import org.springframework.web.servlet.ModelAndView;
public class MyInterceptor1 implements HandlerInterceptor {
    @Override
    public boolean preHandle(HttpServletRequest request,
HttpServletResponse response, Object handler)
                throws Exception {
        System.out.println("第一个拦截器: preHandle 预处理……");
        return true;
    }
    @Override
    public void postHandle(HttpServletRequest request, HttpServletResponse
            response, Object handler,
            ModelAndView modelAndView) throws Exception {
        System.out.println("第一个拦截器: postHandle 后处理……");
    }
    @Override
    public void afterCompletion(HttpServletRequest request,
            HttpServletResponse response, Object handler, Exception ex)
                throws Exception {
        System.out.println("第一个拦截器: afterCompletion 请求结束……");
    }
}
```

接下来再创建第二个拦截器，新建一个 MyInterceptor2 类，创建方式与第一个相同，在方法内部的输出内容上加以区分，具体代码如例 13-14 所示。

【例 13-14】 MyInterceptor2. java

```
package com.qfedu.interceptor;
import javax.servlet.http.HttpServletRequest;
import javax.servlet.http.HttpServletResponse;
import org.springframework.web.servlet.HandlerInterceptor;
import org.springframework.web.servlet.ModelAndView;
public class MyInterceptor2 implements HandlerInterceptor {
    @Override
    public boolean preHandle(HttpServletRequest request,
            HttpServletResponse response, Object handler)
                throws Exception {
```

```java
            System.out.println("第二个拦截器：preHandle 预处理……");
            return true;
        }
        @Override
        public void postHandle(HttpServletRequest request, HttpServletResponse
            response, Object handler,
              ModelAndView modelAndView) throws Exception {
            System.out.println("第二个拦截器：postHandle 后处理……");
        }
        @Override
        public void afterCompletion(HttpServletRequest request,
              HttpServletResponse response, Object handler, Exception ex)
                throws Exception {
            System.out.println("第二个拦截器：afterCompletion 请求结束……");
        }
}
```

注意：这两个拦截器都需要在 preHandle 方法中返回 true，当执行第一个拦截器后，使后面的拦截器或控制器能够继续运行。

springMVC-config. xml 配置文件的内容如例 13-15 所示。

【例 13-15】 springMVC-config. xml

```xml
<?xml version = '1.0' encoding = 'UTF - 8'?>
<beans xmlns = "http://www.springframework.org/schema/beans"
     xmlns:xsi = "http://www.w3.org/2001/XMLSchema - instance"
     xmlns:context = "http://www.springframework.org/schema/context"
     xmlns:mvc = "http://www.springframework.org/schema/mvc"
     xsi:schemaLocation = "http://www.springframework.org/schema/beans
         http://www.springframework.org/schema/beans/spring - beans.xsd
         http://www.springframework.org/schema/context
         http://www.springframework.org/schema/context/spring - context.xsd
         http://www.springframework.org/schema/mvc
         http://www.springframework.org/schema/mvc/spring - mvc.xsd">
    <!-- 扫描 Controller -->
    <context:component - scan base - package = "com.qfedu" />
    <!-- 配置视图解析器 -->
    <bean class = "org.springframework.
           web.servlet.view.InternalResourceViewResolver">

         <property name = "suffix" value = ".jsp" />
    </bean>
    <mvc:annotation - driven/>
    <!-- 配置拦截器 多个组成拦截器链 -->
    <mvc:interceptors>
    <mvc:interceptor>
             <mvc:mapping path = "/*" />
              <bean class = "com.qfedu.interceptor.MyInterceptor1"/>
    </mvc:interceptor>
```

```
        <mvc:interceptor>
            <mvc:mapping path="/*" />
            <bean class="com.qfedu.interceptor.MyInterceptor2"/>
    </mvc:interceptor>
    </mvc:interceptors>
</beans>
```

在浏览器请求任意页面或者是控制器即可得到如图 13.6 所示结果。

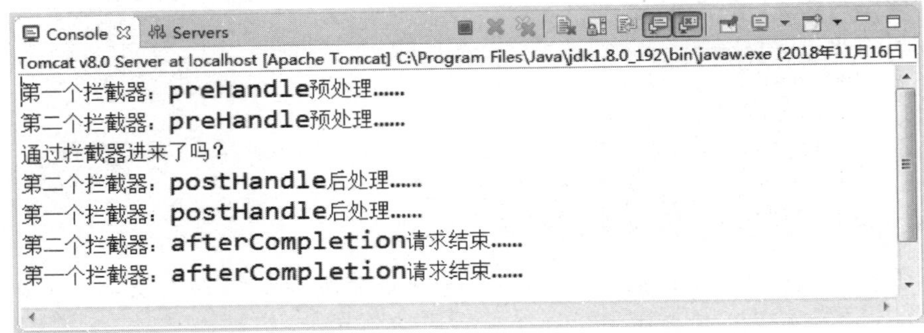

图 13.6　拦截器链处理结果

注意：当多个拦截器一起执行时，则根据拦截器声明的先后顺序执行，直到拦截器的方法都执行完了，才会去执行真正的资源。

13.2.4　拦截器登录控制

拦截器的基本使用大家已经知道了，接下来开始使用拦截器实现登录的拦截控制。如果当前处于未登录状态就跳转到登录页面完成登录，如果已经登录就可以继续访问资源。

1）创建库表

在 MySQL 中创建数据库 chapter13 和数据表 t_user，并新增一条数据，用户名为 qianfeng，密码为 qf6666，对应的 SQL 语句如下所示。

```
DROP DATABASE IF EXISTS chapter13;
CREATE DATABASE chapter13;
use chapter13;
create table t_user(
id int primary key auto_increment,
username varchar(20),password varchar(30));
insert into t_user(username,password) values('qianfeng','qf6666');
```

2）创建对应数据库表的实体类

数据库表中包含的字段，对应的类中就应当有相应的属性，如例 13-16 所示。

【例 13-16】　User.java

```
package com.qfedu.pojo;
public class User {
```

```java
    private int id;
    private String username;
    private String password;
    public int getId() {
        return id;
    }
    public void setId(int id) {
        this.id = id;
    }
    public String getUsername() {
        return username;
    }
    public void setUsername(String username) {
        this.username = username;
    }
    public String getPassword() {
        return password;
    }
    public void setPassword(String password) {
        this.password = password;
    }
}
```

3）创建 dao 层的接口

创建 dao 层就是定义操作数据库的方法，新建 UserDao，如例 13-17 所示。

【例 13-17】 UserDao.java

```java
package com.qfedu.dao;
import com.qfedu.pojo.User;
public interface UserDao {
    User login(String username,String password);
}
```

4）创建 dao 层的接口实现类

接口的实现类就是执行各种 SQL 语句，这里通过 Spring Jdbc 实现对数据库的操作，新建 UserDao 接口的实现类 UserDaoImpl，如例 13-18 所示。

【例 13-18】 UserDaoImpl.java

```java
package com.qfedu.dao.impl;
import org.springframework.jdbc.core.BeanPropertyRowMapper;
import org.springframework.jdbc.core.JdbcTemplate;
import com.qfedu.dao.UserDao;
import com.qfedu.pojo.User;
public class UserDaoImpl implements UserDao{
    private JdbcTemplate jdbcTemplate;
    public void setJdbcTemplate(JdbcTemplate jdbcTemplate) {
            this.jdbcTemplate = jdbcTemplate;
```

```
    }
    @Override
    public User login(String username, String password) {
        // TODO Auto-generated method stub
        return jdbcTemplate.queryForObject(
                "select * from t_user where username = ? and password = ?",
                new Object[] {username,password},
                new BeanPropertyRowMapper<>(User.class));
    }
}
```

5）创建 Service 层的接口

Service 接口是定义业务逻辑层的接口层，方便统一风格约束，新建 UserService 接口，如例 13-19 所示。

【例 13-19】 UserService.java

```
package com.qfedu.service;
import com.qfedu.pojo.User;
public interface UserService {
    User login(String username,String password);
}
```

6）创建 Service 层的接口的实现类

创建 Service 接口的实现类，重写接口中的方法。Service 接口的实现类对象需要依赖 dao 层的对象。新建 UserService 接口的实现类 UserServiceImpl，如例 13-20 所示。

【例 13-20】 UserServiceImpl.java

```
package com.qfedu.service.impl;
import com.qfedu.dao.UserDao;
import com.qfedu.pojo.User;
import com.qfedu.service.UserService;
public class UserServiceImpl implements UserService{
    private UserDao userDao;
    public UserDao getUserDao() {
        return userDao;
    }
    public void setUserDao(UserDao userDao) {
        this.userDao = userDao;
    }
    @Override
    public User login(String username, String password) {
        // TODO Auto-generated method stub
        User user = userDao.login(username, password);
        if(user!= null) {
            if(user.getPassword().equals(password)) {
                return user;
            }
```

```
        }
        return null;
    }
}
```

7）创建控制器实现用户登录的接口

借助 Spring MVC 实现控制器，完成用户登录功能接口的开发，新建一个 UserController 类，具体代码如例 13-21 所示。

【例 13-21】 UserController. java

```
package com.qfedu.service;
package com.qfedu.controller;
import javax.servlet.http.HttpSession;
import org.springframework.beans.factory.annotation.Autowired;
import org.springframework.stereotype.Controller;
import org.springframework.web.bind.annotation.RequestMapping;
import com.qfedu.pojo.User;
import com.qfedu.service.UserService;
@Controller
public class UserController {
    @Autowired
    private UserService userService;
    @RequestMapping("/userlogin")
    public String login(String username,String password,HttpSession session) {
        User user = userService.login(username, password);
        if(user!= null) {
            session.setAttribute("user", user);
            return "index";
        }else {
            return "login";
        }
    }
}
```

8）创建登录拦截器

创建 LoginInterceptor 类实现拦截器接口，在方法内部实现未登录的拦截处理，具体代码如例 13-22 所示。

【例 13-22】 LoginInterceptor. java

```
package com.qfedu.interceptor;
import javax.servlet.http.HttpServletRequest;
import javax.servlet.http.HttpServletResponse;
import javax.servlet.http.HttpSession;
import org.springframework.web.servlet.HandlerInterceptor;
import org.springframework.web.servlet.ModelAndView;
```

```
public class LoginInterceptor implements HandlerInterceptor {
    //需要放行的资源信息
    private String[] urls = {"login.jsp","userLogin"};
    @Override
    public boolean preHandle(HttpServletRequest request,
         HttpServletResponse response, Object handler)
            throws Exception {
        String url = request.getRequestURI();
        if(checkURL(url)){
            //放行
            return true;
        }else {
            HttpSession session = request.getSession();
            if(session.getAttribute("user") == null) {
    request.getRequestDispatcher("login.jsp").forward(request, response);
                return true;
            }else {
                return true;
            }
        }
    }
    @Override
    public void postHandle(HttpServletRequest request, HttpServletResponse
          response, Object handler,
          ModelAndView modelAndView) throws Exception {
    }
    @Override
    public void afterCompletion(HttpServletRequest request,
          HttpServletResponse response, Object handler, Exception ex)
          throws Exception {
    }
    private boolean checkURL(String url) {
        boolean res = false;
        for(String u:urls) {
            if(url.indexOf(u)>-1) {
                res = true;
                break;
            }
        }
        return res;
    }
}
```

在上述代码中，定义了一个方法对当前请求的 url 进行验证，查看是否需要进行未登录拦截处理。

9) 创建登录页面

通过表单标签实现用户登录数据交互，登录页面 login.jsp 的代码如例 13-23 所示。

【例 13-23】 **login.jsp**

```
<%@ page language="java" contentType="text/html; charset=UTF-8"
    pageEncoding="UTF-8"%>
<!DOCTYPE html PUBLIC "-//W3C//DTD HTML 4.01 Transitional//EN"
"http://www.w3.org/TR/html4/loose.dtd">
<html>
<head>
<meta http-equiv="Content-Type" content="text/html; charset=UTF-8">
<title>欢迎登录</title>
</head>
<body>
<form action="userlogin" method="post">
<label>用户名:</label><input name="username"><br/>
<label>密码:</label><input name="password" type="password"><br/>
<input type="submit" value="登录">
</form>
</body>
</html>
```

在上述代码中,需要注意 input 标签的 name 属性的值即为对应控制器中的参数名。

10)创建主页,登录成功之后的跳转页面

通过是否可以跳转到此页面,来验证是否登录成功,创建 index.jsp,具体代码如例 13-24 所示。

【例 13-24】 **index.jsp**

```
<%@ page language="java" contentType="text/html; charset=UTF-8"
    pageEncoding="UTF-8"%>
<!DOCTYPE html PUBLIC "-//W3C//DTD HTML 4.01 Transitional//EN"
"http://www.w3.org/TR/html4/loose.dtd">
<html>
<head>
<meta http-equiv="Content-Type" content="text/html; charset=UTF-8">
<title>主页</title>
</head>
<body>
<h1>欢迎: ${user.username} 登录本系统</h1>
</body>
</html>
```

11)数据库连接信息配置文件

在 src 目录下创建名为 jdbc.properties 的配置文件,注意文件名称拼写不要出错,配置文件的内容如例 13-25 所示。

【例 13-25】 **jdbc.properties**

```
jdbc.driverClass=com.mysql.jdbc.Driver
jdbc.jdbcUrl=jdbc:mysql://localhost:3306/chapter13
```

```
jdbc.user = root
jdbc.password = root
```

注意：在运行的时候需要把该配置文件中的 user 和 password 设置为自己数据库的账号和密码。

12）Spring 的配置文件

在 src 目录下创建配置文件 application.xml 完成 Spring 的标签配置，配置文件内容如例 13-26 所示。

【例 13-26】 application.xml

```
<?xml version = "1.0" encoding = "UTF-8"?>
<beans xmlns = "http://www.springframework.org/schema/beans"
    xmlns:xsi = "http://www.w3.org/2001/XMLSchema-instance"
    xmlns:context = "http://www.springframework.org/schema/context"
    xmlns:aop = "http://www.springframework.org/schema/aop"
    xmlns:tx = "http://www.springframework.org/schema/tx"
    xsi:schemaLocation = "http://www.springframework.org/schema/beans
        http://www.springframework.org/schema/beans/spring-beans.xsd
        http://www.springframework.org/schema/context
        http://www.springframework.org/schema/context/spring-context.xsd
        http://www.springframework.org/schema/aop
        http://www.springframework.org/schema/aop/spring-aop.xsd
        http://www.springframework.org/schema/tx
        http://www.springframework.org/schema/tx/spring-tx.xsd">
<!-- 引入外部 properties 文件 -->
<context:property-placeholder location = "classpath:jdbc.properties"/>
<!-- 注册数据源 -->
<bean name = "dataSource" class = "com.mchange.v2.c3p0.ComboPooledDataSource">
      <property name = "driverClass" value = "${jdbc.driverClass}"/>
      <property name = "jdbcUrl" value = "${jdbc.jdbcUrl}"/>
      <property name = "user" value = "${jdbc.user}"/>
      <property name = "password" value = "${jdbc.password}"/>
</bean>
<!-- 注册 JdbcTemplate 类 -->
<bean name = "jdbcTemplate"
class = "org.springframework.jdbc.core.JdbcTemplate">
      <property name = "dataSource" ref = "dataSource"/>
</bean>
  <bean name = "userDao"  class = "com.qfedu.dao.impl.UserDaoImpl">
      <property name = "jdbcTemplate" ref = "jdbcTemplate"></property>
  </bean>
  <bean name = "accountService"
class = "com.qfedu.service.impl.UserServiceImpl">
      <property name = "userDao" ref = "userDao"></property>
  </bean>
</beans>
```

13）Web.xml 文件

在 Web.xml 文件中需要设置 Spring 加载的配置文件和 Spring MVC 的前端控制器，如例 13-27 所示。

【例 13-27】 Web.xml

```
<?xml version="1.0" encoding="UTF-8"?>
<web-app xmlns:xsi="http://www.w3.org/2001/XMLSchema-instance"
xmlns="http://java.sun.com/xml/ns/javaee"
xsi:schemaLocation="http://java.sun.com/xml/ns/javaee
http://java.sun.com/xml/ns/javaee/web-app_2_5.xsd" version="2.5">
  <display-name>chapter13</display-name>
   <context-param>
        <param-name>contextConfigLocation</param-name>
        <param-value>classpath:application.xml</param-value>
    </context-param>
    <listener>
<listener-class>org.springframework.web.context.ContextLoaderListener</li
stener-class>
    </listener>
  <servlet>
    <servlet-name>springMVC</servlet-name>
    <servlet-class>org.springframework.web.servlet.DispatcherServlet
</servlet-class>
<init-param>
        <param-name>contextConfigLocation</param-name>
        <param-value>/WEB-INF/springMVC-config.xml</param-value>
    </init-param>
    <load-on-startup>1</load-on-startup>
</servlet>
<!-- 访问 DispatcherServlet 对应的路径 -->
<servlet-mapping>
      <servlet-name>springMVC</servlet-name>
      <url-pattern>/</url-pattern>
</servlet-mapping>
  <welcome-file-list>
    <welcome-file>login.jsp</welcome-file>
  </welcome-file-list>
</web-app>
```

完成上述步骤后在浏览器中进行验证，未登录状态下直接访问其他 JSP 页面或控制器时，页面会自动跳转到登录页面，可以看出登录拦截器从中起了作用，如图 13.7 所示。

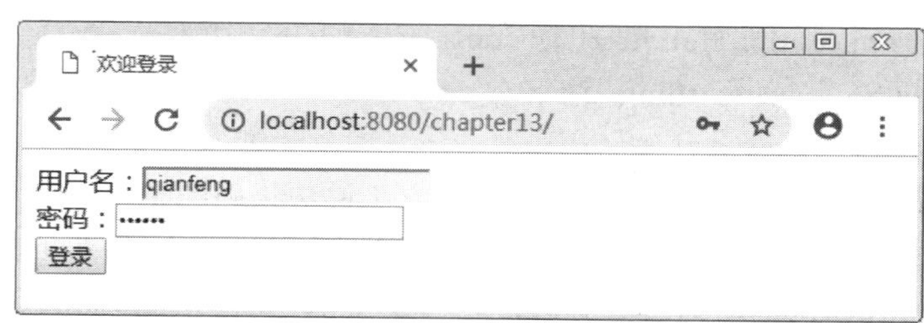

图 13.7 登录

输入正确的用户名和密码，然后单击登录，登录成功即可进入到主页，如果登录失败则继续停留在登录页面,如图 13.8 所示。

图 13.8　登录成功的主页

13.3　本章小结

本章首先介绍了 Spring MVC 的异常处理器,让程序变得更加健壮,并通过代码演示了几种实现方式和效果;其次又介绍了拦截器及其实现;最后通过一个登录拦截器将这几节的内容进行贯穿汇总。在日常使用异常处理器的时候,可以任意选择一种自己习惯的,但要注意不能缺失配置文件中的配置信息。

13.4　习　　题

1. 填空题

(1) HandlerExceptionResolver 接口,它的内部有________方法,提供了对异常进行解析的操作。

(2) 使用________注解可以实现一个专门用于处理异常的 Controller。

(3) 使用________和________注解可以实现全局的异常捕捉。

(4) HandlerInterceptor 接口中的三个方法分别是________方法、postHandle 方法和________方法。

(5) 拦截器的拦截功能是基于 Java 的________来实现的。

2. 选择题

(1) 以下选项可以实现全局异常处理的是(　　)。

A. 使用@ExceptionHandler 注解

B. 使用@ExceptionHandler+@ControllerAdvice 注解

C. 使用@Suppvisewarnings 注解

D. 使用@Deprecated 注解

(2) 下列关于 HandlerInterceptor 中的三个方法说法正确的是(　　)。

A. preHandle 在业务处理器请求前被调用;postHandle 在业务处理器处理请求执行完成后,生成视图之前执行;afterCompletion 在处理完请求后被调用

B. postHandle 在业务处理器请求前被调用;preHandle 在业务处理器处理请求执行完成后,生成视图之前执行;afterCompletion 在处理完请求后被调用

C. postHandle 方法的返回值是 Boolean 类型的

D. preHandle 方法没有返回值

(3) 下列对 HandlerInterceptor 与 WebRequestInterceptor 的说法错误的是(　　)。

A. 两个接口都可用于 Contrller 层请求拦截,接口中定义的方法作用也是一样的

B. SWebRequestInterceptor 的入参 WebRequest 包装了 HttpServletRequest 和 HttpServletResponse

C. WebRequestInterceptor 的 preHandle 是没有返回值的,说明该方法中的逻辑并不影响后续的方法执行

D. HandlerInterceptor 的 preHandle 是没有返回值的,说明该方法中的逻辑并不影响后续的方法执行

(4) 下列选项不是 resolveException 方法中的参数的是(　　)。

A. HttpServletRequest　　B. Object

C. HttpSession　　D. Exception

(5) 下列说法中错误的是(　　)。

A. 拦截器 Interceptor 的拦截功能是基于 Java 的动态代理来实现的

B. Interceptor 拦截器用于拦截 Controller 层接口,表现形式有点像 Spring 的 AOP,但是 AOP 是针对单一的方法

C. WebRequestInterceptor 接口与 HandlerInterceptor 接口的用法相似,WebRequestInterceptor 接口中的 preHandle 方法没有返回值

D. 所有的拦截器都不需要配置在 SpringMVC 的配置文件中

3. 思考题

(1) Spring MVC 框架是怎样处理异常的?

(2) 怎样定义一个拦截器?

4. 编程题

编写一个拦截器,在进入处理器之前记录开始时间,即在拦截器的 preHandle 记录开始时间,在结束请求处理之后记录结束时间,即在拦截器的 afterCompletion 记录结束时间,并用结束时间减去开始时间得到这次请求的处理时间。

第 14 章　Spring MVC 的高级功能

本章学习目标

- 理解文件的上传实现
- 理解文件下载
- 掌握 JSON 格式数据交互
- 掌握 RESTFul 风格接口开发
- 理解静态资源访问的问题

前两章主要讲解了 Spring MVC 的基本应用、异常处理器以及拦截器的使用，本章将讲解开发过程中常用到的文件上传和下载功能，主流的 JSON 格式的数据交互和 RESTFul 风格接口的开发。在接下来的学习中，会详细介绍这些高级功能的应用。

14.1　文件上传下载

14.1.1　利用 Spring MVC 上传文件

Spring MVC 实现文件上传功能是通过 Apache 的 commons-fileupload 实现的，但是 Spring MVC 默认并没有文件上传解析器，因此，需要自己动手配置。

实现文件上传的步骤：

1）准备 jar 包

```
commons-fileupload-1.3.1.jar
commons-io-2.4.jar
```

2）创建控制器

在控制器类中写出映射方法，实现文件上传的保存操作，新建一个 FileUpController 类，具体代码如例 14-1 所示。

【例 14-1】　FileUpController.java

```
package com.qfedu.controller;
import java.io.File;
import java.io.IOException;
import javax.servlet.http.HttpServletRequest;
import javax.servlet.http.HttpServletResponse;
```

```java
import org.springframework.stereotype.Controller;
import org.springframework.web.bind.annotation.RequestMapping;
import org.springframework.web.multipart.MultipartFile;
@Controller
public class FileUpController {
    @RequestMapping("fileup")
    public void file(MultipartFile file,HttpServletRequest request,
        HttpServletResponse response) throws IllegalStateException,
        IOException {
        //获取上传的文件名和后缀名
        String fn = file.getOriginalFilename();
        //创建要保存的文件对象
        File desfile = new
            File(request.getServletContext().getRealPath("/"),fn);
        //保存文件到指定文件中
        file.transferTo(desfile);
        //写出文件存储路径
        response.getWriter().print(desfile.getAbsolutePath());
    }
}
```

上述实现文件上传保存的功能代码中，有几个地方需要注意，分别是：

(1) 请求方式。文件上传的数据接口只能接收 post 请求，否则将获取不到上传文件的内容。因此，在发送文件上传请求时，设置请求方式为 post，否则直接导致请求异常的发生。

(2) 参数类型。Spring MVC 实现文件上传使用的是 Apache 提供的 API，其中把文件上传的部分封装为 MultipartFile，使用时注意参数名称是否与请求上传的 file 标签的 name 名称一致，如果不一致需要使用@RequestParam 进行标记。上传的表单需要设置 enctype="multipart/form-data"，并且发起上传请求的时候需要将文件的内容也上传到服务器，否则将无法读取到文件内容。

3) 文件存储

在方法内部完成上传文件的存储，首先需要获取到上传的文件名称及其后缀名，然后调用 transferTo 方法将文件存储到服务器的某一路径下。这里的路径是指真实路径，也就是项目的发布路径。

4) 创建配置文件

在项目的 WEB-INF 目录下创建 springMVC-config.xml，如例 14-2 所示。

【例 14-2】 springMVC-config.xml

```xml
<?xml version = '1.0' encoding = 'UTF - 8'?>
<beans xmlns = "http://www.springframework.org/schema/beans"
    xmlns:xsi = "http://www.w3.org/2001/XMLSchema - instance"
    xmlns:context = "http://www.springframework.org/schema/context"
    xmlns:mvc = "http://www.springframework.org/schema/mvc"
    xsi:schemaLocation = "http://www.springframework.org/schema/beans
        http://www.springframework.org/schema/beans/spring - beans.xsd
```

```
        http://www.springframework.org/schema/context
        http://www.springframework.org/schema/context/spring-context.xsd
        http://www.springframework.org/schema/mvc
        http://www.springframework.org/schema/mvc/spring-mvc.xsd">
    <!-- 扫描 Controller -->
    <context:component-scan base-package="com.qfedu" />
    <!-- 配置视图解析器 -->
    <bean class="org.springframework.
        web.servlet.view.InternalResourceViewResolver">

        <property name="suffix" value=".jsp" />
    </bean>
    <mvc:annotation-driven/>
     <mvc:default-servlet-handler></mvc:default-servlet-handler>
     <!-- SpringMVC 上传文件时,需要配置 MultipartResolver 处理器 -->
    <bean id="multipartResolver"

class="org.springframework.web.multipart.commons.CommonsMultipartResolver">

        <property name="defaultEncoding" value="UTF-8" />
        <!-- 指定所上传文件的总大小,单位字节.注意 maxUploadSize 属性的限制不是
            针对单个文件,而是所有文件的容量之和 -->
        <property name="maxUploadSize" value="10240000" />
     </bean>
</beans>
```

由于 Spring MVC 默认是没有文件上传解析器的，因此，在上传接口编写完成之后需要在配置文件中配置上传解析器，即添加：

```
<bean id="multipartResolver"
```

class="org.springframework.web.multipart.commons.CommonsMultipartResolver" />这样才能实现对上传文件的类型做自动转换处理。

最后，配置 Web.xml，配置信息的内容如例 14-3 所示。

【例 14-3】 Web.xml

```
<?xml version="1.0" encoding="UTF-8"?>
<web-app xmlns:xsi="http://www.w3.org/2001/XMLSchema-instance"
xmlns="http://java.sun.com/xml/ns/javaee"
xsi:schemaLocation="http://java.sun.com/xml/ns/javaee
http://java.sun.com/xml/ns/javaee/web-app_2_5.xsd" version="2.5">
     <display-name>chapter14</display-name>
     <servlet>
     <servlet-name>springMVC</servlet-name>
     <servlet-class>org.springframework.web.servlet.DispatcherServlet
</servlet-class>
<init-param>
     <param-name>contextConfigLocation</param-name>
```

```
        <param-value>/WEB-INF/springMVC-config.xml</param-value>
        </init-param>
        <load-on-startup>1</load-on-startup>
</servlet>
<!-- 访问 DispatcherServlet 对应的路径 -->
<servlet-mapping>
        <servlet-name>springMVC</servlet-name>
        <url-pattern>/</url-pattern>
</servlet-mapping>
</web-app>
```

5）创建上传页面

在上传页面中使用表单通过<input type="file">完成文件上传，其中，有三个地方需要注意：

（1）请求方式为 post。

（2）设置表单的 enctype 属性。

（3）input 标签的类型为 file。

上传页面 page01.jsp 的具体代码如例 14-4 所示。

【例 14-4】 page01.jsp

```
<%@ page language="java" contentType="text/html; charset=UTF-8"
    pageEncoding="UTF-8"%>
<!DOCTYPE html PUBLIC "-//W3C//DTD HTML 4.01 Transitional//EN"
"http://www.w3.org/TR/html4/loose.dtd">
<html>
<head>
<meta http-equiv="Content-Type" content="text/html; charset=UTF-8">
<title>文件上传</title>
</head>
<body>
<form action="fileup" method="post" enctype="multipart/form-data">
    <input type="file" name="file"><br/>
    <input type="submit" value="上传文件">
</form>
</body>
</html>
```

6）测试

运行项目，在浏览器地址栏中输入上传页面的地址，浏览器中显示的上传页面如图 14.1 所示。

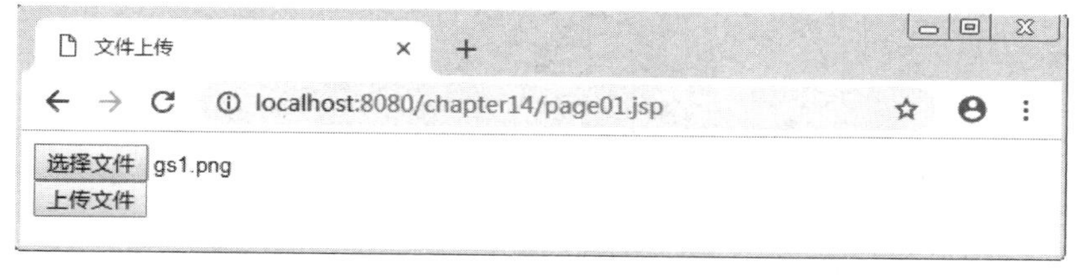

图 14.1 上传页面

在页面中选择要上传的文件，文件上传成功后会在浏览器中显示文件存储的路径，如图 14.2 所示。

图 14.2　上传成功

在浏览器地址栏中输入图 14.2 中显示的文件存储路径，即可查看上传的文件内容，如图 14.3 所示。

图 14.3　访问上传内容

注意：文件上传时，如果 input 标签 file 属性的 name 值和控制器对应的参数名不一致，需要使用@RequestParam 注解。

14.1.2　利用 Spring MVC 下载文件

实现文件的下载功能，实际上就是通过流读取文件的内容，并且通过 response 输出流将文件的内容输出即可。

1）创建控制器

新建文件下载控制器类 FileDownController，代码如例 14-5 所示。

【例 14-5】　FileDownController.java

```
package com.qfedu.controller;
import java.io.File;
import java.io.IOException;
import javax.servlet.http.HttpServletRequest;
import javax.servlet.http.HttpServletResponse;
import org.apache.commons.io.FileUtils;
import org.springframework.stereotype.Controller;
```

```java
import org.springframework.web.bind.annotation.RequestMapping;
@Controller
public class FileDownController {

    @RequestMapping("filedown")
    public void down(HttpServletRequest request,HttpServletResponse
            response) throws IOException {
        response.setContentType("text/html;charset = utf - 8");
        File file = new File(request.getServletContext().
            getRealPath("/"),"gs1.png");
        System.out.println("文件: " + file.getAbsolutePath());
        byte[ ] data = FileUtils.readFileToByteArray(file);
        response.setContentType("application/x - msdownload;");
        response.setHeader("Content - disposition", "attachment; filename = " +
        new String(file.getName().getBytes("utf - 8"), "ISO8859 - 1"));
        response.setHeader("Content - Length",
            String.valueOf(file.length()));
        response.getOutputStream().write(data);
    }
}
```

在上述代码中，首先创建了文件对象，然后通过对上传的文件进行文件内容的读取，最后，将文件的内容通过 response 传递到输出流中。

注意：必须设置消息头，目的是使浏览器获取到这是内容下载。

2）测试

在浏览器地址栏中输入实现下载功能的控制器地址 http://localhost:8080/chapter14/filedown，即可完成下载，如图 14.4 所示。

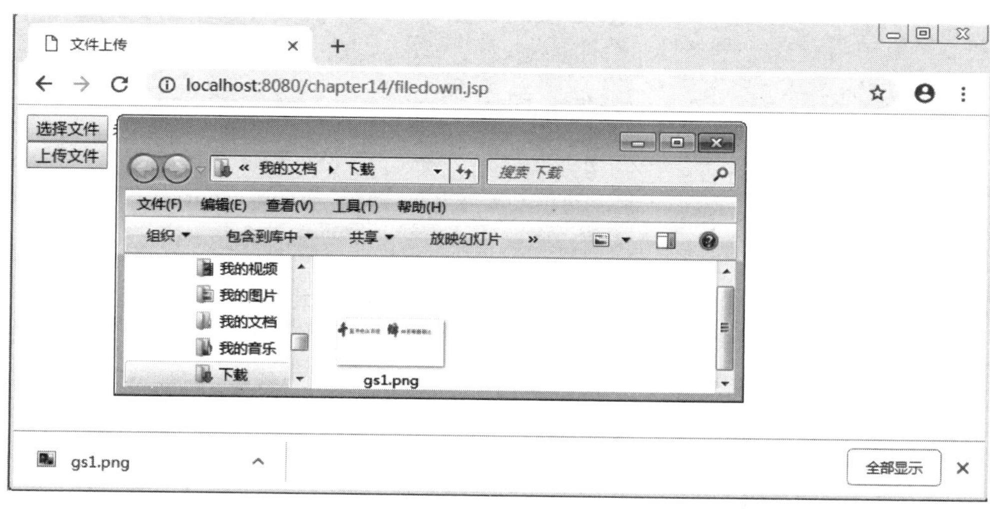

图 14.4　下载展示

14.2 Spring MVC 实现 JSON 交互

JSON(JavaScript Object Notation,JavaScript 对象表示法)是一种存储和交换文本信息的语法。Spring MVC 也支持 JSON 格式的数据交互,其中主要用到的注解为@RequestBody 和@ResponseBody,@RequestBody 注解是解析请求的 JSON 格式的参数,也就是自动解析 json 字符串,而@ResponseBody 注解是将对象转换为 JSON 格式字符串。

在 com.qfedu.controller 包下新建 JsonController 类,具体代码如例 14-6 所示。

【例 14-6】 JsonController.java

```
package com.qfedu.controller;
import java.util.ArrayList;
import java.util.List;
import org.springframework.stereotype.Controller;
import org.springframework.web.bind.annotation.RequestBody;
import org.springframework.web.bind.annotation.RequestMapping;
import org.springframework.web.bind.annotation.ResponseBody;
import com.qfedu.pojo.Page;
import com.qfedu.pojo.Student;
@Controller
public class JsonController {
    @ResponseBody
    @RequestMapping("jsonapp1")
    public List<Student> json1(@RequestBody Page obj){
        List<Student> list = new ArrayList<Student>();
        int start = (obj.getPage() - 1) * obj.getLimit() + 1;
        for(int i = start;i <= obj.getPage() * obj.getLimit();i++) {
            Student stu = new Student();
            stu.setNo("qf-" + i);
            stu.setClassNo("qfjava1801");
            stu.setName("千锋" + i);
            list.add(stu);
        }
        return list;
    }
}
```

在上述代码中,把注解@ResponseBody 标记方法的返回值转换为 json 数据,并在该方法内部创建一个集合,存储对应的数据信息,最后将集合返回至客户端即可。

新建一个 page02.jsp 页面,具体代码如例 14-7 所示。

【例 14-7】 page02.jsp

```
<%@ page language="java" contentType="text/html; charset=UTF-8"
    pageEncoding="UTF-8"%>
<!DOCTYPE html PUBLIC "-//W3C//DTD HTML 4.01 Transitional//EN"
"http://www.w3.org/TR/html4/loose.dtd">
```

```
<html>
<head>
<meta http-equiv="Content-Type" content="text/html; charset=UTF-8">
<title>文件上传</title>
<script type="text/javascript" src="jquery-2.1.0.min.js"></script>
<script type="text/javascript">
function getJson() {
    $.ajax({
        url:"jsonapp1",
        method:"get",
        data:{"page":1,"limit":10},
        contentType:"application/json",
        success:function(arr){
            var s = "";
            $("#sl1").html("");
            for(var i = 0;i < arr.length;i++){
                s += "<li>学号: " + arr[i].no + ",姓名: " + arr[i].name + "</li>";
            }
            $("#sl1").append(s);
        }
    });
}
</script>
</head>
<body>
<h1><input type="button" value="请求" onclick="getJson()"></h1>
<ol id="sl1"></ol>
</body>
</html>
```

在上述代码中,通过 Ajax 请求后台接口,再遍历返回值(数组),把内容拼接到一个字符串中,最后再匹配到对应的标签上。

14.3 Spring MVC 实现 RESTful 风格

14.3.1 REST

REST(Representational State Transfer)翻译成中文是“表现层状态转化”,是所有 Web 应用都应该遵守的架构设计指导原则。面向资源是 REST 的核心,同一个资源可以完成一组不同的操作。资源是服务器上一个可命名的抽象概念,资源是以名词为核心来组织的,首先关注的是名词。REST 要求,必须通过统一的接口来对资源执行各种操作。对每个资源只能执行一组有限的操作(7 个 HTTP 方法: GET/POST/PUT/DELETE/PATCH/HEAD/OPTIONS)。符合 REST 设计标准的 API,即 RESTful API。REST 架构设计,遵循的各项标准和准则,就是 HTTP 的表现,换句话说,HTTP 就是属于 REST 架构的设计模式。

14.3.2 使用 Spring MVC 实现 RESTful 风格

Spring MVC 的控制器支持各种请求方式，也有注解@PathVariable 配合使用，可以实现 RESTful 的风格的数据接口。在 Spring-mvc 中实现 RESTful 风格的控制器代码如例 14-8 所示。

【例 14-8】 RestContrller.java

```java
package com.qfedu.controller;
import org.springframework.stereotype.Controller;
import org.springframework.web.bind.annotation.PathVariable;
import org.springframework.web.bind.annotation.RequestMapping;
import org.springframework.web.bind.annotation.RequestMethod;
import org.springframework.web.bind.annotation.ResponseBody;
@Controller
public class RestContrller {
    //新增
    @RequestMapping(value = "user/{name}/{pass}",method = RequestMethod.POST)
    @ResponseBody
    public String add(@PathVariable String name,@PathVariable String pass) {
        System.out.println("Post 请求 -- 新增：" + name + "_" + pass);
        return "新增成功";
    }
    //修改
    @RequestMapping(value = "user/{id}/{name}",method = RequestMethod.PUT)
    @ResponseBody
    public String update(@PathVariable int id,@PathVariable String name) {
        System.out.println("Put 请求 -- 修改：" + id + "_" + name);
        return "修改成功";
    }
    //删除
    @RequestMapping(value = "user/{id}",method = RequestMethod.DELETE)
    @ResponseBody
    public String del(@PathVariable int id,@PathVariable String name) {
        System.out.println("Delete 请求 -- 删除：" + id);
        return "删除成功";
    }
    //修改
    @RequestMapping(value = "user/{id}",method = RequestMethod.GET)
    @ResponseBody
    public String select(@PathVariable int id) {
        System.out.println("GET 请求 -- 查询：" + id);
        return "查询成功";
    }
}
```

在上述代码中，定义了四个映射方法实现 CRUD 操作，这四个映射方法分别对应四种请求方式：GET\POST\PUT\DELETE。例如，通过浏览器地址栏输入以下信息：

http://localhost:8080/User? _method=post&id=001&name=zhangsan，该信息可

以向数据库 user 表里面插入一条 id 为 001,name 为 zhangsan 的用户记录。

http://localhost:8080/User? _method=put&id=001&name=lisi,该信息可以将 user 表里面 id=001 的用户名改为 lisi。

http://localhost:8080/User? _method=delete&id=001,该信息可以将数据库 user 表里面的 id=001 的信息删除。

http://localhost:8080/User? _method=get&id=001,该信息可以通过 get 请求获取到数据库 user 表里面 id=001 的用户信息。

14.3.3 静态资源访问问题

当使用 Spring MVC 的时候,需要在 web. xml 中配置前端控制器,而且还要配置匹配规则。在实际应用中往往会出现静态资源无法访问的现象,这是因为 DispatchServlet 会把所有的请求都进行拦截,静态资源也不例外,从而导致资源无法访问(js 文件、css 文件、图片等),这样就必须要配置允许访问静态变量的方法。

静态资源无法访问,根据文件位置的不同又可以划分为以下两种情况。

1) < mvc:default-servlet-handler >

项目类似下面的结构,如图 14.5 所示。

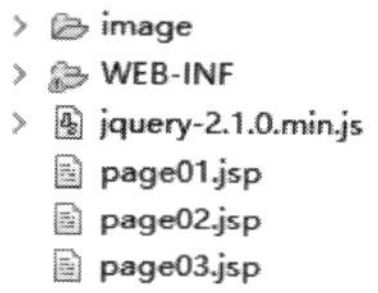

在 webContent 目录下,创建包含图片内容的文件夹 image,然后再创建一个页面引用 image 中的图片,代码如例 14-9 所示。

> image
> WEB-INF
> jquery-2.1.0.min.js
page01.jsp
page02.jsp
page03.jsp

图 14.5 项目结构

【例 14-9】 page03. jsp

```
<%@ page language="java" contentType="text/html; charset=UTF-8"
    pageEncoding="UTF-8"%>
<!DOCTYPE html PUBLIC "-//W3C//DTD HTML 4.01 Transitional//EN"
"http://www.w3.org/TR/html4/loose.dtd">
<html>
<head>
<meta http-equiv="Content-Type" content="text/html; charset=UTF-8">
<title>静态资源无法访问</title>
</head>
<body>
<h1>可以看到图片吗</h1>
<img alt="图片在哪里" src="image/gs1.png"/>
</body>
</html>
```

在上述代码中,通过一个 img 标签的 src 属性引用刚刚准备的图片,然后运行项目访问 page03. jsp 页面,结果如图 14.6 所示。

发现图片无法显示,原因是 web. xml 文件中,配置了前端控制器使用的匹配规则为 < url-pattern >/</url-pattern >,而"/"的意思就是除了 JSP 页面放行,剩下的都匹配,可是匹配完之后发现没有这样的控制器和对应的页面,所以图片就无法显示,那么该如何解决呢?

这种情况下只要在 SpringMVC 的配置文件中添加如下标签即可。

图 14.6 静态资源无法访问

<mvc:default-servlet-handler></mvc:default-servlet-handler>这个标签的意思是放行静态资源，比如 js、css、图片等，配置文件信息如例 14-10 所示。

【例 14-10】 springMVC-config.xml

```
<?xml version='1.0' encoding='UTF-8'?>
<beans xmlns="http://www.springframework.org/schema/beans"
    xmlns:xsi="http://www.w3.org/2001/XMLSchema-instance"
    xmlns:context="http://www.springframework.org/schema/context"
    xmlns:mvc="http://www.springframework.org/schema/mvc"
    xsi:schemaLocation="http://www.springframework.org/schema/beans
        http://www.springframework.org/schema/beans/spring-beans.xsd
        http://www.springframework.org/schema/context
        http://www.springframework.org/schema/context/spring-context.xsd
        http://www.springframework.org/schema/mvc http://www.springframework.org/
            schema/mvc/spring-mvc.xsd">
    <!-- 扫描 Controller -->
    <context:component-scan base-package="com.qfedu" />
    <mvc:annotation-driven/>
    <!-- 配置视图解析器 -->
    <bean class="org.springframework.
            web.servlet.view.InternalResourceViewResolver">
        <property name="suffix" value=".jsp" />
    </bean>
    <!-- SpringMVC 上传文件时,需要配置 MultipartResolver 处理器 -->
    <bean id="multipartResolver"class="org.springframework.
            web.multipart.commons.CommonsMultipartResolver">
        <property name="defaultEncoding" value="UTF-8" />
        <!-- 指定所上传文件的总大小,单位字节.注意 maxUploadSize 属性的限制不是
        针对单个文件,而是所有文件的容量之和 -->
        <property name="maxUploadSize" value="10240000" />
    </bean>
    <mvc:default-servlet-handler></mvc:default-servlet-handler>
</beans>
```

在完成上述配置后，再重新访问 page03.jsp 页面，即可解决静态资源无法加载的问题，浏览器中的展示效果如图 14.7 所示。

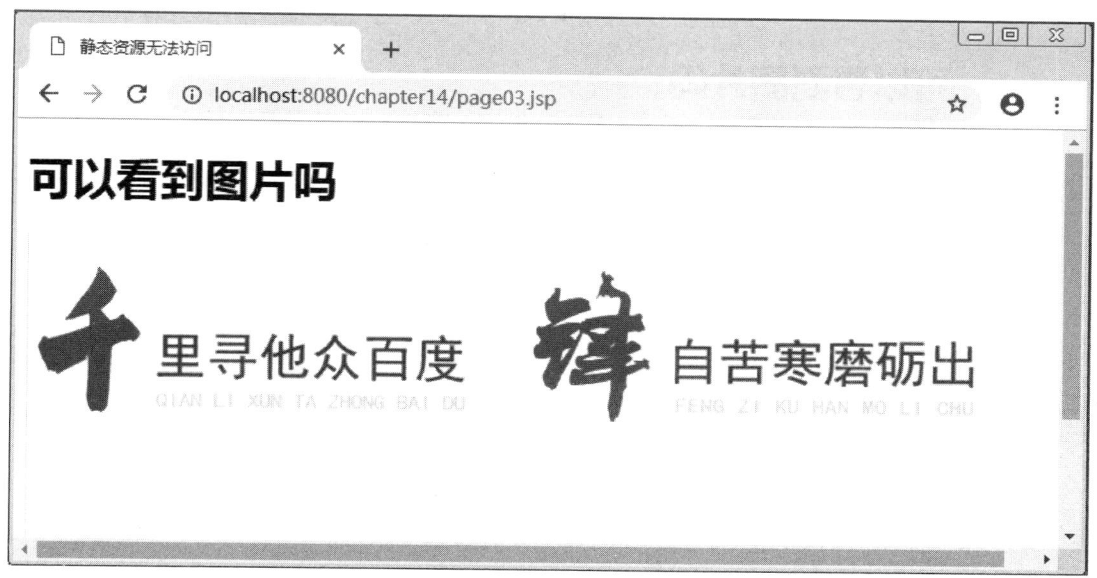

图 14.7　演示静态资源访问

2) < mvc:resources >

创建 page04.jsp 页面，并在页面中加载静态资源信息，代码如例 14-11 所示。

【例 14-11】　page04.jsp

```
<%@ page language = "java" contentType = "text/html; charset = UTF - 8"
    pageEncoding = "UTF - 8" %>
<!DOCTYPE html PUBLIC " - //W3C//DTD HTML 4.01 Transitional//EN"
"http://www.w3.org/TR/html4/loose.dtd">
<html>
<head>
<meta http - equiv = "Content - Type" content = "text/html; charset = UTF - 8">
<title>静态资源无法访问</title>
</head>
<body>
<h1>可以看到图片吗</h1>
<img alt = "图片在哪里" src = "image/gs5.jpg"/>
</body>
</html>
```

当运行上述代码，在浏览器中访问页面时，发现出现静态资源无法加载的问题，如图 14.8 所示。

要解决此类问题，可以通过路径映射的方式，在 Spring MVC 的配置文件中设置静态资源映射即可，配置文件如例 14-12 所示。

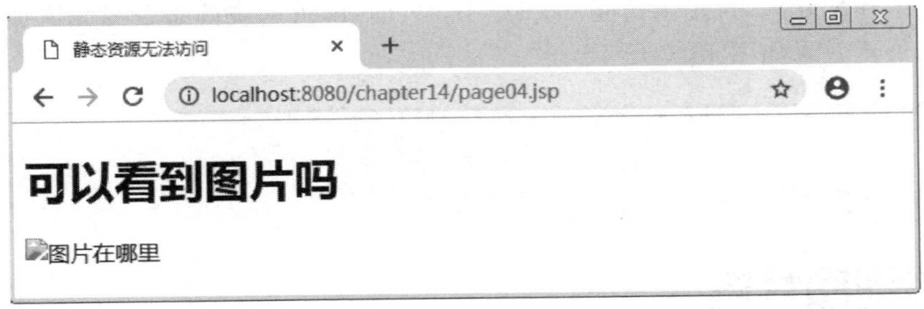

图 14.8　静态资源无法加载

【例 14-12】　springMVC-config. xml

```
<?xml version='1.0' encoding='UTF-8'?>
<beans xmlns="http://www.springframework.org/schema/beans"
    xmlns:xsi="http://www.w3.org/2001/XMLSchema-instance"
    xmlns:context="http://www.springframework.org/schema/context"
    xmlns:mvc="http://www.springframework.org/schema/mvc"
    xsi:schemaLocation="http://www.springframework.org/schema/beans
        http://www.springframework.org/schema/beans/spring-beans.xsd
        http://www.springframework.org/schema/context
        http://www.springframework.org/schema/context/spring-context.xsd
        http://www.springframework.org/schema/mvc
        http://www.springframework.org/schema/mvc/spring-mvc.xsd">
    <!-- 扫描 Controller -->
    <context:component-scan base-package="com.qfedu" />
    <!-- 配置视图解析器 -->
    <bean class="org.springframework.
        web.servlet.view.InternalResourceViewResolver">

        <property name="suffix" value=".jsp" />
    </bean>
    <!-- SpringMVC 上传文件时,需要配置 MultipartResolver 处理器 -->
    <bean id="multipartResolver"class="org.springframework.
        web.multipart.commons.CommonsMultipartResolver">

        <property name="defaultEncoding" value="UTF-8" />
        <!-- 指定所上传文件的总大小,单位字节.注意 maxUploadSize 属性的限制不是
       针对单个文件,而是所有文件的容量之和 -->
        <property name="maxUploadSize" value="10240000" />
    </bean>
    <mvc:annotation-driven/>
    <!-- <mvc:default-servlet-handler></mvc:default-servlet-handler> -->
    <mvc:resources location="/image/" mapping="/image/**"/>
</beans>
```

< mvc:resources >标签配置说明：

location 元素表示 webapp 目录下的 image 包下的所有文件。

mapping 元素表示以/image 开头的所有请求路径，如/image/qf.jpg。

此时，再次访问 page04.jsp 页面，发现图片可以看到了，如图 14.9 所示。

图 14.9　加载静态资源

以上就是对静态资源无法访问的处理。

14.4　本章小结

本章首先介绍了 Spring MVC 的文件上传和下载，包括 Spring MVC 对文件上传信息的设置；其次通过实例演示了 Spring MVC 的 JSON 数据格式的转换；最后详细讲解了 Spring MVC 的 RESTful 风格的接口实现和静态资源访问常见的问题及其解决方案。通过本章知识的学习，应当能理解 Spring MVC 的一些高级操作方式，掌握通过 Spring 实现文件上传和下载的功能、通过 Spring MVC 实现 JSON 数据的交互、通过 Spring MVC 实现 RESTful 接口的开发以及解决静态资源无法访问的具体方法。

14.5　习　　题

1. 填空题

(1) 文件上传时，应设置请求方式为________。

(2) 上传的表单需要设置________属性。

(3) 在方法内部完成上传文件的存储，首先需要获取到上传的文件名称及其后缀名，然后调用________方法将文件存储到服务器的路径下。

(4) Spring MVC 实现 JSON 格式的数据交互，主要用到的注解为________注解和________注解。

(5) REST 是一种架构风格，其核心是________。

2. 选择题

(1) 下列关于文件上传说法不正确的是(　　)。

A. Spring MVC 默认并没有文件上传解析器，因此，需要自己动手配置

B. 文件上传的数据接口只能接收 post 请求，否则将获取不到上传文件的内容

C. 上传的表单可以不用设置 enctype="multipart/form-data"

D. 在方法内部完成上传文件的存储，首先需要获取到上传的文件名称及其后缀名，然后调用 transferTo 方法将文件存储到服务器的某一路径下

(2)（　　）注解用于将 controller 类的方法返回对象转换为 JSON 响应给客户端。

A. @ResponseBody　　B. @RequestBody

C. @RequestMapping　　D. @Controller

(3) 以下（　　）不是 HTTP 的请求方式。

A. put　　B. head

C. options　　D. rest

(4) 当 DispatchServlet 拦截请求导致静态资源无法访问时，在配置文件中添加（　　）标签。

A. <mvc:annotation-driven/>　　B. <mvc:default-servlet-handler>

C. <context:annotation-config>　　D. <context:component-scan/>

(5)（　　）注解是解析请求的 JSON 格式的参数，也就是自动解析 JSON 字符串。

A. @ResponseBody　　B. @RequestBody

C. @RequestMapping　　D. @Controller

3. 思考题

(1) 静态资源该如何配置才能够在浏览器中加载成功？

(2) Spring MVC 和 AJAX 之间是怎么完成调用的？

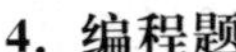

4. 编程题

仿照本章的案例试着编写一个项目能够实现文件的上传下载的功能。

第 15 章 SSM 框架整合

本章学习目标

- 了解整合环境的搭建
- 掌握整合思路
- 了解整合的 jar 包
- 理解整合的代码编写
- 理解整合的配置文件

前面的章节中通过对 Mybatis、Spring 和 Spring MVC 框架的学习，可以了解到这三种框架在项目开发的过程中分别充当了持久层(Mybatis)、业务层(Spring)和表现层(Spring MVC)的角色。Mybatis+ Spring+Spring MVC 的 SSM 框架整合也是目前比较主流的开发方式，适用于搭建各种大型的企业级应用系统。本章节将对 SSM 框架整合的开发步骤以及开发过程中的注意事项进行详细讲解。

15.1 整合环境搭建

项目的开发通常采用三层结构，即界面层(User Interface Layer)、业务逻辑层(Business Logic Layer)和数据访问层(Data Access Layer)，区分层次的目的是为了满足“高内聚低耦合”的编程思想。在 SSM 的整合中需要配备如表 15.1 所示的开发环境。

表 15.1 SSM 整合的开发环境

名称	系统配置条件
硬件环境	一般 PC，4GB 以上内存
操作系统	Windows 7、Windows 10
语言	Java、JavaScript、HTML、CSS 等
开发工具	Eclipse
服务器软件	Tomcat 7
数据库	MySQL 5.5
浏览器	FireFox、Chrome

为了使整合能够顺利完成，大家在学习时应当尽量按照表 15.1 所列举的条件配置开发环境，以保证系统运行平稳、响应及时。

15.2 整合思路

本次 SSM 的整合，在开发过程中整体采用三层模型的思路，即从数据库表设计、实体的创建开始到应用层代码的实现，最后再编写展示信息的页面。

下面将通过 SSM 整合来实现一个学校信息的统计管理。首先，在数据库的选择上，本次开发使用 MySQL 作为数据库。数据库文件信息如例 15-1 所示。

【例 15-1】 chapter15. sql

```
create database chapter15;
use chapter15;
create table t_school(id int primary key auto_increment,name
varchar(50),persons int);
```

在上述代码中实现了数据库的创建和数据库表的创建，其中数据表中的 id 属性作为自增主键。

数据库表创建完成之后就开始准备 jar 包，编写项目所需的配置文件，然后编写数据层到业务逻辑层再到显示层的代码，通过 jsp 实现所需的页面操作。最后进行项目整体的运行测试就可以了。

15.3 准备所需 jar 包

在整理所需的 jar 包之前，首先创建本次的整合项目，如图 15.1 所示。

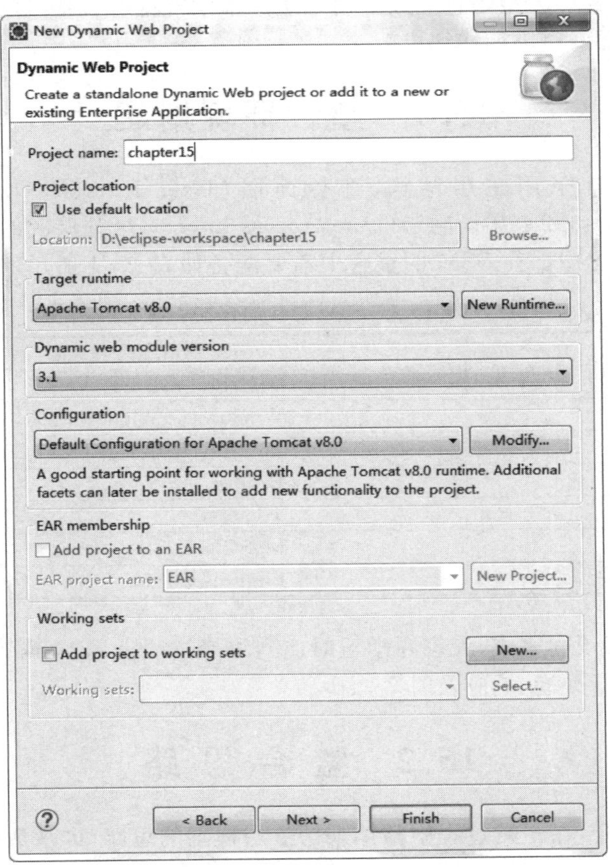

图 15.1 创建项目

注意：在创建 Web 项目时，需要拥有 web.xml 文件进行整个项目的全局设置。项目创建完成之后，接下来根据整合的框架整理所需要的 jar 包。

Spring 框架对应的 jar 包如图 15.2 所示。

spring-aop-5.0.8.RELEASE.jar	2018/7/26 07:23	Executable Jar File	358 KB
spring-aspects-5.0.8.RELEASE.jar	2018/7/26 07:26	Executable Jar File	46 KB
spring-beans-5.0.8.RELEASE.jar	2018/7/26 07:23	Executable Jar File	645 KB
spring-context-5.0.8.RELEASE.jar	2018/7/26 07:23	Executable Jar File	1,066 KB
spring-core-5.0.8.RELEASE.jar	2018/7/26 07:23	Executable Jar File	1,199 KB
spring-jdbc-5.0.8.RELEASE.jar	2018/7/26 07:24	Executable Jar File	392 KB
spring-test-5.0.8.RELEASE.jar	2018/7/26 07:25	Executable Jar File	592 KB
spring-tx-5.0.8.RELEASE.jar	2018/7/26 07:24	Executable Jar File	250 KB
spring-web-5.0.8.RELEASE.jar	2018/7/26 07:24	Executable Jar File	1,234 KB

图 15.2　Spring 所需 jar

Spring MVC 框架所需的 jar 是 spring-webmvc-5.0.8.RELEASE.jar，如图 15.3 所示。

spring-webmvc-5.0.8.RELEASE.jar	2018/7/26 07:25	Executable Jar File	773 KB

图 15.3　spring-webmvc jar 包

Spring 框架内部有日志记录，采用的是 common-logging 日志工具，需要的 jar 如图 15.4 所示。

commons-logging-1.2.jar	2018/10/28 14:55	Executable Jar File	61 KB

图 15.4　common-logging jar 包

整合 Mybatis 持久化操作，少不了 Mysql 的驱动和数据库连接池，数据库连接池还是选用 C3P0 技术。本次整合的完整 jar 如图 15.5 所示。

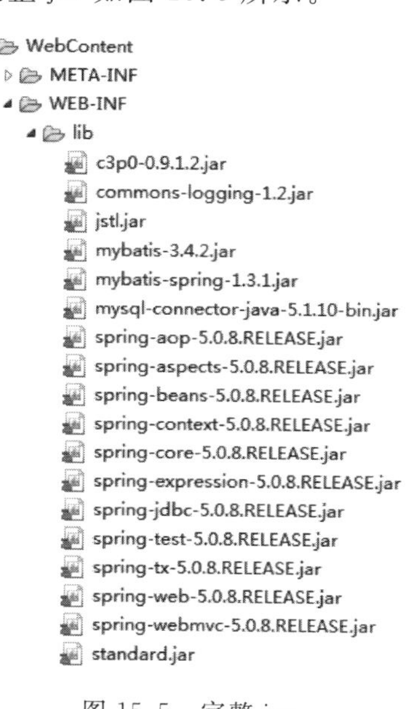

图 15.5　完整 jar

15.4 编写配置文件

项目创建完成并导入相关 jar 包后，进行项目配置文件的编写。SSM 整合中的配置文件主要有四个部分，第一个是数据库连接信息文件，用来标记数据库的地址和用户密码等信息；第二个是 Spring 框架的配置文件，用来实现数据库连接信息的配置和数据库连接池MyBatis 的一些配置；第三个是 Spring MVC 框架的配置文件，用来实现控制器的一些配置信息；第四个是在项目启动的时候完成 Spring 监听器的监听和 Spring MVC 前端控制器的配置实现。

首先在项目的 src 下右击 New Source Folder 创建资源文件夹（在编译的时候会编译到classpath 路径下面），用于存储本次整合的配置文件，如图 15.6 所示。

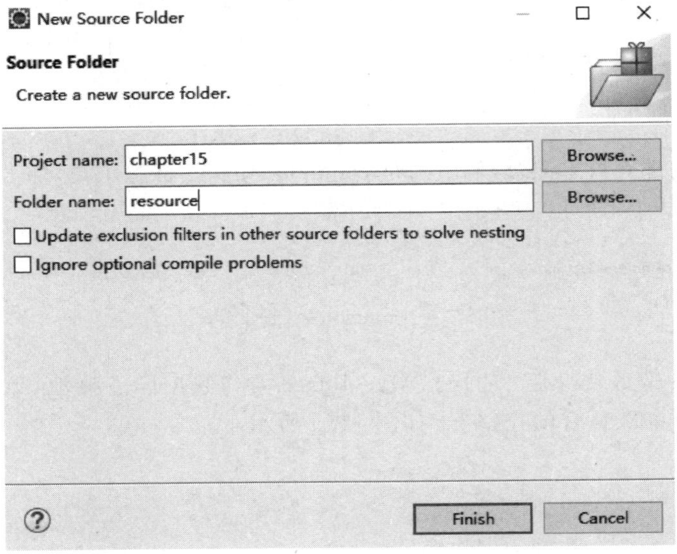

图 15.6　资源文件夹

接下来在该文件夹下创建数据库的配置文件和 Spring 的配置文件。数据库配置文件名称叫 jdbc. properties，注意：后缀名不要写错了。数据库配置文件信息如例 15-2 所示。

【例 15-2】 jdbc. properties

```
jdbc.driverClass = com.mysql.jdbc.Driver
jdbc.jdbcUrl = jdbc:mysql://localhost:3306/chapter15?characterEncoding = UTF -
8
 jdbc.user = root
 jdbc.password = root
```

上述文件主要实现了数据库的连接信息的设置，当然还可以进行数据库连接池的信息设置，只是笔者这里没有配置那么复杂。

application. xml 配置文件的内容，如例 15-3 所示。

【例 15-3】 application.xml

```
<?xml version="1.0" encoding="UTF-8"?>
<beans xmlns="http://www.springframework.org/schema/beans"
     xmlns:xsi="http://www.w3.org/2001/XMLSchema-instance"
     xmlns:context="http://www.springframework.org/schema/context"
     xmlns:aop="http://www.springframework.org/schema/aop"
     xmlns:tx="http://www.springframework.org/schema/tx"
     xsi:schemaLocation="http://www.springframework.org/schema/beans
          http://www.springframework.org/schema/beans/spring-beans.xsd
          http://www.springframework.org/schema/context
          http://www.springframework.org/schema/context/spring-context.xsd
          http://www.springframework.org/schema/aop
          http://www.springframework.org/schema/aop/spring-aop.xsd
          http://www.springframework.org/schema/tx
          http://www.springframework.org/schema/tx/spring-tx.xsd">
    <!-- 1、引入外部 properties 文件 -->
    <context:property-placeholder location="classpath:jdbc.properties" />
    <!-- 2、注册数据源 -->
    <bean name="dataSource"
          class="com.mchange.v2.c3p0.ComboPooledDataSource">
        <property name="driverClass" value="${jdbc.driverClass}" />
        <property name="jdbcUrl" value="${jdbc.jdbcUrl}" />
        <property name="user" value="${jdbc.user}" />
        <property name="password" value="${jdbc.password}" />
    </bean>
    <!-- 3、配置 Session 工厂对象 -->
    <bean id="sessionFactory"
          class="org.mybatis.spring.SqlSessionFactoryBean">
        <property name="dataSource" ref="dataSource"></property>
    </bean>
    <!-- 4、配置扫描的包 -->
    <bean id="mapperScannerConfigurer"
          class="org.mybatis.spring.mapper.MapperScannerConfigurer">
        <property name="basePackage" value="com.qfedu.dao"></property>
    </bean>
    <!-- 5、扫描对应的类 IOC 创建对象   -->
    <!-- 扫描注解的类 -->
    <context:component-scan
base-package="com.qfedu.service.impl"></context:component-scan>
</beans>
```

上述文件实现了 Spring 的配置，借助 IOC 创建对象，设置数据库连接信息，采用数据库连接池 C3P0 进行数据库连接，并且实现 MyBatis 的映射文件的扫描。要注意 MyBatis 映射文件扫描的配置信息，对应的路径信息不能写错，否则就会导致 dao 层接口无法使用。

数据库和 Spring 的配置文件都编写完成后，接下来在项目的 WEB-INF 目录下编写 Spring MVC 的配置文件，实现控制器的扫描，统一视图后缀名的设置等操作。

springMVC-config.xml 的配置文件内容如例 15-4 所示。

【例 15-4】 springMVC-config.xml

```
<?xml version='1.0' encoding='UTF-8'?>
<beans xmlns="http://www.springframework.org/schema/beans"
    xmlns:xsi="http://www.w3.org/2001/XMLSchema-instance"
    xmlns:context="http://www.springframework.org/schema/context"
    xmlns:mvc="http://www.springframework.org/schema/mvc"
    xsi:schemaLocation="http://www.springframework.org/schema/beans
        http://www.springframework.org/schema/beans/spring-beans.xsd
        http://www.springframework.org/schema/context
        http://www.springframework.org/schema/context/spring-context.xsd
        http://www.springframework.org/schema/mvc
        http://www.springframework.org/schema/mvc/spring-mvc.xsd">
    <!-- 扫描 Controller -->
    <context:component-scan base-package="com.qfedu.controller" />
    <!-- 配置视图解析器 -->
    <bean class="org.springframework.
          web.servlet.view.InternalResourceViewResolver">
        <property name="suffix" value=".jsp" />
    </bean>
    <mvc:annotation-driven/>
    <mvc:default-servlet-handler></mvc:default-servlet-handler>
</beans>
```

注意：这个配置文件需要在项目的WebContent目录的WEB-INF下创建Spring MVC的配置文件，用来实现控制器的扫描和视图解析器的配置。例如在视图解析器中标记视图的后缀名称，静态资源默认放行等配置。

最后配置web.xml文件，web.xml配置文件的内容如例15-5所示。

【例 15-5】 web.xml

```
<?xml version="1.0" encoding="UTF-8"?>
<web-app xmlns:xsi="http://www.w3.org/2001/XMLSchema-instance"
xmlns="http://xmlns.jcp.org/xml/ns/javaee"
xsi:schemaLocation="http://xmlns.jcp.org/xml/ns/javaee
http://xmlns.jcp.org/xml/ns/javaee/web-app_3_1.xsd" version="3.1">
  <display-name>chapter15</display-name>
  <!--Spring 的配置文件-->
   <context-param>
        <param-name>contextConfigLocation</param-name>
        <param-value>classpath:application.xml</param-value>
    </context-param>
    <!--Spring 的监听器-->
    <listener>
        <listener-class>
org.springframework.web.context.ContextLoaderListener
</listener-class>
    </listener>
    <!--SpringMVC 的前端控制器-->
```

```xml
    <servlet>
        <servlet-name>springMVC</servlet-name>
        <servlet-class>
org.springframework.web.servlet.DispatcherServlet
</servlet-class>
        <init-param>
            <param-name>contextConfigLocation</param-name>
            <param-value>/WEB-INF/springMVC-config.xml</param-value>
        </init-param>
        <load-on-startup>1</load-on-startup>
    </servlet>
    <!-- 访问 DispatcherServlet 对应的路径 -->
    <servlet-mapping>
        <servlet-name>springMVC</servlet-name>
        <url-pattern>/</url-pattern>
    </servlet-mapping>
    <!-- 编码格式过滤器 -->
     <filter>
        <filter-name>encoding</filter-name>
        <filter-class>org.springframework.web.filter.CharacterEncodingFilter
        </filter-class>
        <init-param>
            <param-name>encoding</param-name>
            <param-value>UTF-8</param-value>
        </init-param>
    </filter>
    <filter-mapping>
        <filter-name>encoding</filter-name>
        <url-pattern>/*</url-pattern>
    </filter-mapping>
    <!-- 引导页 -->
    <welcome-file-list>
        <welcome-file>page01.jsp</welcome-file>
     </welcome-file-list>
</web-app>
```

注意：在整个项目中需要实现对 Spring 的配置和对 Spring MVC 的配置，在 web.xml 文件中实现 Spring 的加载和 Spring MVC 前端控制器的配置，还有编码自动转换过滤器等的配置。切记这个配置文件在整个项目中只有一个。

15.5 编写项目代码

无论是环境的搭建还配置文件的编写，都是准备工作。在上述的步骤中完成了数据库表的创建和配置文件的编写，那么接下来需要实现代码的编写。整个代码逻辑上采用三层架构来完成。也就是从数据库表的映射类到数据库操作层再到业务逻辑处理层最后到控制器编写映射接口，供页面访问。

1）数据库表的映射类编写

根据数据库对应的表，编写对应的映射类，类中的属性和数据库表中的字段要求一一对应。如果类的属性名称和数据库标的字段名称不一致，则需要在 dao 的映射配置文件中使用< resultMap >进行标记起别名，实现不同的名称映射。创建实体类 School 的代码如例 15-6 所示。

【例 15-6】 School. java

```
package com.qfedu.pojo;
public class School {
    private int id;
    private String name;
    private int persons;
    public int getId() {
        return id;
    }
    public void setId(int id) {
        this.id = id;
    }
    public String getName() {
        return name;
    }
    public void setName(String name) {
        this.name = name;
    }
    public int getPersons() {
        return persons;
    }
    public void setPersons(int persons) {
        this.persons = persons;
    }
}
```

注意：上述的类在包 com. qfedu. pojo 内部，该包表示简单对象类，它是数据库的表的映射类。如果还有其他的映射类都写到该包下。然后编写数据持久化操作接口，这里只需要写出操作接口即可，实现类通过 MyBatis 框架映射生成。创建 dao 的代码如例 15-7 所示。

【例 15-7】 SchoolDao. java

```
package com.qfedu.dao;
import java.util.List;
import org.apache.ibatis.annotations.Insert;
import org.apache.ibatis.annotations.ResultType;
import org.apache.ibatis.annotations.Select;
import com.qfedu.pojo.School;
public interface SchoolDao {
    //新增
```

```
    @Insert("insert into t_school(name,persons)
values(#{name},#{persons})")
    int insert(School school);
    //查询
    @Select("select * from t_school")
    @ResultType(School.class)
    List<School> selectAll();
}
```

上述代码中的接口就是操作数据库的接口，位置在 com. qfedu. dao 包下。这里采用的是对 MyBatis 中注解的应用，具体请参照前面章节中对 MyBatis 的讲解。该段代码实现了对数据库的新增和查询的方法以及对应的 SQL 语句。

注意：使用@ResultType 注解的前提是数据库中的表的字段名称和对应类的属性名称都一致，而且对应的类内部也没有复杂的嵌套关系。

2）业务逻辑层接口

Dao 层接口编写完成之后，接下来就是编写业务逻辑层的接口，完成新增和查询方法的定义，代码如例 15-8 所示。

【例 15-8】 SchoolService. java

```
package com.qfedu.service;
import java.util.List;
import com.qfedu.pojo.School;
public interface SchoolService {
    int save(School school);
    List<School> queryAll();
}
```

该接口位于 com. qfedu. service 包下，实现了添加和查询的方法。参考阿里巴巴公司的技术开发规范，接口中的方法一般不用声明 public，因为接口中的方法默认修饰符就是 public。

该接口的实现类 SchoolServiceImpl. java 的代码如例 15-9 所示。

【例 15-9】 SchoolServiceImpl. java

```
package com.qfedu.service.impl;
import java.util.List;
import org.springframework.beans.factory.annotation.Autowired;
import org.springframework.stereotype.Service;
import com.qfedu.dao.SchoolDao;
import com.qfedu.pojo.School;
import com.qfedu.service.SchoolService;
@Service
public class SchoolServiceImpl implements SchoolService{
    @Autowired
    private SchoolDao dao;
    @Override
    public int save(School school) {
```

```
        return dao.insert(school);
    }
    @Override
    public List<School> queryAll() {
        return dao.selectAll();
    }
}
```

上述代码就是业务逻辑接口层对应的实现类，该类需要通过 Spring 的 IOC 创建对象，并且依赖 dao 中的对象。需要注意的是这里用到了 SchoolDao 对象，而这个对象要注册在 Spring 的 IOC 容器中，当前这个类的对象也要注册在 Spring 的 IOC 容器中，所以使用@Service 注解。

接下来实现控制器类的编写，代码如例 15-10 所示。

【例 15-10】 SchoolController.java

```
package com.qfedu.controller;
import org.springframework.beans.factory.annotation.Autowired;
import org.springframework.stereotype.Controller;
import org.springframework.ui.Model;
import org.springframework.web.bind.annotation.RequestMapping;
import org.springframework.web.bind.annotation.RequestMethod;
import com.qfedu.pojo.School;
import com.qfedu.service.SchoolService;
@Controller
public class SchoolController {
    @Autowired
    private SchoolService schoolService;
    //新增
    @RequestMapping(value = "/schooladd",method = RequestMethod.POST)
    public String save(School school,Model model) {
        if(schoolService.save(school)> 0) {
            return "page02";
        }else {
            model.addAttribute("msg", "服务器异常,新增学校信息有误!");
            return "page01";
        }
    }
    //查询
    @RequestMapping(value = "/schoolall",method = RequestMethod.GET)
    public String all(Model model) {
        model.addAttribute("schools", schoolService.queryAll());
        return "page02";
    }
}
```

注意：SchoolController 类位于 com.qfedu.controller 包下，这个包也就是控制器类所在包。类是基于 Spring MVC 实现的控制器，创建对应的数据接口，让页面可以通过 url 进行访问，达到操作数据库的目的。在新增的方法中，要注意参数为School，那么前台页面在传递的时候就需要写出对应的请求参数和值，而且参数名称就是School类中的属性名称，必须要一致，否则就会出现报数据库中有空数据的错误。在查询的方法中，切记需要将查询的结果存储到Model中，也可以用HttpServletRequest的setAttribute方法，只要保证名称为schools即可，这样在前台页面中通过EL表达式就能够直接获取到数据。

Page01.jsp的代码如例15-11所示。

【例15-11】 page01.jsp

```
<%@ page language = "java" contentType = "text/html; charset = UTF - 8"
     pageEncoding = "UTF - 8" %>
<!DOCTYPE html PUBLIC " - //W3C//DTD HTML 4.01 Transitional//EN"
"http://www.w3.org/TR/html4/loose.dtd">
<html>
<head>
<meta http - equiv = "Content - Type" content = "text/html; charset = UTF - 8">
<title>新增学校</title>
</head>
<body>
<div>
<label style = "color: red;">${msg}</label>
<form action = "schooladd" method = "post">
     <label>校名: </label><input name = "name"><br/>
     <label>人数: </label><input name = "persons"><br/>
     <input type = "submit" value = "新增学校">
</form>
</div>
</body>
</html>
```

上述代码实现了新增学校信息，并通过表单标签实现数据交互，注意input标签的name属性名称需要和对应类的属性名称一致，否则无法接收到数据。如果新增失败，可以获取错误信息，在css中把错误信息显示到页面，并且把字体样式标记为红色。请求的方式需要和控制器的映射方法中的请求方式保持一致，否则请求异常。

Page02.jsp的代码如例15-12所示。

【例15-12】 page02.jsp

```
<%@ page language = "java" contentType = "text/html; charset = UTF - 8"
     pageEncoding = "UTF - 8" %>
     <%@ taglib prefix = "c" uri = "http://java.sun.com/jsp/jstl/core" %>
<!DOCTYPE html PUBLIC " - //W3C//DTD HTML 4.01 Transitional//EN"
"http://www.w3.org/TR/html4/loose.dtd">
<html>
<head>
<meta http - equiv = "Content - Type" content = "text/html; charset = UTF - 8">
```

```
<title>学校列表</title>
</head>
<body>
<div>
<h2>学校列表</h2>
<h6><a href = "schoolall">刷新</a></h6>
<table border = "1">
<tr>
    <th>序号</th>
    <th>校名</th>
    <th>人数</th>
</tr>
<c:forEach items = "${schools}" var = "s">
    <tr>
        <td align = "center">${s.id}</td>
        <td align = "center">${s.name}</td>
        <td align = "center">${s.persons}</td>
    </tr>
</c:forEach>
</table>
</div>
</body>
</html>
```

该页面实现了查询功能，通过 JSTL 的循环标签来实现数据的显示，可以查询并展示出所有的学校信息。

15.6 整合应用测试

整个代码编写完成后进行测试，运行项目，等待启动成功之后，在浏览器地址栏中输入 http://localhost:8080/chapter15/，访问成功后在页面上输入要添加的信息，如图 15.7 所示。

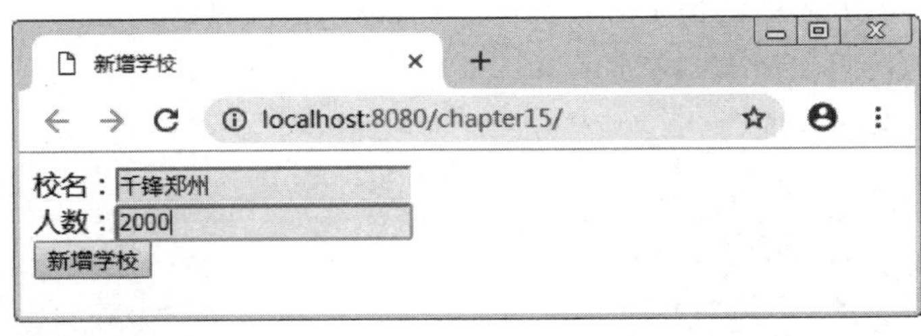

图 15.7 新增学校页面

单击新增学校按钮后，自动跳转至列表页面，然后单击刷新按钮，就可以发现数据库中的学校信息显示在网页上了。如果刷新之后发现数据没有显示，可以到数据库进行查询，查看数据是否添加成功。浏览器中的显示效果如图 15.8 所示。

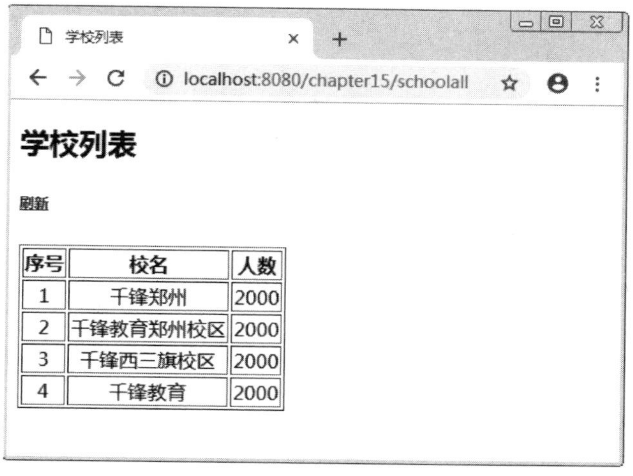

图 15.8　学校列表

15.7　本章小结

本章完成了 Spring＋Spring MVC＋Mybatis 的三大框架的整合开发。通过本章学习，大家应该能了解 SSM 框架整合的开发步骤，整个流程从数据库表设计开始到项目的创建、配置文件的编写以及展示页面的开发。特别要注意在 SSM 整合过程中数据库连接信息和包扫描位置的准确性。采用 MyBatis 注解的方式进行数据库持久化操作，只需要写出对应的 dao 层接口和接口中的 sql 语句，至于 dao 层接口的实现类，则不再进行编写。

15.8　习　　题

1. 填空题

(1) 本章节学习的 SSM 框架整合是指________、________、________这三个框架的整合。

(2) ________框架使用简单的 xml 或注解用于配置和原始映射，将接口和 java 的 POJO 映射成数据库中的记录。

(3) SSM 是基于 mvc 结构的开发模式，________充当了控制层，________充当了业务逻辑层，________充当了持久层。

(4) 将一个请求 url 指向一个类的方法的注解是________。

(5) Spring MVC 中有一个 servlet 将前端的请求分发到各个控制器，这个 servlet 的名字是________。

2. 选择题

(1) 声明控制器类的注解是(　　)。

A. @autowire　　　　B. @Controller

C. @RequestMapping　　　　D. @RequestParam

(2) ModelAndView 类中 addObject 方法和 Model 类中的 addAttribute 方法相当于执行了(　　)对象中的 setAttribute 方法。

A. response　　B. request
C. session　　D. document

(3) 一个 SSM 项目中有以下四个配置文件，哪个配置文件中配置了数据源和数据库连接池(　　)。

A. log4j. properties　　B. web. xml
C. applicationContext. xml　　D. springmvc. xml

(4) 以下(　　)标签用于开启注解扫描。

A. <context-param>　　B. <context:component-scan>
C. <mvc:annotation-driven/>　　D. <aop:config>

(5) 以下说法中错误的是(　　)。

A. Mybatis 的核心是 sqlsessionfactory
B. 控制器类中有一个成员变量，已经在 spring 配置文件中声明，可用@autowire 注解将该成员变量注入
C. Spring MVC 中，controller 执行完成返回的 ModelAndView 即为最终响应用户的视图
D. spring 上下文是一个配置文件，向 spring 框架提供上下文信息

3. 思考题

(1) SSM(Spring、Spring MVC、MyBatis)在项目开发过程中各有什么作用?

(2) 使用 SSM 框架开发项目有什么好处?

4. 编程题

使用 ssm 整合框架编写一个能够实现学生信息的注册和登录功能的代码。在数据库中新建 Student 表，通过浏览器向表中插入学生信息内容，插入成功后，再实现登录功能。

第16章 SSM整合开发案例——锋迷网

本章学习目标

- 整合环境搭建
- 整合思路
- 准备所需jar包
- 编写配置文件
- 编写代码
- 整合应用测试

本书在之前的章节中，详细讲解了Java常用框架在开发过程中所涉及的基础知识与核心技术，其中包括Spring、Spring MVC、MyBatis三大框架等。为了帮助大家能够串联和巩固之前所学的知识，本章将通过一个项目案例来讲解这些知识点在实际开发中的应用，让读者真正理解Java常用框架中的精髓并做到融会贯通、学以致用。

16.1 项目背景及系统架构

16.1.1 应用背景

近几年电子商务的兴起，在一定程度上改变了人们的生活方式。购物网站更是为人们的日常生活带来了极大的便利，让消费者足不出户便可以购买到来自全国各地的商品。这种网上商城的模式不仅仅为卖家节省了大量的成本，也为买家节省了时间和开销。随着电子商务的发展，作为程序开发工作者，了解和学习网上商城的开发势在必行，是"顺势而为"的一件事。因此，本章节将通过实现一个电子商务网站，来巩固SSM框架整合技术以及重要知识点在项目中的实际应用。

16.1.2 系统架构介绍

本系统将采用Java语言进行开发，通过Spring、Spring MVC和MyBatis框架的整合来实现整个项目，并使用MySQL作为数据库。

16.1.3 功能模块介绍

项目主要由五大模块组成，分别是：用户模块，该模块包含了用户的注册、登录功能，以及收货地址的添加、修改等功能；商品模块，该模块包含了商品列表的展示以及单个商品详细信息的展示功能；购物车模块，该模块包含了用户向购物车中添加和删除商品的功能；

订单模块,该模块包含订单预览、订单生成、历史订单查询、删除订单等功能;后台管理系统模块,该模块包含了网上商城内部工作人员对整个网站的管理,例如会员管理、商品管理、订单管理等功能。

16.1.4 运行效果

整个网站分为两个部分,第一部分是普通用户能够正常访问的系统,该系统可以实现普通用户注册、会员登录、商品浏览、商品详情、购物车、订单等功能。第二部分是后台管理系统,该系统可以由系统管理员实现会员管理、商品管理、订单管理等功能。

首先在浏览器的地址栏中输入锋迷网的后台登录地址,可以进入锋迷网的后台登录页面,在该页面输入创建数据库时添加的管理员账号、密码进行登录,如图 16.1 所示。

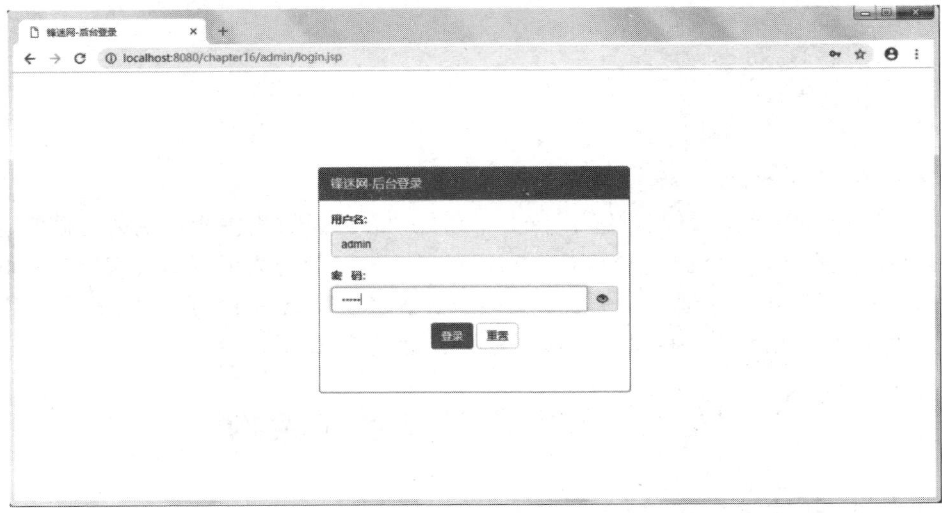

图 16.1 后台登录页面

用户名和密码输入无误,单击登录按钮,即可进入锋迷网后台模块的首页,如图 16.2 所示。

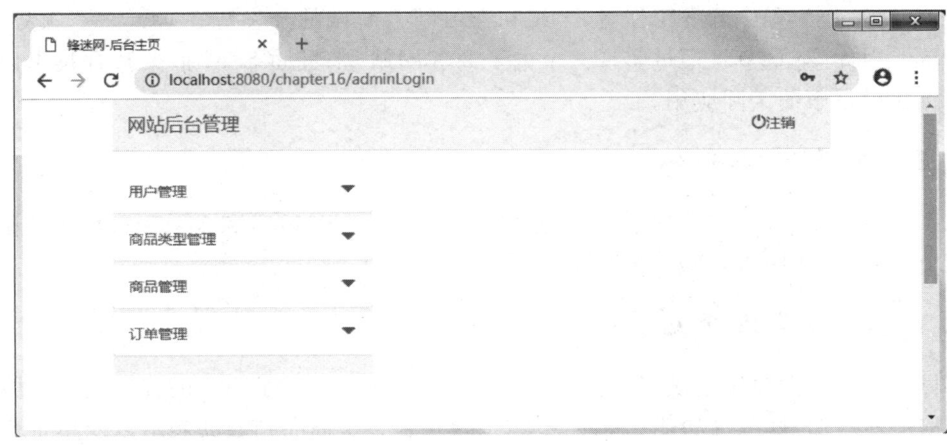

图 16.2 后台主页

首页展示了锋迷网的后台管理功能，左侧是管理菜单，管理员可以通过这些管理菜单实现对应的后台管理操作。

分别单击首页左侧管理菜单下的用户管理、商品类型管理、商品管理、订单管理等菜单项，可以将这些菜单项下的对应子菜单展开或隐藏，当单击子菜单时，右侧会弹出与之对应的功能页面。

以用户管理下的子菜单为例，当单击会员管理菜单项时，右侧会弹出所有注册的会员信息并以列表的形式展现，具体如图16.3所示。

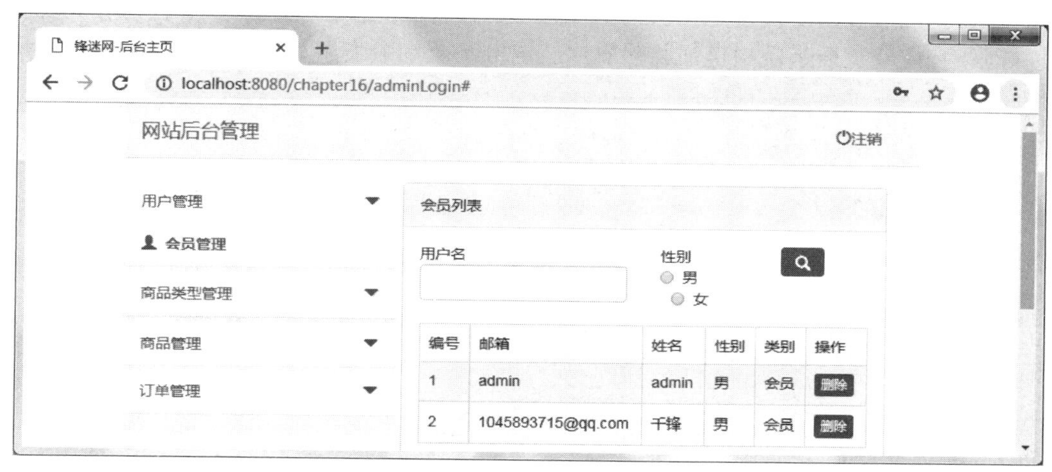

图16.3 会员列表页面

该列表页面可以实现会员管理的一些基本操作。管理员可以在用户名后的输入框中输入用户名并勾选性别，从数据库中查询到符合条件的会员，还可以在操作栏中单击“删除”按钮，将该行会员信息从数据库用户表中移除。在用户名输入框中输入“admin”，性别选项勾选“男”，可以查询到用户名为admin的所有男性用户信息。查询结果如图16.4所示。

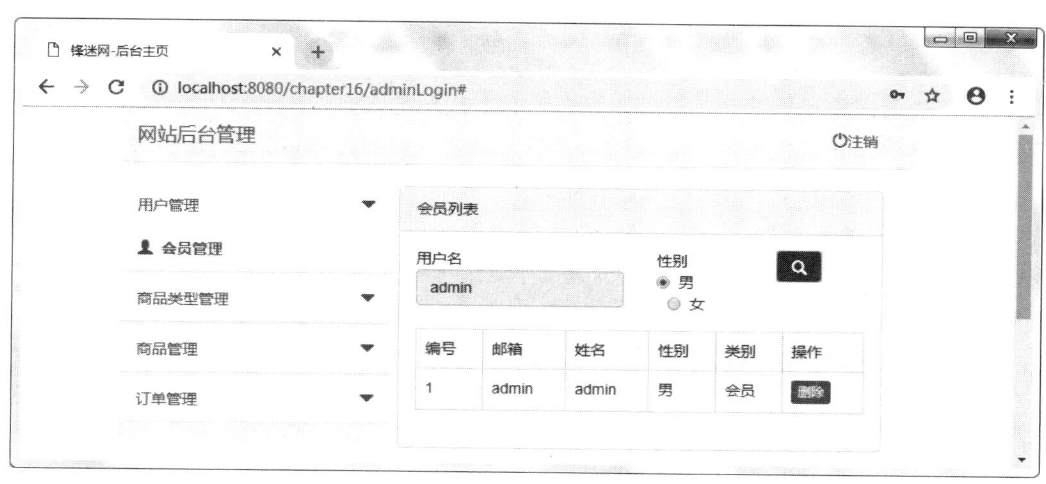

图16.4 会员查询

当单击商品类型管理下的查看商品分类子菜单时，网页右侧将查询到已存在的所有商品类型，并以列表形式展示出来。在该列表中，可以看到每个商品类型的级别以及该类型的父类型(所属类型)，如果类型为一级，则父类型为空。通过输入商品类型等级和商品类型名称可以查询到该商品类型的详细信息，并且在操作栏中可以单击"删除"按钮完成对该行类型的删除功能。

商品类型的管理是为了方便普通用户在登录网上商城时快速定位到自己所要选购的商品。例如用户想要购买橙汁，那么用户可以首先进入酒水饮料这个类型所包含的商品栏中寻找，这样就能方便用户快速定位选购。而且，通过将商品进行分类也可以对网站的页面商品进行归纳，使网页展示做到更加井然有序。商品类型列表展示如图 16.5 所示。

图 16.5　商品类型页面

同样在输入框中输入等级和类型名称就可以搜索到该类型的详细信息，然后单击"删除"按钮，该类型随之从数据库 t_goodstype 表中删除。类型搜索效果如图 16.6 所示。

当单击"商品类型"管理下的"添加商品分类"子菜单时，在网页右侧将完成商品类型信息的添加，如果添加的类型为一级类型，则所属类型选框中不再选择，直接在种类名称中填写要添加的类型即可，若添加的类型为子类型，则在所属类型的选框中选中该类型的所属类型，即父类型之后再完成商品类型的添加。商品类型添加页面如图 16.7 所示。

当单击商品管理下的查看商品子菜单时，网页右侧将会查询到数据库中已存在的所有商品信息，并以列表形式展示出来。管理员可以通过输入商品名称和商品上架的时间搜索到该商品的信息，也可以查看到该商品的所属类型，还可以在操作栏中单击"描述"按钮查看到关于该商品的一些详细信息。最后管理员可以通过单击"删除"按钮完成选中商品从数据库中删除的操作。商品列表展示如图 16.8 所示。

商品的查询和删除功能的效果图与商品类型类似，此处不再演示。

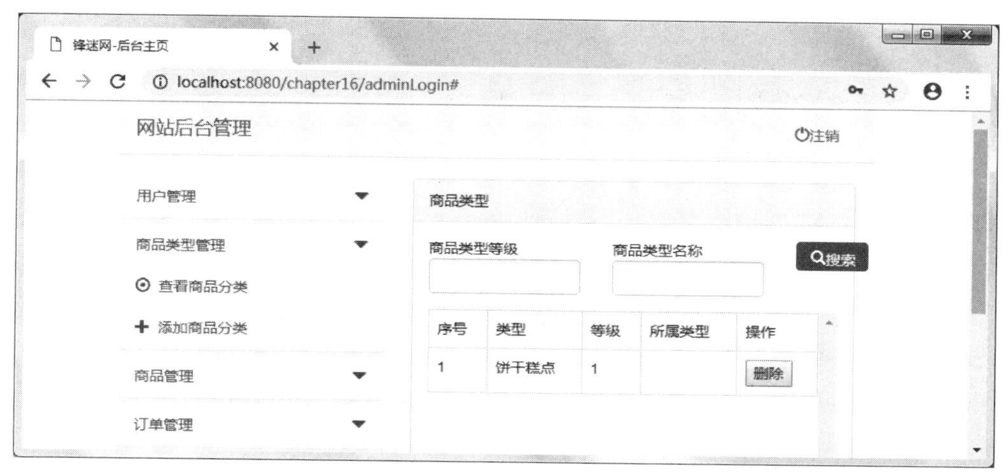

图 16.6　商品类型查询

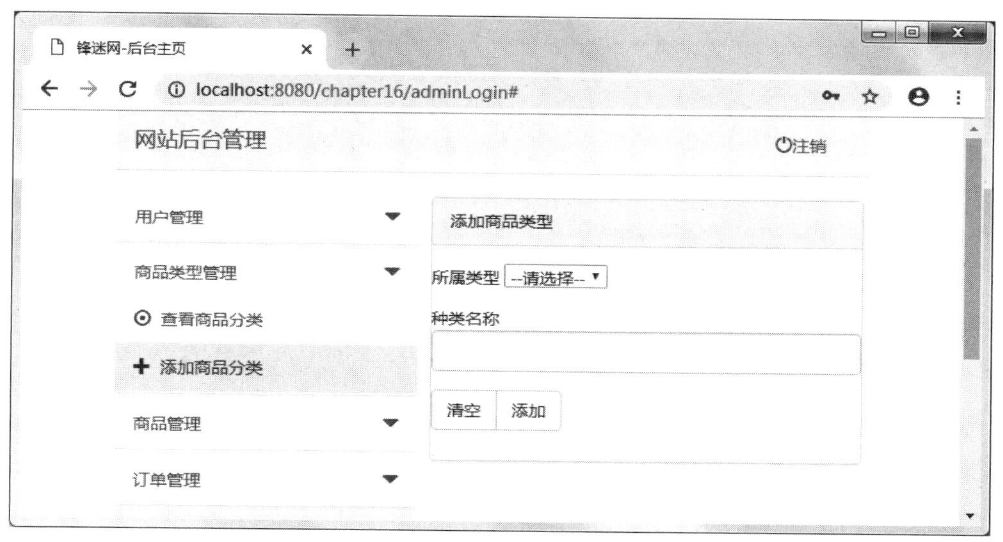

图 16.7　添加商品类型页面

当单击“商品管理”下的添加商品子菜单时，在网页右侧将完成商品信息的添加，在添加商品信息时需要注意选择商品类型，上传图片时注意图片存储的路径，否则图片显示不出来。页面中还添加了信息重置按钮。新增商品的页面如图 16.9 所示。

当单击订单管理下的查看订单子菜单时，右侧将会展示出所有用户的下单信息，可以通过输入用户名和选择订单状态查询到该用户不同状态下的订单信息，一共将订单分为未支付、已支付，待发货、已发货待收货，已收货，未评价，完成订单这些状态。在操作栏中可以选择对该订单完成发货处理，订单管理页面如图 16.10 所示。

最后单击页面右上角的注销按钮即可退出该后台管理系统。

后台管理系统的内容大致如上所述，接下来将进行网上商城系统相关页面的展示。该系统有用户模块、商品模块、购物车模块、订单模块等，访问网站的主页，如图 16.11 所示。

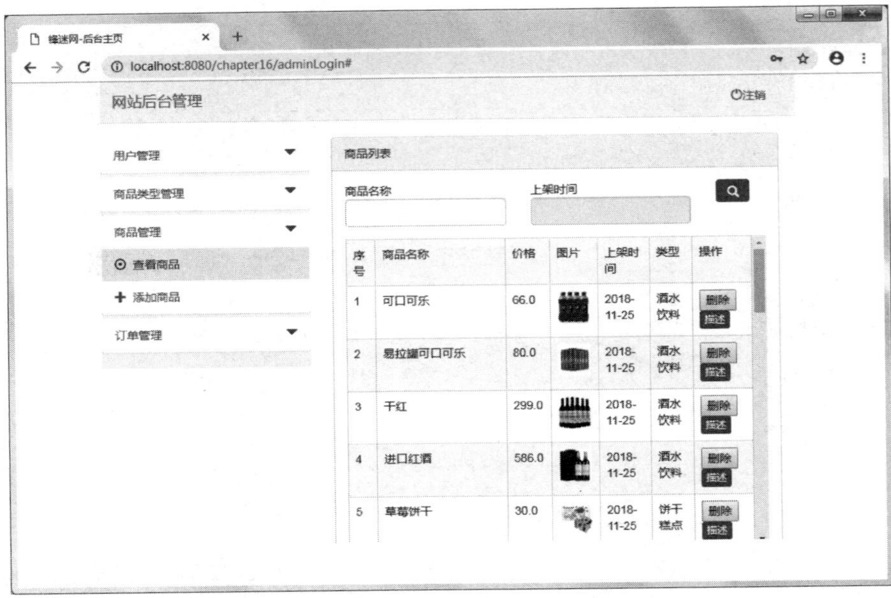

图 16.8　商品列表

图 16.9　添加商品页面

图 16.10　订单管理页面

图 16.11　锋迷网主页

在网站主页的右上角单击登录链接即可跳转至会员登录页面，如果之前没有注册过账号可以单击注册链接，完成注册，具体页面如图 16.12 和图 16.13 所示。

锋迷网-登陆
localhost:8080/chapter16/login.jsp
欢迎登录
锋迷网
用户名：qianfeng
密码：
请不输入6位以上字符
登录　重置

图 16.12　登录页面

登录成功后可以看到网站的主页，该页面的上中部位有一行分类标签，该标签就是根据商品的类型进行分类的，单击标签可以展示该标签类型下的商品列表。在该页面的右上部位还有一个搜索框，可以实现对商品的模糊查询功能。此外，页面中还有推广活动的轮播图和一些热销商品的展示。在搜索框中输入关键字“酒”，即可查询到所有名字中带有“酒”字样的商品，如图 16.14 所示。

单击网页上方酒水饮料的类型链接，就可以浏览当前网站的酒水饮料对应的商品列表，具体如图 16.15 所示。

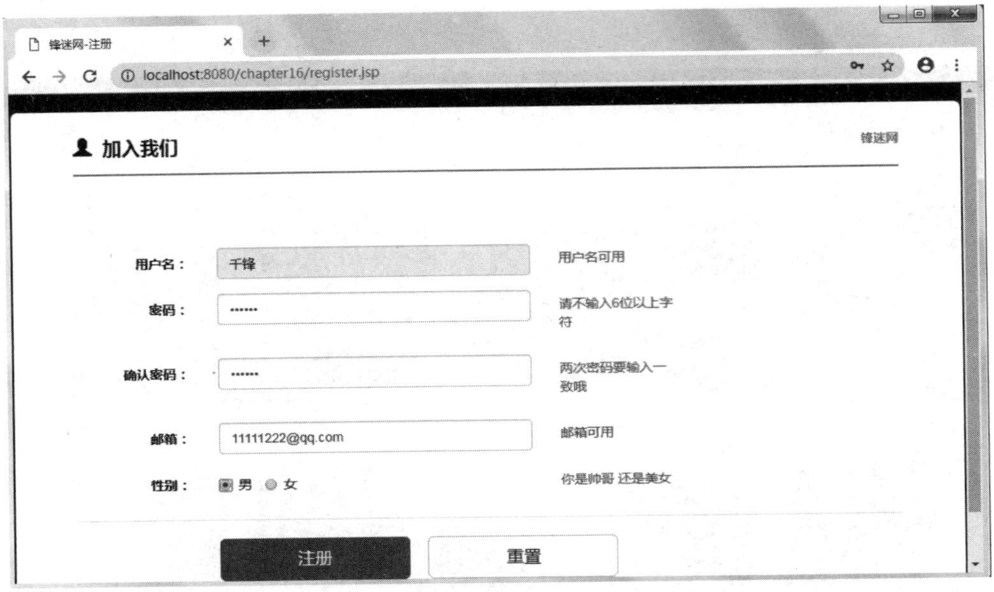

图 16.13　注册页面

图 16.14　商品查询

在图 16.15 中可以看到商品的详细信息，如商品名称、热销指数、上架日期以及对应的价格等，在这些详细信息的上面有与之对应的商品图片。单击商品的名称可以看到该商品的详情信息，如图 16.16 所示。

通过该页面可以实现向购物车中添加商品的操作，此时，需要注意，如果目前用户处于未登录状态，则系统会跳转至登录页面，要求用户完成登录，如果用户有账号就使用账号密码进行登录，如果没有账号就先注册然后再登录。登录成功之后才能选择需要加入购物车的商品，完成购物车的添加，查看购物车的内容，如图 16.17 所示。

图 16.15　商品列表

图 16.16　商品详情

图 16.17 中在“数量”栏中单击“－”或者“＋”可以减少或者增加购买商品的数量，商品的总金额也会随之改变。

在购物车页面中单击“结算”就进入订单预览页面，注意该页面需要设置收货地址，如果尚未添加收货地址，则需要先添加收货地址，然后再完成下单，否则订单无法生成，如图 16.18 所示。

单击“收货地址”框下的链接，添加收货地址。可以依次填写联系人、手机号和收货地址的详细信息，填写完成后单击“添加地址”按钮即可，如图 16.19 所示。

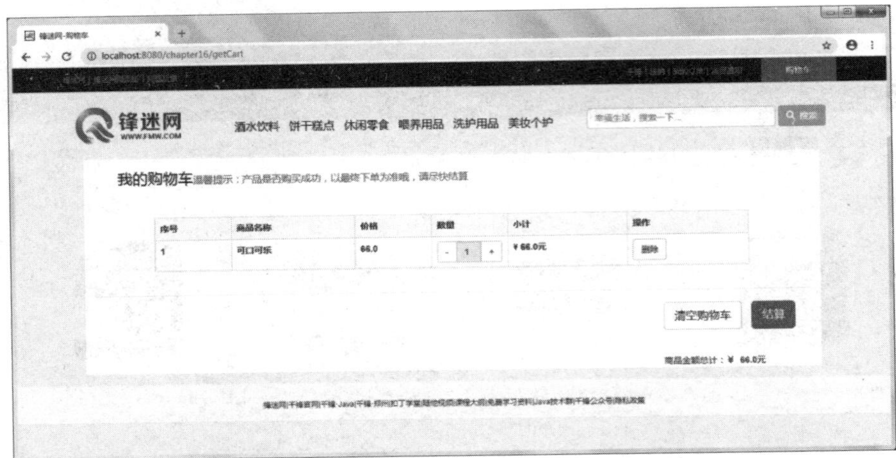

图 16.17　购物车列表

图 16.18　订单预览页面

图 16.19　添加收货地址

收货地址添加成功后可以在当前页面浏览到添加的地址信息，而且可以在操作栏中单击“修改”按钮，完成错误地址信息的修改，如图 16.20 所示。

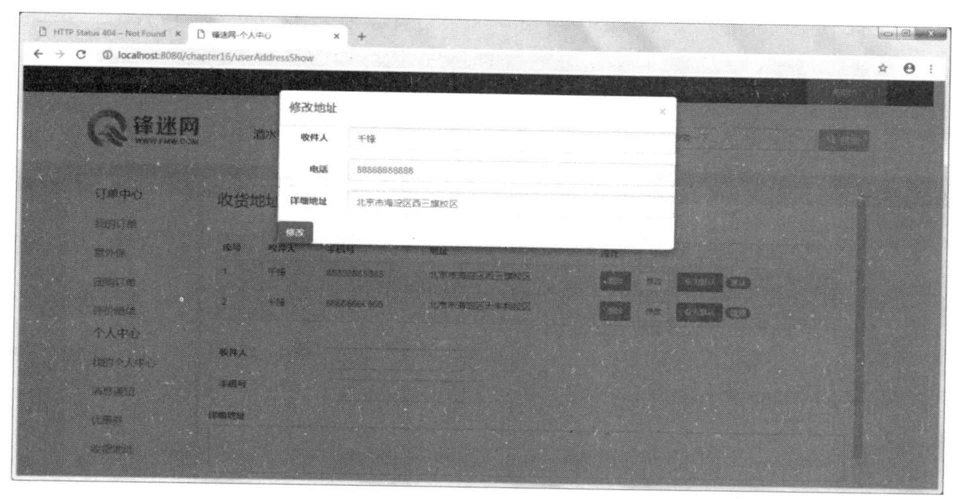

图 16.20　修改地址信息

在图 6.20 中可以看到，用户单击“修改”按钮后，页面会弹出一个信息修改的对话框，用户可以在对话框中填写想要修改的内容，然后单击左下方的“修改”按钮，即可完成收货信息的修改。

一个用户可以添加多条收货地址，还可以挑选其中一条设为默认地址，被设为默认的地址会在操作栏中标注出来，以后每次提交订单时，系统会为该订单自动匹配该默认地址。如图 16.21 所示。

图 16.21　设置默认地址

地址设置完成后单击页面左下方的“返回购物车”即可重新提交订单。

用户可以在自己的订单中心查看到自己未支付、已支付、派送中等各种状态的订单信息，还可以在操作栏中完成查看订单详情、删除订单、确认收货一系列操作，如图 16.22 所示。

图 16.22　订单列表

最后单击页面右上方的“注销”即可退出登录。上述就是锋迷网系统一些主要页面的运行展示，通过这些页面功能的展示，可以更直观地了解一个电商网站的业务逻辑和需求，接下来将开始锋迷网的代码编写工作。

16.2　SSM 框架整合

16.2.1　配置 SSM 开发环境

整个项目的框架是基于 Spring＋Spring MVC＋Mybatis 实现的，因此，首先完成这三种框架整合所需要的 jar 包，详情可以参考本书的第 15 章。项目所需 jar 包如图 16.23 所示：

将 jar 包复制到项目的 WebContent 目录的 WEB-INF 下的 lib 文件夹中，这样项目就可以自动进行依赖了。

16.2.2　相关的配置文件

通过前面的学习，可以知道框架整合除了所需要的 jar 包之外，还需要进行相关的配置。接下来将完成所需配置文件的编写，主要包括数据库的配置文件、Spring 的配置文件、Spring MVC 的配置文件，最后还要在 web.xml 的配置文件中添加实现 Spring 和 Spring

- WEB-INF
  - lib
    - c3p0-0.9.1.2.jar
    - commons-logging-1.2.jar
    - druid-1.1.2.jar
    - fastjson-1.1.22.jar
    - jstl.jar
    - mail.jar
    - mybatis-3.4.2.jar
    - mybatis-spring-1.3.1.jar
    - mysql-connector-java-5.1.10-bin.jar
    - spring-aop-5.0.8.RELEASE.jar
    - spring-aspects-5.0.8.RELEASE.jar
    - spring-beans-5.0.8.RELEASE.jar
    - spring-context-5.0.8.RELEASE.jar
    - spring-core-5.0.8.RELEASE.jar
    - spring-expression-5.0.8.RELEASE.jar
    - spring-jdbc-5.0.8.RELEASE.jar
    - spring-test-5.0.8.RELEASE.jar
    - spring-tx-5.0.8.RELEASE.jar
    - spring-web-5.0.8.RELEASE.jar
    - spring-webmvc-5.0.8.RELEASE.jar
    - standard.jar
    - ValidateCode.jar

图 16.23　图 16.13 所需 jar

MVC 的相关配置。

数据库连接配置文件 jdbc.properties 的内容如例 16-1 所示。

【例 16-1】 jdbc.properties

```
jdbc.driverClass = com.mysql.jdbc.Driver
jdbc.jdbcUrl =
jdbc:mysql://localhost:3306/chapter16?characterEncoding = UTF - 8
jdbc.user = root
jdbc.password = root
```

上述代码主要实现 MySql 数据库的连接配置，该配置是笔者自己的数据库账号和密码，读者需要更改为自己的。

接下来需要实现 Spring 配置文件中的相关的配置，具体实现如例 16-2 所示。

【例 16-2】 application.xml

```
<! -- 1、引入外部 properties 文件 -->
<context:property - placeholder location = "classpath:jdbc.properties" />
<! -- 2、注册数据源 -->
<bean name = "dataSource" class = "com.mchange.v2.c3p0.ComboPooledDataSource">
<property name = "driverClass" value = " $ {jdbc.driverClass}" />
<property name = "jdbcUrl" value = " $ {jdbc.jdbcUrl}" />
<property name = "user" value = " $ {jdbc.user}" />
<property name = "password" value = " $ {jdbc.password}" />
</bean>
<! -- 3、配置 Session 工厂对象 -->
<bean id = "sessionFactory"
```

```
class = "org.mybatis.spring.SqlSessionFactoryBean">
<property name = "dataSource" ref = "dataSource"></property>
</bean>
<! -- 4、配置扫描的包 -->
<bean id = "mapperScannerConfigurer"
class = "org.mybatis.spring.mapper.MapperScannerConfigurer">
<property name = "basePackage" value = "com.qfedu.dao"></property>
</bean>
```

在上述的配置中，配置了数据库的连接信息和数据库连接池对象，并且设置了 MyBatis 的扫描信息。由于篇幅有限，所以并没有给出完整的代码，具体代码可以参考本书的源码下载链接。

在 WEB-INF 下创建 Spring MVC 的配置文件 springMVC-config. xml，配置扫描的控制器类和文件上传等配置信息，如例 16-3 所示。

【例 16-3】 springMVC-config. xml

```
<?xml version = '1.0' encoding = 'UTF - 8'?>
<beans xmlns = "http://www.springframework.org/schema/beans"
xmlns:xsi = "http://www.w3.org/2001/XMLSchema - instance"
xmlns:context = "http://www.springframework.org/schema/context"
xmlns:mvc = "http://www.springframework.org/schema/mvc"
xsi:schemaLocation = "http://www.springframework.org/schema/beans
http://www.springframework.org/schema/beans/spring - beans.xsd
http://www.springframework.org/schema/context
http://www.springframework.org/schema/context/spring - context.xsd
http://www.springframework.org/schema/mvc
http://www.springframework.org/schema/mvc/spring - mvc.xsd">
<! -- 扫描 Controller -->
<context:component - scan base - package = "com.qfedu.controller" />
<! -- 配置视图解析器 -->
<bean
class = "org.springframework.web.servlet.view.InternalResourceViewResol
ver">
<property name = "suffix" value = ".jsp" />
</bean>
<mvc:annotation - driven/>
<mvc:default - servlet - handler></mvc:default - servlet - handler>
<! -- Spring MVC 上传文件时,需要配置 MultipartResolver 处理器 -->
<bean id = "multipartResolver"
class = "org.springframework.web.multipart.commons.CommonsMultipartReso
lver">
<property name = "defaultEncoding" value = "UTF - 8" />
<! -- 指定所上传文件的总大小,单位为字节.注意 maxUploadSize 属性的限制不是针对单个
文件,而是所有文件的容量之和 -->
<property name = "maxUploadSize" value = "10240000" />
</bean>
</beans>
```

本系统中由于有大量商品信息的展示，所以少不了使用商品的图片信息。因此，在上述配置中，需要注意的是不要忘记文件上传的配置信息。

16.3 锋迷网数据库设计

16.3.1 用户表

本章实现的是一个在线商务网站，根据需求分析设计如下的数据库和对应的数据表：

创建本章对应的数据库 chapter16 和数据表。实现用户表的 SQL 语句如例 16-4 所示。

【例 16-4】 数据库表结构

```
create database chapter16;
use chapter16;
#创建用户表
create table t_user(
    id int primary key auto_increment,
    role int comment '角色类型 0 管理员 1 会员',
    username varchar(50),password varchar(50),
    email varchar(30),
    gender varchar(2),
    createtime datetime,
    flag int comment '标记位  1 未激活  2 有效  3 临时无效 4 永久无效',
    activatecode varchar(50)
);
#添加管理员账号
INSERT INTO
t_user(role ,username,PASSWORD,email,gender,createtime ,flag ,activatecode )
VALUES(1,'admin','21232f297a57a5a743894a0e4a801fc3','admin','男',
NOW() ,2 ,NULL );
```

该系统为了节约时间，将会员信息和管理员信息存储到一张表中，在 user 表中通过 role 字段来标记区分。默认新增一个管理员账号，注意，本系统密码采用的是密文存储，所以 admin 对应的密码是 admin，转换为密文存储即可。

16.3.2 购物车相关表

购物车表信息，一个用户只能有一个购物车，可是一个购物车中却可以有多个商品。此处首先设计两个数据库表，SQL 如例 16-5 所示。

【例 16-5】 购物车相关表结构

```
#创建购物车表
create table t_cart(
    id int primary key auto_increment,
    uid int,
    money bigint
);
```

```
#创建购物车详情表
create table t_cartdetail(
     id int primary key auto_increment,
     cid int,
     gid int,
     num int,
     money bigint
);
#创建触发器,实现用户新增的时候自动创建购物车
DELIMITER $
CREATE TRIGGER tri_usercart AFTER INSERT ON t_user
FOR EACH ROW
BEGIN
     INSERT INTO t_cart(uid,money) VALUES(new.id,0);
END $
```

在上述 SQL 语句中,首先实现购物车表和购物车详情表的创建,因为每个用户都有一个购物车,为方便起见这里采用触发器实现监听用户表的 insert 命令,如果用户表执行了 insert 语句,就进行触发,自动在购物车表中新增一条数据。

16.3.3 商品相关表

根据业务的需要,设计商品类型和商品表,对应的 SQL 语句如例 16-6 所示:

【例 16-6】 商品相关表结构

```
#创建商品类型表
create table t_goodstype(
     id int primary key auto_increment,
     name varchar(20),
     level int comment '1 或 2',
     parentName varchar(20),
     flag int comment '1 有效 2 无效'
);
#新增默认商品类型数据
insert into t_goodstype(name,level,parentName,flag) values('酒水饮料',
1,null,1);
insert into t_goodstype(name,level,parentName,flag) values('饼干糕点',
1,null,1);
insert into t_goodstype(name,level,parentName,flag) values('休闲零食',
1,null,1);
insert into t_goodstype(name,level,parentName,flag) values('喂养用品',
1,null,1);
insert into t_goodstype(name,level,parentName,flag) values('洗护用品',
1,null,1);
insert into t_goodstype(name,level,parentName,flag) values('美妆个护',
1,null,1);
#创建商品表
create table t_goods(
```

```
    id int primary key auto_increment,
    name varchar(50),
    price bigint,
    pubdate date,
    typeName varchar(20),
    intro varchar(200),
    picture    varchar(150),
    flag int comment '1 上架 2 下架'
);
```

在上述的 SQL 语句中，实现了商品类型表和商品表的创建。注意：商品类型只记录最多 2 级类型，没有记录太深层次的商品类型。创建表之后，在表中添加商品的类型信息，也可以以管理员的身份登录系统后，通过网页菜单工具栏添加新增的商品类型信息。

16.3.4 订单相关表

创建与订单有关的表，订单和商品之间存在多对多的关系，而且订单也需要收货地址，所以对应的 SQL 语句如例 16-7 所示。

【例 16-7】 订单相关表结构

```
#创建用户收货地址表
create table t_useraddress(
    id int primary key auto_increment,
    name varchar(50),
    phone varchar(20),
    detail varchar(150),
    uid int,
    flag int
);
#创建订单表
create table t_order(
    id varchar(100) primary key,
    uaid int,
    uid int,
    createtime datetime,
    money bigint,
    flag int
);
#创建订单详情表
create table t_orderdetail(
    id int primary key auto_increment,
    gid int,
    oid varchar(100),
    money bigint,
    num int
);
```

在上述的 SQL 语句中，主要就是实现用户收货地址表、订单表和订单详情表的创建，注意：用户和用户收货地址表存在一对多的关系，订单表和订单详情表存在一对多的关系，用户与订单表存在一对多的关系。对于锋迷网的开发而言，首要任务就是理清网站中对应的数据库的表间关系。

16.4 完成通用模块

在锋迷网的开发中，需要用到一些工具类，例如使用工具类解决邮箱中的编码格式问题、密码加密问题、订单号的生成需要的签名问题等。实现 Base64 格式的相互转换，具体代码如例 16-8 所示。

【例 16-8】 Base64Utils.java

```
package com.qfedu.common.utils;
import java.util.Base64;
public class Base64Utils {
    //编码
    public static String encode(String msg){
        return Base64.getEncoder().encodeToString(msg.getBytes());
    }
    //解码
    public static String decode(String msg){
        return new String(Base64.getDecoder().decode(msg));
    }
}
```

密码的加密算法使用 MD5 加密实现，具体代码如例 16-9 所示。

【例 16-9】 MD5Utils.java

```
package com.qfedu.common.utils;
package com.qfedu.common.utils;
import java.math.BigInteger;
import java.security.MessageDigest;
import java.security.NoSuchAlgorithmException;
  //MD5 摘要算法 签名或者简易加解密
public class MD5Utils {
    public static String md5(String password){
        try {
            //获取摘要对象
            MessageDigest md = MessageDigest.getInstance("MD5");
            //设置要签名的内容
            md.update(password.getBytes());
            //获取摘要结果
            return new BigInteger(1, md.digest()).toString(16);
        } catch (NoSuchAlgorithmException e) {
            // TODO Auto - generated catch block
```

```
                e.printStackTrace();
            }
            return null;
        }
}
```

当用户需要注册会员账号时将会用到验证码验证，本系统是采用邮箱发送验证码的方式来实现的，另外邮箱还具备推送日常活动信息的功能等，邮箱工具类实现的具体代码，如例 16-10 所示。

【例 16-10】 EmailUtils.java

```
public class EmailUtils {
public static void sendEmail(User user){
    //邮箱
    String myAccount = "lx_teach@163.com";
    //授权码
    String myPass = "java168";
    //邮箱服务器
    String SMTPHost = "smtp.163.com";
    //设置属性信息
    Properties prop = new Properties();
    //设置协议
    prop.setProperty("mail.transport.protocol", "smtp");
    //邮件服务器
    prop.setProperty("mail.smtp.host", SMTPHost);
    //认证
    prop.setProperty("mail.smtp.auth", "true");
    //1.创建会话
    Session session = Session.getDefaultInstance(prop);
    //设置是否需要调试
    session.setDebug(false);
    //2.创建发送信息
    MimeMessage message = createMsg(session,myAccount,user);
    //3.发送信息操作
    try {
        Transport tran = session.getTransport();
        //连接
        tran.connect(myAccount, myPass);
        //发送消息
        tran.sendMessage(message, message.getAllRecipients());
        //关闭
        tran.close();
    } catch (MessagingException e) {
        // TODO Auto - generated catch block
        e.printStackTrace();
    }
}
    //生成邮件消息
```

```
    private static MimeMessage createMsg(Session session, String myAccount, User user) {
        //创建消息对象
        MimeMessage message = new MimeMessage(session);
        //设置
        try {
            //3.1 发送方
            message.setFrom(new InternetAddress(myAccount, "锋迷网官方邮件",
                    "utf-8"));
            //3.2 设置接收方
            message.setRecipient(MimeMessage.RecipientType.TO, new InternetAddress
                    (user.getEmail(), user.getUsername(), "utf-8"));
            //3.3 设置主题
            message.setSubject("锋迷网激活码","utf-8");
            //获取本机的 ip 地址
            String ip = Inet4Address.getLocalHost().getHostAddress();
            String url = "http://" + ip + ":8080/chapter16/activate?e=" +
                    Base64Utils.encode(user.getEmail()) + "&c=" +
                    Base64Utils.encode(user.getActivatecode());
            //设置正文信息
            message.setContent(user.getUsername() + ",欢迎你加入我们<br>为
                    了更好体验我们的产品,请<a href='" + url + "'>单击激活
                    " + url + "</a>","text/html;charset=utf-8");
            //设置日期
            message.setSentDate(new Date());
            //保存
            message.saveChanges();
        } catch (UnsupportedEncodingException | MessagingException |
UnknownHostException e) {
            // TODO Auto-generated catch block
            e.printStackTrace();
        }
        return message;
    }
}
```

本系统经常还会用到随机数的生成,例如激活码、订单编号等随机数的生成实现。随机数生成的工具类的具体代码如例 16-11 所示。

【例 16-11】 RandomUtils.java

```
package com.qfedu.common.utils;
import java.text.SimpleDateFormat;
import java.util.Calendar;
import java.util.Random;
import java.util.UUID;
//随机数
public class RandomUtils {
    //生成激活码
    public static String createActive(){
        return getTime() + Integer.toHexString(new
```

```
Random().nextInt(900) + 100);
    }
    //设置时间戳
    public static String getTime(){
        return new SimpleDateFormat
            ("yyyyMMddHHmmssSSS").format(Calendar.getInstance().getTime());
    }
    //生成订单编号
    public static String createOrderId(){
        return getTime() + UUID.randomUUID().toString();
    }
}
```

16.5 用户模块

用户模块中需要实现两种功能，首先是用户在系统中的注册、登录、退出功能，其次是用户的查询功能，接下来，将对用户模块的代码编写步骤做出详细讲解。

1）编写 JavaBean 类 User

在工程的 src 目录下新建 com.qfedu.domain 包，在该包下新建类 User，该类用于封装用户信息，其主要代码如例 16-12 所示。

【例 16-12】 User.java

```
package com.qfedu.domain;
public class User {
    private int id;
    private String username;
    private String password;
    private String createtime;
    private int flag;
    private String email;
    private String gender;
    private String activatecode;
    private int role;
    ……
}
```

以上代码中列举了 User 类的属性，作为一个完整的 JavaBean，User 类还需要有无参构造方法和 getter/setter 方法，这些需要大家在开发过程中自行编写。

2）编写 dao 层

在项目的 src 目录下新建 com.qfedu.dao 包，在该包下新建接口 UserDao，该接口用于定义用户模块的持久层方法，具体代码如例 16-13 所示。

【例 16-13】 UserDao.java

```
public interface UserDao {
//新增
```

```
@Insert("insert into
t_user(role ,username,password,email,gender,createtime ,flag ,activatecode)
values(1,#{username},#{password},#{email},#{gender},now(),1,#{activatecode})")
int insert(User user);
//登录 用户名或密码都可以使用
@Select("select * from t_user where flag=2 and (username=#{name} or
email=#{email})")
@ResultType(User.class)
User select(String name);
//检查用户名是否存在
@Select("select count(*) from t_user where username=#{username}")
int checkName(String name);
//检查用户名是否存在
@Select("select count(*) from t_user where email=#{email}")
int checkEmail(String email);
//查询全部
@Select("select * from t_user where flag=2")
@ResultType(User.class)
List<User> selectAll();
//激活
@Update("update t_user set flag=2 where email=#{email} and
activatecode=#{code}")
int updateAcode(@Param("email")String email,@Param("code")String code);
//删除用户
@Delete("delete from t_user where id=#{id} ")
int deleteByid(int id);
//根据用户名和性别搜索用户
@Select("select * from t_user where username like CONCAT('%',#{username},'%')
and gender=#{gender}")
@ResultType(User.class)
List<User> selectsearch(@Param("username")String username ,
@Param("gender")String gender);
}
```

在上述代码中，通过使用 Mybatis 注解开发的方式实现了用户模块中对数据库表中信息的增加、删除、修改、查询等基础操作。

3）编写 service 层

在工程的 src 目录下新建 com.qfedu.service 包，在该包下新建接口 UserService，该接口用于定义用户模块的业务层方法，具体代码如例 16-14 所示。

【例 16-14】 UserService.java

```
public interface UserService {
    //新增
    boolean save(User user);
    //根据名称 用户名或者邮箱
    User getUserByName(String name);
    //查询全部
```

```
    List<User> selectAll();
    //校验用户名是否存在  注册页面
    boolean checkName(String name);
    //校验邮箱是否存在    注册页面
    boolean checkEmail(String email);
    //检查登录用户是否存在  可能是邮箱 也可能是用户名
    boolean checkLogin(String name);
    //激活
    boolean activateUser(String email,String code);
    //删除用户
    int deleteByid(int id);
    //用户搜索
    List<User> usersearch(String username,String gender);
}
```

在项目的 src 目录下新建 com.qfedu.service.impl 包，在该包下新建 UserService 接口的实现类 UserServiceImpl，该类用于实现用户模块的业务层方法，具体代码如例 16-15 所示。

【例 16-15】 UserServiceImpl.java

```
@Service
public class UserServiceImpl implements UserService{
    @Autowired
    private UserDao dao;
    @Override
    public boolean save(User user) {
        // TODO Auto-generated method stub
        //密码使用密文
        user.setPassword(MD5Utils.md5(user.getPassword()));
        return dao.insert(user)>0;
    }
    @Override
    public User getUserByName(String name) {
        // TODO Auto-generated method stub
        return dao.select(name);
    }
    @Override
    public List<User> selectAll() {
        // TODO Auto-generated method stub
        return dao.selectAll();
    }
    /* 检查用户名
     * true 表示存在,false 表示不存在 */
    @Override
    public boolean checkName(String name) {
        return dao.checkName(name)>0;
    }
    /* 检查邮箱
```

```
     *   true 表示存在,false 表示不存在 */
    @Override
    public boolean checkEmail(String email) {
        return dao.checkEmail(email)> 0;
    }
    @Override
    public boolean activateUser(String email, String code) {
        // TODO Auto - generated method stub
        if(!StrUtils.empty(email,code)){
            return dao.updateAcode(Base64Utils.decode(email),
            Base64Utils.decode(code))> 0;
        }else {
            return false;
        }
    }
    /* @Return true 表示存在,false 表示不存在
     */
    @Override
    public boolean checkLogin(String name) {
        // TODO Auto - generated method stub
        return dao.select(name)!= null;
    }
    @Override
    public int deleteByid(int id) {
        return dao.deleteByid(id);
    }
    //搜索用户(根据 username 和 gender)
    @Override
    public List < User > usersearch(String username, String gender) {
        return dao.selectsearch(username, gender);
    }
}
```

4）编写 Web 层

在 src 目录下的 com.qfedu.controller 包下新建 Servlet 类 UserController,该类主要用于实现用户模块的 Web 层方法,具体代码如例 16-16 所示。

【例 16-16】 UserController.java

```
@Controller
public class UserController {
    @Autowired
    private UserService userService;
    @Autowired
    private CartService cartService;
    //登录
    @RequestMapping("userlogin")
    public String login(String username, String password,Model model,
          HttpSession session) {
```

```
        if (!StrUtils.empty(username, password)) {
            User user = userService.getUserByName(username);
            if (user != null) {
                //校验密码
                if (user.getPassword().equals(MD5Utils.md5(password))){
                    //正确
                    //记录登录信息到会话中
                    session.setAttribute("user", user);
                    Cart cart = cartService.queryByUid(user.getId());
                    if(cart == null) {
                        cart = new Cart();
                        cart.setUid(user.getId());
                        cart.setMoney(0);
                        cartService.createCart(cart);
                    }
                    session.setAttribute("cart",
                        cartService.queryByUid(user.getId()));
                    //页面跳转
                    return "index";
                }
            }
        }
        model.addAttribute("loginMsg", "用户名或密码错误");
        return "login";
    }
    //注册
    @RequestMapping("userregister")
    public String register(User user, Model model, HttpSession session) {
        //创建激活码
        String acode = RandomUtils.createActive();
        user.setActivatecode(acode);
        if (userService.save(user)) {
            //新增成功
            session.setAttribute("acode", acode);
            //发送激活码
            EmailUtils.sendEmail(user);
            return "registerSuccess";
        } else {
            model.addAttribute("registerMsg", "服务器开小差,请稍后再来");
            return "register";
        }
    }
    //注销
    @RequestMapping("userloginout")
    public String loginout(String t, HttpSession session) {
        if (t != null) {
            session.removeAttribute("adminuser");
            return "admin/login";
        } else {
            session.removeAttribute("user");
```

```
                return "index";
            }
        }
        //校验用户名是否可用
        @RequestMapping("usercheckname")
        @ResponseBody
        public ResultBean checkname(String name) {
            //创建激活码
            if (userService.checkName(name)) {
                return ResultBean.setSuccess("OK");
            } else {
                return ResultBean.setError("ERROR");
            }
        }
        //删除用户
        @RequestMapping("userdel")
        @ResponseBody
        public int checkname(int id) {
            return userService.deleteByid(id);
        }
    //用户搜索
    @RequestMapping("usersearch")
    @ResponseBody
    public List<User> usersearch(String username,String gender) {
        List<User> users = userService.usersearch(username, gender);
        return users;
    }
    //校验邮箱是否可用
    @RequestMapping("usercheckemail")
    @ResponseBody
    public ResultBean chakeemail(String email) {
        //创建激活码
        if (userService.checkEmail(email)) {
            return ResultBean.setSuccess("OK");
        } else {
            return ResultBean.setError("ERROR");
        }
    }
    //用户列表
    @RequestMapping("userlist")
    @ResponseBody
    public List<User> list() {
        return userService.selectAll();
    }
    //管理员登录
    @RequestMapping("adminLogin")
    public String adminLogin(String username, String password,
    HttpServletRequest request) {
        if (!StrUtils.empty(username, password)) {
            User user = userService.getUserByName(username);
```

```
            if (user != null) {
                //校验密码
                if (user.getPassword().equals(MD5Utils.md5(password))) {
                        //正确
                        //记录登录信息到会话中
                    request.getSession().setAttribute("adminuser", user);
                        //页面跳转
                        return "admin/admin";
                    }
                }
            }
            request.setAttribute("loginMsg", "用户名或密码错误");
            return "admin/login";
        }
//激活激活码
@RequestMapping("/activate")
public String checkCode(String e, String c, HttpSession session) {
    if (userService.activateUser(e, c)) {
        //激活成功
            return "login";
        } else {
            return "index";
        }
    }
}
```

5）编写注册页面

register.jsp 是本系统的注册页面，由于搭建开发环境时已经导入所有的 JSP 页面文件，因此，书中不再给出 JSP 页面文件的全部代码，同时，大家在学习编写本系统时无须自行编写 JSP 页面文件，只需理解 JSP 页面文件中的主要代码即可。

注册页面的核心代码，通过表单标签实现数据的收集和提交，如例 16-17 所示。

【例 16-17】 register.jsp

```
< form class = "form - horizontal" action = "userregister" method = "post">
< div class = "form - group">
< label class = "col - sm - 2 control - label">用户名: </label >
< div class = "col - sm - 8" style = "width: 40 % ">
< input type = "text" id = "username" name = "username" class = "form - control
        col - sm - 10"placeholder = "请输入用户名……" />
</div >
< div class = "col - sm - 2">
< p class = "text - danger">< span class = "help - block "
id = "usernameMsg"></span ></p >
</div >
</div >
```

```
<div class = "form - group">
<label class = "col - sm - 2 control - label">密码: </label>
<div class = "col - sm - 8" style = "width: 40 % ">
<input type = "password" name = "password"
class = "form - control col - sm - 10" placeholder = "请输入密码…… " />
</div>
<div class = "col - sm - 2">
<p class = "text - danger"><span id = "helpBlock" class = "help - block ">
请不要输入 6 位以上字符</span></p>
</div>
</div>
<div class = "form - group">
<label class = "col - sm - 2 control - label">确认密码: </label>
<div class = "col - sm - 8" style = "width: 40 % ">
<input type = "password" class = "form - control col - sm - 10"
placeholder = "请确认密码…… " />
</div>
<div class = "col - sm - 2">
<p class = "text - danger"><span id = "helpBlock" class = "help - block ">
两次密码要输入一致哦</span></p>
</div>
</div>
<div class = "form - group">
<label class = "col - sm - 2 control - label">邮箱: </label>
<div class = "col - sm - 8" style = "width: 40 % ">
<input type = "text" name = "email" id = "email" class = "form - control
col - sm - 10"
placeholder = "请输入邮箱…… " />
</div>
<div class = "col - sm - 2">
<p class = "text - danger"><span id = "emailMsg" class = "help - block ">填
写正确邮箱格式</span></p>
</div>
</div>
<div class = "form - group">
<label class = "col - sm - 2 control - label">性别: </label>
<div class = "col - sm - 8" style = "width: 40 % ">
<label class = "radio - inline">
<input type = "radio" name = "gender" checked = "checked" value = "男">男
</label>
<label class = "radio - inline">
<input type = "radio" name = "gender" value = "女">女
</label>
</div>
<div class = "col - sm - 2">
<p class = "text - danger"><span id = "helpBlock" class = "help - block ">
你是帅哥 还是美女</span></p>
</div>
</div>
<hr>
```

```
<div class = "form - group">
<div class = "col - sm - 7 col - sm - push - 2">
< input id = "registerBtn" type = "submit" value = "注册" class = "btn
btn - primary  btn - lg" style = "width: 200px;" disabled/>    
< input type = "reset" value = "重置" class = "btn btn - default  btn - lg"
style = "width: 200px;"  />
</div>
</div>
<div> $ {registerMsg}</div>
</form>
```

6）编写登录页面

login.jsp 是本系统的登录页面，用来实现用户的登录跳转。由于角色不同，因此本系统中需要有会员登录和后台模块的管理员登录这两个不同的登录页面。会员登录页面的核心代码如例 16-18 所示。

【例 16-18】 login.jsp

```
<form class = "form - horizontal" action = "userlogin" method = "post">
<div class = "form - group">
<label class = "col - sm - 2 control - label">用户名: </label>
<div class = "col - sm - 8" style = "width: 40 % ">
< input type = "text" id = "username" name = "username" class = "form - control
col - sm - 10" placeholder = "请输入用户名……" />
</div>
<div class = "col - sm - 2">
<p class = "text - danger">< span class = "help - block "
id = "usernameMsg"></span></p>
</div>
</div>
<div class = "form - group">
<label class = "col - sm - 2 control - label">密码: </label>
<div class = "col - sm - 8" style = "width: 40 % ">
< input type = "password" name = "password" class = "form - control col - sm - 10"
placeholder = "请输入密码……" />
</div>
<div class = "col - sm - 2">
<p class = "text - danger">< span id = "helpBlock" class = "help - block ">请不要输
入 6 位以上字符</span></p>
</div>
</div>
<hr>
<div class = "form - group">
<div class = "col - sm - 7 col - sm - push - 2">
< input id = "registerBtn" type = "submit" value = "登录" class = "btn btn - primary btn - lg"
      style = "width: 200px;"/>    
< input type = "reset" value = "重置" class = "btn btn - default  btn - lg"
style = "width: 200px;"  />
</div>
```

```
</div>
<div>${registerMsg}</div>
</form>
```

后台模块的管理员登录页面,在 admin 文件夹下,该页面主要是对锋迷网的数据进行管理操作,代码如例 16-19 所示。

【例 16-19】 login.jsp

```
<form action="${pageContext.request.contextPath}/adminLogin"
method="post">
<div class="form-group">
<label>用户名:</label>
<input type="text" name="username" id="" class="form-control" placeholder="请输入用户名……"/>
</div>
<div class="form-group">
<label>密    码:</label>
<div class="input-group">
<input type="password" name="password" id="password" class="form-control"
    placeholder="请输入密码……"/>
<span class="input-group-addon">
<span class="glyphicon glyphicon-eye-open" id="eye"></span>
</span>
</div>
</div>
<div class="form-group" style="text-align: center;">
<input type="submit" value="登录" class="btn btn-primary">
<input type="reset" value="重置" class="btn btn-default">
</div>
</form>
```

7) 编写用户列表页面

在 amdin 文件夹下,需要一个用户列表页面来展示当前系统中已存在的所有用户信息,并且可以对这些用户进行操作,代码如例 16-20 所示。

【例 16-20】 UserList.jsp

```
<script type="text/javascript">
    $(function(){
        loadUser();
    })
    //连接 servlet 获取数据
    function loadUser(){
        $.ajax({
            url:"${pageContext.request.contextPath}/userlist",
            method:"get",
            success:function(data){
                showMsg(data);
```

```
                },
                error:function(XMLHttpRequest,textStatus,errorThrown){
                    alert("失败" + XMLHttpRequest.status + ":" +
                        textStatus + ":" + errorThrown);
                }
            });
        }
        //显示用户信息
        function showMsg(list){
            $("#tb_list").html("<tr class = 'tr_head'><td>编号</td><td>邮箱</td>
                <td>姓名</td><td>性别</td><td>类别</td><td>操作</td></tr>");
            var i = 1;
            for(var u in list){
                //声明 tr  td  对象
                var tr = $("<tr></tr>");
                var td1 = $("<td>" + (i++) + "</td>");
                var td2 = $("<td>" + list[u].email + "</td>");
                var td3 = $("<td>" + list[u].username + "</td>");
                var td4 = $("<td>" + list[u].gender + "</td>");
                var td5 = $("<td>" + (list[u].role == 0?"管理员":"会员") + "</td>");
                var td6 = $("<td><a href = 'javascript:delUser(" + list[u].id + ")'
                    class = 'btn btn - primary btn - xs'>删除</a></td>");
                //将 td 添加到 tr 中
                tr.append(td1);
                tr.append(td2);
                tr.append(td3);
                tr.append(td4);
                tr.append(td5);
                tr.append(td6);
                $("#tb_list").append(tr);
            }
        }
        //删除用户
        function delUser(id){
            if(confirm("确认要删除吗?")){
                $.ajax({
                url:"${pageContext.request.contextPath}/userdel?id = " + id,
                    method:"get",
                    success:function(data){
                        if(data > 0){
                            loadUser();
                        }
                    },
                error:function(XMLHttpRequest,textStatus,errorThrown){
                        alert("失败" + XMLHttpRequest.status + ":" +
                            textStatus + ":" + errorThrown);
                    }
                })
            }
        }
```

```
    //条件查询
    $ (function(){
        //给查询按钮添加单击事件
        $ ("#search").click(function(){
            var username = $ ("input[name = 'username']").val();
            var genders = $ ("input[name = 'gender']");
            var gender = "";
            for(var i = 0;i < genders.length;i++){
                if(genders[i].checked){
                    gender += genders[i].value;
                }
            }
            //使用 ajax 进行异步交互
            $ .ajax({
    url:" $ {pageContext.request.contextPath}/usersearch?username = " + username + "
                &gender = " + gender,method:"post",
                success:function(data){
                    if(data == 0){
                        alert("未找到指定内容");
                        $ ("input[name = 'username']").val("");
                        $ ("input[name = 'gender']").removeAttr("checked");
                    }else{
                        showMsg(data);
                    }
                },
                error:function(XMLHttpRequest,textStatus,errorThrown){
                    alert("失败" + XMLHttpRequest.status + ":" +
                        textStatus + ":" + errorThrown);
                }
            })
        })
    })
</script>
</head>
<body>
    <div class = "row" style = "width: 100 % ;">
        <div class = "col - md - 12">
            <div class = "panel panel - default">
                <div class = "panel - heading">会员列表</div>
                <div class = "panel - body">
                    <! -- 条件查询 -->
                    <div class = "row">
                        <div class = "col - xs - 6 col - sm - 6 col - md - 6
                            col - lg - 6">
                            <div class = "form - group form - inline">
                                <span>用户名</span>
                                <input type = "text" name = "username"
                                    class = "form - control">
                            </div>
```

```
                        </div>
        <div class = "col - xs - 3 col - sm - 3 col - md - 3 col - lg - 3">
                             <div class = "form - group form - inline">
                                 < span >性别</span >
                                     
                                 < label class = "radio - inline">
                                   < input type = "radio" name = "gender" value =
                                        "男"> 男     
                                 </label >
                                 < label class = "radio - inline">
                                   < input type = "radio"name = "gender"
                                        value = "女"> 女
                                 </label >
                             </div >
                        </div >
          < div class = "col - xs - 3 col - sm - 3 col - md - 3 col - lg - 3">
                   < button type = "button" class = "btn
                        btn - primary" id = "search">< span class = "glyphicon
                        glyphicon - search"></span ></button >
                             </div >
                        </div >
                        <! -- 列表显示 -->
                        < table id = "tb_list" class = "table table - striped
                             table - hover table - bordered">
                        </table >
                   </div >
              </div >
          </div >
     </div >
</body >
```

16.6 商品模块

16.6.1 商品类型

每种商品都有所属的类型，通过商品类型的不同可以对商品进行分类，以便更好地管理商品。商品类型的操作主要涉及对商品类型的增加、删除和查询等功能。接下来本节将对商品类型的编写步骤做详细的讲解。

1）编写 JavaBean 类 GoodsType

在 src 目录的 com. qfedu. domain 包下新建类 GoodsType，该类用于封装商品信息，其主要代码如例 16-21 所示。

【例 16-21】 GoodsType. java

```
package com.qfedu.domain;
public class GoodsType {
```

```
    private int id;
    private String name;
    private int level;
    private int flag;
    private String parentName;
    ...
}
```

以上代码块列举了 GoodsType 类的属性，作为一个完整的 JavaBean，GoodsType 类还需要有无参构造方法和 getter/setter 方法，这些需要读者在开发过程中自行编写。

2）编写 dao 层

在 src 目录的 com. qfedu. dao 包下新建接口 GoodsTypeDao，该接口用于定义商品类型的持久层方法，本章代码中都提供了注释，以方便读者理解和学习。具体代码如例 16-22 所示。

【例 16-22】 GoodsTypeDao. java

```
public interface GoodsTypeDao {
    //新增
    @Insert("insert into t_goodstype(name,level,parentName,flag)
    values(#{name},#{level},#{parentName},1)")
    int save(GoodsType gt);
    //查询一级类型
    @Select("select * from t_goodstype where level = 1")
    @ResultType(GoodsType.class)
    List<GoodsType> queryByLevel();
    //查询全部
    @Select("select * from t_goodstype")
    @ResultType(GoodsType.class)
    List<GoodsType> queryAll();
    //根据商品等级和商品名称查询商品(admin)
    @Select("select * from t_goodstype where level = #{flag} and name = #{name}")
    @ResultType(GoodsType.class)
    public List<GoodsType> queryNameAndFlag(@Param("name")String name,
    @Param("flag")int flag);
    //根据 id 删除商品类型
    @Delete("delete from t_goodstype where id = #{id}")
    public int deleteType(@Param("id")int tid);
}
```

该部分使用的是 Mybatis 注解开发的方式完成对数据库的操作，因此，不再需要创建 dao 对应的实现类。

3）编写 service 层

在 src 目录的 com. qfedu. service 包下新建接口 GoodsTypeService，该接口用于定义商品类型模块的业务层方法，具体代码如例 16-23 所示。

【例 16-23】 **GoodsTypeService. java**

```
public interface GoodsTypeService {
    //新增
    boolean save(GoodsType gt);
    //查询一级类型
    List<GoodsType> queryByLevel();
    //查询全部
    List<GoodsType> queryAll();
    //删除
    int deleteType(int tid);
    //根据商品等级和商品名称查询类型(admin)
    List<GoodsType> queryNameAndFlag(String name, int flag);
}
```

在 src 目录的 com. qfedu. service. impl 包下新建接口 GoodsTypeService 的实现类 GoodsTypeServiceImpl,该类用于实现管理模块的业务层方法,核心代码如例 16-24 所示。

【例 16-24】 **GoodsTypeServiceImpl. java**

```
@Service
public class GoodsTypeServiceImpl implements GoodsTypeService {
    @Autowired
    private GoodsTypeDao dao;
    @Override
    public boolean save(GoodsType gt) {
        //添加商品类型
        return dao.save(gt)>0;
    }
    @Override
    public List<GoodsType> queryByLevel() {
        //通过 level 查询商品类型
        return dao.queryByLevel();
    }
    @Override
    public List<GoodsType> queryAll() {
        //查询所有商品类型
        return dao.queryAll();
    }
    @Override
    public List<GoodsType> queryNameAndFlag(String name, int flag) {
        //根据商品等级和商品名称查询商品(admin)
        return dao.queryNameAndFlag(name, flag);
    }
    @Override
    public int deleteType(int tid) {
        //根据 id 删除商品类型
        return dao.deleteType(tid);
    }
}
```

4）编写 controller 层

在 src 目录的 com. qfedu. controller 包下新建类 GoodsTypeController，该类主要用于实现商品类型管理的控制器层方法，具体代码如例 16-25 所示。

【例 16-25】 GoodsTypeController. java

```
@Controller
public class GoodsTypeController {
    @Autowired
    private GoodsTypeService service;
    //跳转到新增页面
    @RequestMapping("goodstypeshowadd")
    public String showadd(HttpServletRequest request, Model model) {
        model.addAttribute("gtlist", service.queryByLevel());
        return "addGoodsType";
    }
    //实现新增类型
    @RequestMapping("goodstypeadd")
    public String add(GoodsType goodsType, HttpServletRequest request, Model model) {
        if("1".equals(goodsType.getParentName())) {
            goodsType.setLevel(1);
            goodsType.setParentName(null);
        }else {
            goodsType.setLevel(2);
        }
        goodsType.setFlag(1);
        if (service.save(goodsType)) {
            return "redirect:getGoodsType";
        } else {
            model.addAttribute("msg", "服务器异常,请稍后再来");
            return "redirect:goodstypeshowadd";
        }
    }
    //实现删除商品类型
    @RequestMapping("deleteGoodsType")
    @ResponseBody
    public String deleteGoodsType(HttpServletRequest req, int count) {
        service.deleteType(count);
        return "success";
    }
    //显示类型列表
    @RequestMapping("getGoodsType")
    public String show(HttpServletRequest request, Model model) {
        model.addAttribute("gtlist", service.queryAll());
        return "/admin/showGoodsType";
    }
    //goodstypejson
    //显示类型
    @RequestMapping("goodstypejson")
    @ResponseBody
```

```
    public List < GoodsType > showjson() {
        return service.queryByLevel();
    }
    //根据商品等级和商品名称查询类型(admin)
    @RequestMapping("selectByNameAndFlag")
    public String queryNameAndFlag(String name, int flag, Model model) {
        model.addAttribute("gtlist", service.queryNameAndFlag(name,
            flag));
        return "/admin/showGoodsType";
        }
}
```

5）编写页面 addGoodsType.jsp

addGoodsType.jsp 用于显示添加商品类型的页面，它提供了提交商品类型信息的表单，主要代码如例 16-26 所示。

【例 16-26】 addGoodsType.jsp

```
<form action = "${pageContext.request.contextPath}/goodstypeadd"
method = "post">
<div class = "row">
<div class = "form-group form-inline">
<span>所属类型</span>
<select name = "parentName">
<option value = "1">-- 请选择 --</option>
<c:forEach items = "${gtlist}" var = "type">
<c:if test = "${type.level <= 2}">
<option value = "${type.name}">${type.name}</option>
</c:if>
</c:forEach>
</select>
</div>
</div>
<div class = "row">
<div class = "form-group form-inline">
<span>种类名称</span>
<input type = "text" name = "name" class = "form-control">
</div>
</div>
<div class = "row">
<div class = "btn-group">
<button type = "reset" class = "btn btn-default">清空</button>
<button type = "submit" class = "btn btn-default">添加</button>
</div>
</div>
</form>
<p style = "color:red">${msg}</p>
```

从以上代码可以看出，表单信息将被提交给 GoodsTypeController 的 goodstypeadd 方法处理。

6）编写页面 showGoodsType.jsp

showGoodsType.jsp 用于展示商品类型列表，主要代码如例 16-27 所示。

【例 16-27】 showGoodsType.jsp

```
<body>
<div class="row" style="width:98%;margin-left: 1%;">
   <div class="col-xs-12 col-sm-12 col-md-12 col-lg-12">
      <div class="panel panel-default">
         <div class="panel-heading">商品类型</div>
            <div class="panel-body">
               <div class="row">
                  <div class="col-xs-5 col-sm-5 col-md-5 col-lg-5">
                     <div class="form-group form-inline">
                     <span>商品类型等级</span>
                     <input type="text" name="flag" class="form-control">
                     </div>
                  </div>
               <div class="col-xs-5 col-sm-5 col-md-5 col-lg-5">
               <div class="form-group form-inline">
                 <span>商品类型名称</span>
               <input type="text" name="name" class="form-control">
               </div>
               </div>
<div class="col-xs-2 col-sm-2 col-md-2 col-lg-2">
<button type="button" class="btn btn-primary" id="search"><span
class="glyphicon glyphicon-search">搜索</span></button>
</div>
</div>
<div style="height: 400px;overflow: scroll;">
<table id="tb_list" class="table table-striped table-hover table-bordered">
     <tr>
          <td>序号</td>
          <td>类型</td>
          <td>等级</td>
          <td>所属类型</td>
          <td>操作</td>
     </tr>
<c:forEach items="${gtlist}" var="gtype" varStatus="i">
     <tr>
          <td>${i.count}</td>
          <td>${gtype.name}</td>
          <td>${gtype.level}</td>
          <td>${gtype.parentName}</td>
          <td>
     <!-- <button>修改</button>   --><button
onclick="deleteGoodsType(${gtype.id})">删除</button></td>
```

```
        </tr>
</c:forEach>
</table>
</div>
</div>
</div>
</div>
</div>
<script type="text/javascript">
    //条件查询
$(function(){
    //给查询按钮添加单击事件
    $("#search").click(function(){
        var flag = $("input[name='flag']").val();
        var name = $("input[name='name']").val();
        location.href = "${pageContext.request.contextPath}/
            selectByNameAndFlag?flag=" + flag + "&name=" + name;
    })
})
    //连接 servlet 获取数据
    function deleteGoodsType(count){
        $.ajax({
            url:"${pageContext.request.contextPath}/deleteGoodsType",
            method:"get",
            data:{"count":count},
            success:function(data){
                location.href =
                    "${pageContext.request.contextPath}/getGoodsType";
            },
            error:function(XMLHttpRequest,textStatus,errorThrown){
                alert("失败" + XMLHttpRequest.status +
                    ":" + textStatus + ":" + errorThrown);
            }
        });
    }
</script>
</body>
```

16.6.2 商品

商品模块用于实现对整个网站的商品的管理，其中包括对后台模块的商品管理，如增加、查询、修改等操作以及实现对应商品的列表展示、详情显示等功能，接下来本节将对商品模块的编写步骤做详细讲解。

1）编写 JavaBean 类 Goods

在 src 目录的 com.qfedu.domain 包下新建类 Goods，该类用于封装商品信息，其主要代码如例 16-28 所示。

【例 16-28】 Goods. java

```
package com.qfedu.domain;
public class Goods {
    private int id;
    private String name;
    private double price;
    private String pubdate;
    private String typeName;
    private String intro;
    private String picture;
    private int flag;
    private int star;
        ……
}
```

以上代码块中列举了 Goods 类的属性，作为一个完整的 JavaBean，Goods 类还需要有无参构造方法和 getter/setter 方法，这些需要大家在开发过程中自行编写。

2）编写 dao 层

在 src 目录下的 com. qfedu. dao 包下新建接口 GoodsDao，该接口用于定义商品的持久层方法，具体代码如例 16-29 所示。

【例 16-29】 GoodsDao. java

```
public interface GoodsDao {
    //新增
    @Insert("insert into t_goods(name,price,pubdate,typeName,
        intro,picture,flag,star) values(#{name},#{price},#{pubdate},
        #{typeName},#{intro},#{picture},1,#{star})")
    public int save(Goods goods);
    //查询
    @Select("select * from t_goods")
    @ResultType(Goods.class)
    public List<Goods> queryAll();
    //查询单个
    @Select("select * from t_goods where id = #{id}")
    @ResultType(Goods.class)
    public Goods querySingle(int id);
    //根据商品类型查询
    @Select("select * from t_goods where flag = 1 and typename = #{type}")
    @ResultType(Goods.class)
    public List<Goods> queryByType(String type);
    //查询主页的商品信息
    @Select("select * from t_goods where typename = #{type} order by star desclimit 5")
    @ResultType(Goods.class)
    public List<Goods> queryIndex(String type);
    //根据商品名模糊查询
```

```
    @Select("select * from t_goods where name LIKE CONCAT(CONCAT('%',
          #{name}),'%');")
    @ResultType(Goods.class)
    public List<Goods> queryName(String name);
    //根据商品名称和上架时间查询商品(admin)
    @Select("<script>" +
            "select * from t_goods" +
            "<where>" +
                "<if test='name != null'>" +
                    "and name like concat('%', #{name}, '%')" +
                "</if>" +
                "<if test='pubdate != null'>" +
                    "and pubdate = #{pubdate}" +
                "</if>" +
            "</where>" +
            "</script>")
    @ResultType(Goods.class)
    public List<Goods> queryNameAndPub(@Param("name")String name,
          @Param("pubdate")String pubdate);
    //删除商品
    @Delete("delete from t_goods where id= #{id} ")
    int deleteById(int id);
}
```

该部分同样使用 Mybatis 注解开发的方式完成对数据库的操作，因此，不再需要创建 dao 对应的实现类。

3）编写 service 层

在 src 目录的 com.qfedu.service 包下新建接口 GoodsService，该接口用于定义商品模块的业务层方法，具体代码如例 16-30 所示。

【例 16-30】 GoodsService.java

```
public interface GoodsService {
    //新增
    boolean save(Goods goods);
    //查询
    List<Goods> queryAll();
    //模糊查询
    List<Goods> queryByName(String name);
    //根据商品类型查询
    List<Goods> queryByType(String type);
    //查询单个
    Goods querySingle(int id);
    List<List<Goods>> queryIndex();
    //根据商品名称和上架时间查询商品(admin)
    List<Goods> queryNameAndPub(String name,String pubdate);
    //删除商品
```

```
    int deleteById(int id);
}
```


在 src 目录的 com. qfedu. service. impl 包下新建接口 GoodsTypeService 的实现类 GoodsTypeServiceImpl，该类用于实现招聘管理模块的业务层方法，具体代码如例 16-31 所示。

【例 16-31】 GoodsServiceImpl. java

```
@Service
public class GoodsServiceImpl implements GoodsService{
    @Autowired
    private GoodsDao dao;
    @Override
    public boolean save(Goods goods) {
        //添加商品
        goods.setPrice(goods.getPrice());
        return dao.save(goods)>0;
    }
    @Override
    public List<Goods> queryAll() {
        //查询所有商品
        return dao.queryAll();
    }
    @Override
    public Goods querySingle(int id) {
        //根据商品 id 查询商品
        return dao.querySingle(id);
    }
    @Override
    public List<Goods> queryByType(String type) {
        //根据商品类型查询商品
        return dao.queryByType(type);
    }
    @Override
    public List<List<Goods>> queryIndex() {
        List<List<Goods>> list = new ArrayList<List<Goods>>();
        list.add(dao.queryIndex("酒水饮料"));
        list.add(dao.queryIndex("饼干糕点"));
        list.add(dao.queryIndex("休闲零食"));
        return list;
    }
    @Override
    public List<Goods> queryByName(String name) {
        //模糊查询
        return dao.queryName(name);
    }
    @Override
    public List<Goods> queryNameAndPub(String name, String pubdate) {
```

```java
        if(name != null && "".equals(name)) {
            name = null;
        }
        if(pubdate != null && "".equals(pubdate)) {
            pubdate = null;
        }
        //根据商品名称和上架时间查询商品(admin)
        return dao.queryNameAndPub(name, pubdate);
    }
    @Override
    public int deleteById(int id) {
        //删除商品
        return dao.deleteById(id);
    }
}
```

4）编写 controller 层

在 src 目录的 com.qfedu.controller 包下新建类 GoodsController，该类主要用于实现商品类型管理的控制器层方法，具体代码如例 16-32 所示。

【例 16-32】 GoodsController.java

```java
@Controller
public class GoodsController {
    @Autowired
    private GoodsService goodsService;
    @Autowired
    private GoodsTypeService service;

    //跳转到新增页面
    @RequestMapping("toAddGoods")
    public String showadd(HttpServletRequest request, Model model) {
        model.addAttribute("gtlist", service.queryByLevel());
        return "addGoods";
    }
    //商品新增
    @RequestMapping("addGoods")
    public String save(MultipartFile file, Goods goods, HttpServletRequest request)
            throws IllegalStateException, IOException {
        File dir = FileUtils.createDir
            (request.getServletContext().getRealPath("/"));
        File desFile = new File(dir,
            FileUtils.createFileName(file.getOriginalFilename()));
        file.transferTo(desFile);
        goods.setPicture(dir.getName() + "/" + desFile.getName());
        goods.setFlag(1);  //上架
        if (goodsService.save(goods)) {
            return "redirect:toAddGoods";
        } else {
```

```
                request.setAttribute("msg", "添加失败,重新再来");
                return "redirect:toAddGoods";
            }
        }
        //查看商品详情
        @RequestMapping("getGoodsById")
        public String goodsbyid(int id, Model model) {
            model.addAttribute("goods", goodsService.querySingle(id));
            return "goodsDetail";
        }
        //查看商品详情
        @RequestMapping("getGoodsIndex")
        @ResponseBody
        public List<List<Goods>> goodsindex(HttpServletResponse response)
        throws IOException {
            return goodsService.queryIndex();
        }
        //查看商品列表
        @RequestMapping("getGoodsListByTn")
        public String goodsbytn(String tn, Model model) {
            model.addAttribute("glist", goodsService.queryByType(tn));
            return "goodsList";
        }
        //查看商品列表
        @RequestMapping("getGoodsList")
        public String goodslist(Model model) {
            model.addAttribute("goodsList", goodsService.queryAll());
            return "admin/showGoods";
        }
        //模糊查询商品列表
        @RequestMapping("selectByName")
        public String selectByName(String name, Model model) {
            model.addAttribute("glist", goodsService.queryByName(name));
            return "goodsList";
            }
        //根据商品名称和上架时间查询商品(admin)
        @RequestMapping("selectByNameAndPub")
        public String queryNameAndPub(String name,String pubdate, Model model)
        {
            List<Goods> goods =  goodsService.queryNameAndPub(name,pubdate);
            model.addAttribute("goodsList",goods );
            return "admin/showGoods";
        }
        //删除商品
        @RequestMapping("goodsDeleteById")
        public String goodsDeleteById(Integer id ,Model model) {
            goodsService.deleteById(id);
            return "redirect:getGoodsList";
        }
}
```

5）编写页面 index.jsp

index.jsp 用于显示整个项目的主页，它包含了轮播图、热销商品等样式的展示，主要代码如例 16-33 所示。

【例 16-33】 index.jsp

```
<body>
    <%@ include file="header.jsp" %>
    <div id="thred" style="height:310px">
        <div id="myCarousel" class="carousel slide" >
            <!-- 轮播(Carousel)指标 -->
            <ol class="carousel-indicators">
                <li data-target="#myCarousel" data-slide-to="0"
                      class="active"></li>
                <li data-target="#myCarousel" data-slide-to="1"></li>
                <li data-target="#myCarousel" data-slide-to="2"></li>
            </ol>
            <!-- 轮播(Carousel)项目 -->
            <div class="carousel-inner">
                <div class="item active">
                    <img src="image/b1.JPG" alt="First slide">
                </div>
                <div class="item">
                    <img src="image/b2.JPG" alt="Second slide">
                </div>
                <div class="item">
                    <img src="image/b3.JPG" alt="Third slide">
                </div>
            </div>
            <!-- 轮播(Carousel)导航 -->
            <a class="left carousel-control" href="#myCarousel"
                    role="button"
                data-slide="prev"><span
                class="glyphicon glyphicon-chevron-left"
                    aria-hidden="true"></span>
                <span class="sr-only">上一张</span>
            </a><a class="right carousel-control" href="#myCarousel"
                    role="button"
                data-slide="next"><span
                class="glyphicon glyphicon-chevron-right"
                    aria-hidden="true"></span>
                <span class="sr-only">下一张</span>
            </a>
        </div>
    </div>
    <div class="fifth">
        <span class="fif_text">热销饮料</span>
    </div>
    <div class="sixth" id="data1">

```

```
    </div>
    <div class = "fifth">
        <span class = "fif_text">热销酒类</span>
    </div>
    <div class = "sixth" id = "data2">

    </div>
    <div class = "fifth">
        <span class = "fif_text">热销零食</span>
    </div>
    <div class = "sixth" id = "data3">

    </div>
    <script type = "text/javascript">
     $(function() {
         $.get("getGoodsIndex",null,function(arr){
             var s = "";
             for(var i = 0;i < arr[0].length;i++){
                 var g = arr[0][i];
                 s += "<span class = 'sindex'><a class = 'siximg' href = '$
                        {pageContext.request.contextPath}/getGoodsById?id = " +
                        g.id + "'><img src = './fmwimages/" + g.picture + "' width =
                        '234px' height = '234px' /></a><a class = 'na'>" +
                        g.name + "</a><p class = 'chip'>" + g.intro + "</p><p
                        class = 'pri'>¥" + g.price + "元</p></span>";
             }
             $("#data1").html(s);
             s = "";
             for(var i = 0;i < arr[1].length;i++){
                 var g = arr[1][i];
                 s += "<span class = 'sindex'><a class = 'siximg'
                        href = '${pageContext.request.contextPath}/getGoodsById?id = " +
                        g.id + "'><img src = './fmwimages/" + g.picture + "'
                        width = '234px' height = '234px' /></a><a class = 'na'>" +
                        g.name + "</a><p class = 'chip'>" + g.intro + "</p><p
                        class = 'pri'>¥" + g.price + "元</p></span>";
             }
             $("#data2").html(s);
             s = "";
             for(var i = 0;i < arr[2].length;i++){
                 var g = arr[2][i];
                 s += "<span class = 'sindex'><a class = 'siximg'
                        href = '${pageContext.request.contextPath}/getGoodsById?id = " +
                        g.id + "'><img src = './fmwimages/" + g.picture + "'
                        width = '234px' height = '234px' /></a><a class = 'na'>" +
                        g.name + "</a><p class = 'chip'>" + g.intro + "</p><p
                        class = 'pri'>¥" + g.price + "元</p></span>";
             }
             $("#data3").html(s);

```

```
        });
    });
    </script>
    <!-- 底部 -->
    <%@ include file="footer.jsp" %>
    </body>
```

从以上代码可以看出，表单信息将被提交给 GoodsTypeController 的 goodstypeadd 方法处理。

6）编写页面 goodsList.jsp

goodsList.jsp 是用来展示商品列表的页面，该页面通过传递的商品类型来获取此类型下的所有商品并完成商品展示功能，主要代码如例 16-34 所示。

【例 16-34】 goodsList.jsp

```
<body>
    <%@ include file="header.jsp" %>
<div class="panel panel-default" style="margin: 0 auto;width: 95%;">
    <div class="panel-heading">
        <h3 class="panel-title"><span class="glyphicon
              glyphicon-th-list"></span>  商品列表</h3>
    </div>
    <div class="panel-body">
            <!--列表开始-->
        <div class="row" style="margin: 0 auto;">
            <c:forEach items="${glist}" var="g" varStatus="i">
                <div class="col-sm-3">
                    <div class="thumbnail">
                      <img src="./fmwimages/${g.picture}" width="180"
                          height="180"  alt="小米 6" />
                      <div class="caption">
                        <h4>商品名称:<a href="${pageContext.request.
                            contextPath}/getGoodsById?id=${g.id}">${g.name}</a></h4>
                        <p>热销指数:
                            <c:forEach begin="1" end="${g.star}">
                                <img src="image/star_red.gif" alt="star"/>
                            </c:forEach>
                        </p>
                         <p>上架日期: ${g.pubdate}</p>
                         <p style="color:orange">价格: ${g.price} 元</p>
                      </div>
                    </div>
                </div>
            </c:forEach>
        </div>
      </div>
</div>
  <!-- 底部 -->
```

```
<%@ include file = "footer.jsp" %>
</body>
```


7）编写页面 goodsDetail.jsp

goodsDetail.jsp 用于展示商品详情，通过传递的商品主键值获取该商品的详细信息进行展示，主要代码如例 16-35 所示。

【例 16-35】 goodsDetail.jsp

```
<body>
    <%@ include file = "header.jsp" %>
    <div style = "margin: 0 auto; width: 90%;">
        <ol class = "breadcrumb">
            <li><a href = "#">锋迷网</a></li>
            <li class = "active"><a href = "getGoodsListByTn?tn = $
                    {goods.typeName}">${goods.typeName}</a></li>
        </ol>
    </div>
    <div class = "container">
        <div class = "row">
            <div class = "col-xs-6 col-md-6">
                <a href = "#" class = "thumbnail"><img src = "./fmwimages/$
                        {goods.picture}" width = "560" height = "560"
                    alt = "${goods.name}" />
                </a>
            </div>
            <div class = "col-xs-6 col-md-6">
                <div class = "panel panel-default" style = "height: 560px">
                    <div class = "panel-heading">商品详情</div>
                    <div class = "panel-body">
                        <h3>
                            产品名称:<small>${goods.name}</small>
                        </h3>
                        <div style = "margin-left: 10px;">
<p>市场价格:   <span class = "text-danger"style = "font-size:
15px;">${goods.price}</span>   <span class = "glyphicon
glyphicon-yen"></span></p>
<p>上市时间:    ${goods.pubdate}</p>
<p>热销指数:   
<c:forEach begin = "1" end = "${goods.star}">
    <img src = "image/star_red.gif" alt = "star" />
</c:forEach>
</p>
<p>详细介绍:</p>
<p>   ${goods.intro }</p>
<a
href = "${pageContext.request.contextPath}/addCart?gid = ${goods.id}&price = ${
goods.price}" class = "btn btn-warning">加入购物车    <span
class = "glyphicon glyphicon-shopping-cart"></span></a>   
```

```
<a
href = " ${pageContext.request.contextPath}/getDirectOrder?id = ${goods.id}&pr
ice = ${goods.price}&name = ${goods.name}" class = "btn btn - warning">直接购
买    < span
class = "glyphicon glyphicon - shopping - cart"></span></a>
</div>
</div>
</div>
</div>
</div>
</div>
    <! -- 底部 -->
    <%@ include file = "footer.jsp" %>
</body>
```

8）编写页面 addGoods.jsp

addGoods.jsp 是用于完成新增商品信息的页面，在 admin 路径下面，通过 form 表单实现商品信息的添加，主要代码如例 16-36 所示。

【例 16-36】 addGoods.jsp

```
< form action = " ${pageContext.request.contextPath }/addGoods" method = "post"
enctype = "multipart/form - data">
< div >< h3 >新增商品</h3 ></div>
< hr />< div class = "row">
< div class = "col - sm - 6">
    < p style = "color:red"> ${msg}</p>
        < div class = "form - group form - inline">
            < label >名称:</label >
            < input type = "text" name = "name" class = "form - control" />
        </div>
< div class = "form - group form - inline">
    < label >分类:</label>
    < select name = "typeName" class = "form - control">
    < option value = "0">--- 请选择商品类型 ---</option>
        < c:forEach items = " ${gtlist}" var = "t">
        < option value = " ${t.name}"> ${t.name}</option>
    </c:forEach>
    </select>
</div>
< div class = "form - group form - inline">
        < label >时间:</label>
< input type = "text" name = "pubdate" readonly = "readonly" class = "form - control"
onclick = "setday(this)" />
</div>
</div>
< div class = "col - sm - 6">
    < div class = "form - group form - inline">
        < label >价格:</label>
```

```
    < input type = "text" name = "price" class = "form - control" /> 元
    </div >
< div class = "form - group form - inline">
    < label >评分 :</label >
    < input type = "text" name = "star" class = "form - control" />
     </div >
    </div >
</div >
< div class = "row">
< div class = "col - sm - 10">
  < div class = "form - group form - inline">
    < label >商品图片</label >
    < input type = "file" name = "file" />
   </div >
< div class = "form - group ">
            < label >商品简介</label >
    < textarea  name = "intro" class = "form - control" rows = "5"></textarea >
    </div >
    < div class = "form - group form - inline">
    < input type = "submit" value = "添加" class = "btn btn - primary" />
    < input type = "reset" value = "重置" class = "btn btn - default" />
    </div >
  </div >
</div >
</form >
```

9）编写页面 showGoods. jsp

showGoods. jsp 是商品的列表展示页面，该页面在 admin 路径下，只有后台管理员才能看到，主要实现整个网站的商品显示、查询等操作，主要代码如例 16-37 所示。

【例 16-37】 showGoods. jsp

```
<% @ taglib uri = "http://java. sun. com/jsp/jstl/core" prefix = "c" %>
<% @ taglib uri = "http://java. sun. com/jsp/jstl/functions" prefix = "fn" %>
<% @ page language = "java" contentType = "text/html; charset = UTF - 8"
    pageEncoding = "UTF - 8" %>
<! DOCTYPE html PUBLIC " - //W3C//DTD HTML 4. 01 Transitional//EN"
"http://www. w3. org/TR/html4/loose. dtd">
< html >
< head >
< meta http - equiv = "Content - Type" content = "text/html; charset = UTF - 8">
< meta http - equiv = "X - UA - Compatible" content = "IE = edge">
< meta name = "viewport" content = "width = device - width, initial - scale = 1">
< link rel = "stylesheet"
href = " $ {pageContext. request. contextPath }/css/bootstrap. min. css" />
< script
src = " $ {pageContext. request. contextPath }/js/jquery. min. js"></script >
< script
src = " $ {pageContext. request. contextPath }/js/bootstrap. min. js"></script >
```

```
< script
src = " $ {pageContext. request. contextPath }/js/DatePicker. js"></script >
< script type = "text/javascript">
function deleteGoods(id){
     location. href  =
" $ {pageContext. request. contextPath}/goodsDeleteById? id = " + id;
}
</script >
< title>锋迷网 - 商品列表</title >
</head >
< body >
<div class = "row" style = "width:98 % ;margin - left: 1 % ;">
     < div class = "col - xs - 12 col - sm - 12 col - md - 12 col - lg - 12">
          < div class = "panel panel - default">
               < div class = "panel - heading">
                    商品列表
               </div >
               < div class = "panel - body">
                    < div class = "row">
                         < div class = "col - xs - 5 col - sm - 5 col - md - 5 col - lg - 5">
                              < div class = "form - group form - inline">
                                   < span >商品名称</span >
                                   < input type = "text" name = "name"
                                        class = "form - control">
                              </div >
                         </div >
                         < div class = "col - xs - 5 col - sm - 5 col - md - 5 col - lg - 5">
                              < div class = "form - group form - inline">
                                   < span >上架时间</span >
                                   < input type = "text" readonly = "readonly"
                                        name = "pubdate" class = "form - control"
                                        onclick = "setday(this)">
                              </div >
                         </div >
                         < div class = "col - xs - 2 col - sm - 2 col - md - 2 col - lg - 2">
                              < button type = "button" class = "btn btn - primary"
                                        id = "search">< span class = "glyphicon
                                   glyphicon - search"></span ></button >
                         </div >
                    </div >
                    < div style = "height: 400px;overflow: scroll;">
                         < table id = "tb_list" class = "table table - striped
                                   table - hover table - bordered">
                              < tr >
                                   < td >序号</td >
                                   < td >商品名称</td >
                                   < td >价格</td >
                                   < td >图片</td >
                                   < td >上架时间</td >
                                   < td >类型</td >
                                   < td >操作</td >
                              </tr >
```

```
                    <c:forEach items="${goodsList}" var="goods"
                        varStatus="i">
                        <tr>
                            <td>${i.count}</td>
                            <td>${goods.name}</td>
                            <td>${goods.price}</td>
                            <td><img src="./fmwimages/${goods.picture}"
                                width="40px" height="40px"/></td>
                            <td>${goods.pubdate}</td>
                            <td>${goods.typeName}</td>
                            <td><button onclick="deleteGoods
                                ('${goods.id}')">删除</button>
                                <a tabindex="0" id="example${goods.id}"
                                class="btn btn-primary btn-xs"
                                role="button" data-toggle="popover"
                                data-trigger="focus"
                                data-placement="left"
                                data-content="${goods.intro}">描述
                                </a>
                                <script type="text/javascript">
                                    $(function(){
                            $("#example${goods.id}").popover();
                                    })
                                </script>
                            </td>
                        </tr>
                    </c:forEach>
                </table>
            </div>
        </div>
    </div>
  </div>
</div>
<script type="text/javascript">
//条件查询
$(function(){
   //给查询按钮添加单击事件
   $("#search").click(function(){
      var pubdate = $("input[name='pubdate']").val();
      var name = $("input[name='name']").val();
      location.href =
         "${pageContext.request.contextPath}/selectByNameAndPub?pubdate=
         " + pubdate + "&name=" + name;
   })
})
</script>
</body>
</html>
```

16.7 购物车模块

传统的购物车一般指超市中顾客去结算前暂时存放所选商品的一种手推车。该项目中的购物车模块用来实现用户对商品的操作，用户可以将自己想要购买的商品添加进购物车，也可以将购物车中的商品移除，还可以修改购物车中商品的数量、查询购物车详情等，接下来本节将对购物车的编写步骤做详细讲解。

每个用户都拥有一辆购物车，而且购物车中可以添加多个商品，所以这里采用两个类来描述用户的购物车信息。第一个类是购物车类，该类和用户是一对一的关系；第二个类是购物车详情类，主要是描述购物车和商品的对应关系。

1）编写 JavaBean 类

在 src 目录的 com.qfedu.domain 包下新建类 Cart，该类用于封装购物车信息，主要代码如例 16-38 所示。

【例 16-38】 Cart.java

```
package com.qfedu.domain;
public class Cart {
    private int id;
    private int uid;
    private double money;
    … …
}
```

以上代码块列举了 Cart 类的属性，作为一个完整的 JavaBean，Cart 类还需要有无参构造方法和 getter/setter 方法，这些需要大家在开发过程中自行编写。

在 src 目录的 com.qfedu.domain 包下新建类 CartDetail，该类用于封装购物车和订单整合信息，其主要代码如例 16-39 所示。

【例 16-39】 CartDetail.java

```
package com.qfedu.domain;
public class CartDetail {
    private int id;
    private int cid;
    private int gid;
    private int num;
    private double money;
    … …
}
```

以上代码块列举了 CartDetail 类的属性，作为一个完整的 JavaBean，CartDetail 类还需要有无参构造方法和 getter/setter 方法，这些需要大家在开发过程中自行编写。

2）编写 dao 层

在 src 目录的 com.qfedu.dao 包下新建接口 CartDao，该接口用于定义购物车和购物车详情的持久层方法，具体代码如例 16-40 所示。

【例 16-40】 CartDao.java

```
public interface CartDao {
    //新增－创建购物车
    @Insert("insert into t_cart(uid,money) values(#{uid},0)")
    public int insert(Cart cart);
    //购物车添加商品
    @Insert("insert into t_cartdetail(cid,gid,num,money)
        values(#{cid},#{gid},#{num},#{money})")
    public int insertDetail(CartDetail cd);
    //修改购物车商品
    @Update("update t_cart set money=#{money} where id=#{id}")
    public int update(Cart cd);
    //修改购物车中数量
    @Update("update t_cartdetail set num=${num},money=${money} where
        cid=#{cid} and gid=#{gid}")
    public int updateDetail(CartDetail cartdetail);
    //清空购物车
    @Update("update t_cart set money=0 where id=#{id}")
    public int updateEmpty(int id);
    //获取用户的购物车
    @Select("select * from t_cart where uid=#{uid}")
    @ResultType(Cart.class)
    public Cart queryByUid(int uid);
    //获取用户的购物车详情
    @Select("select cd.* from t_cartdetail cd left join t_cart c on
        cd.cid=c.id where c.uid=#{uid}")
    @ResultType(CartDetail.class)
    public List<CartDetail> queryByDetail(int uid);
    //删除购物车中的商品
    @Delete("delete from  t_cartdetail where cid=#{cid} and gid=#{gid}")
    public int deleteDetail(@Param("cid")int cid,@Param("gid")int gid);
    //清空购物车
    @Delete("delete from  t_cartdetail where cid=#{cid} ")
    public int deleteDetailByCid(int cid);
    //获取详情对象
    @Select("select * from t_cartdetail where cid=#{cid} and gid=#{gid}")
    @ResultType(CartDetail.class)
    public CartDetail queryByCdid(@Param("cid")int cid,@Param("gid")int gid);
    //购物车的数据
    @Select("SELECT cd.num,cd.money,cd.gid,g.name,g.price FROM t_cartdetail
        cd LEFT JOIN t_goods g ON cd.gid=g.id  WHERE cd.cid=#{cid}")
    @ResultType(ViewCart.class)
    public List<ViewCart> queryCart(int cid);
}
```

这里用的是 Mybatis 对数据库的操作，所以不用创建对应的实现类，用 MyBatis 的注解实现 SQL 语句的编写。注意：这里把提供对购物车表的操作和购物车详情表的操作都放在了一起，而且为了页面的数据需要，这里也定义了一个视图类（View Object 简称 vo 类是针对页面创建的类），在项目的 com. qfedu. common. vo 下创建类 ViewCart，代码如例 16-41 所示。

【例 16-41】 ViewCart. java

```
package com.qfedu.common.vo;
public class ViewCart {
    private String name;
    private double price;
    private int num;
    private double money;
    private int gid;   //商品主键
     … …
}
```

以上代码块列举了 ViewCart 类的属性，作为一个完整的 JavaBean，ViewCart 类还需要有无参构造方法和 getter/setter 方法，这些需要大家在开发过程中自行编写。

3）编写 service 层

在 src 目录的 com. qfedu. service 包下新建接口 CartService，该接口用于定义购物车模块的业务层方法，具体代码如例 16-42 所示。

【例 16-42】 CartService. java

```
public interface CartService {
    // 创建购物车
    public boolean createCart(Cart cart);
    // 加入购物车详情页
    public boolean add(int cid, Goods gds, int num);
    // 修改数量 购物车页面
    public boolean changeNum(int cid, Goods gds, int num);
    // 获取购物车对象
    public Cart queryByUid(int uid);
    // 购物车的数据
    public List<ViewCart> queryCart(int cid);
    // 删除购物车的商品
    public int deleteDetail(int cid, int gid);
}
```

在 src 目录的 com. qfedu. service. impl 包下新建接口 CartService 的实现类 CartServiceImpl，该类用于实现购物车模块的业务层方法，关键代码如例 16-43 所示。

【例 16-43】 CartServiceImpl. java

```
@Service
public class CartServiceImpl implements CartService{
    @Autowired
    private CartDao dao;
    @Override
```

```
public boolean createCart(Cart cart) {
    // TODO Auto - generated method stub
    return dao. insert(cart)> 0;
}
/ * 在商品详情页面 加入购物车 * /
@Override
public boolean add(int cid,Goods gds,int num) {
    // TODO Auto - generated method stub
    CartDetail detail = dao. queryByCdid(cid,gds. getId());
    //判断是新增还是修改
    if(detail == null) {
        //第一次
        CartDetail cd = new CartDetail();
        cd. setCid(cid);
        cd. setGid(gds. getId());
        cd. setNum(num);
        cd. setMoney(num * gds. getPrice());
        return dao. insertDetail(cd)> 0;
    }else {
        //N 次
        detail. setMoney(detail. getMoney() + gds. getPrice());
        detail. setNum(detail. getNum() + 1);
        return dao. updateDetail(detail)> 0;
    }
}
//更改数量
@Override
public boolean changeNum(int cid, Goods gds, int num) {
    // TODO Auto - generated method stub
    CartDetail detail = dao. queryByCdid(cid,gds. getId());
    if(num == - 1) {
        gds. setPrice( - gds. getPrice());
        detail. setNum(detail. getNum() - 1);
    }else {
        detail. setNum(detail. getNum() + 1);
    }
    detail. setMoney(detail. getMoney() + gds. getPrice());
    return dao. updateDetail(detail)> 0;
}
//查询
@Override
public Cart queryByUid(int uid) {
    // TODO Auto - generated method stub
    return dao. queryByUid(uid);
}
//详情查询
@Override
public List < ViewCart > queryCart(int cid) {
```

```
            // TODO Auto - generated method stub
            return dao.queryCart(cid);
        }
        //删除购物车中的商品
        @Override
        public int deleteDetail(int cid, int gid) {
            if(gid == 0) {//清空购物车
                return dao.deleteDetailByCid(cid);
            }else {
                return dao.deleteDetail(cid,gid);
            }
        }
}
```

在上述代码中,需要注意的是更改购物车中的商品数量,如果传递的数据为-1,约定是删除购物车中的商品。在商品详情页面,加入购物车的时候,需要验证之前购物车中是否有该商品,如果有就会修改数量,如果之前没有就是新增商品数据。

4) 编写 controller 层

在 src 目录的 com.qfedu.controller 包下新建类 CartController,该类用于实现商品类型管理的控制器层方法,具体代码如例 16-44 所示。

【例 16-44】 CartController.java

```
@Controller
public class CartController {
@Autowired
private CartService cartService;
//新增购物车
@RequestMapping("addCart")
public String add(int gid,double price,HttpSession session) {
    Goods goods = new Goods();
    goods.setId(gid);
    goods.setPrice(price);
if(cartService.add(((Cart)session.getAttribute("cart")).getId(), goods, 1)) {
        //成功
        //跳转到购物车页面
        return "cartSuccess";
    }else {
        return "index";
    }
}
//查询购物车
@RequestMapping("getCart")
public String get(HttpServletRequest req) {
    Cart cart = (Cart)req.getSession().getAttribute("cart");
    req.setAttribute("carts", cartService.queryCart(cart.getId()));
    //转发
```

```
        return "cart";
    }
    //购物车删除商品
    @RequestMapping("clearCart")
    public String clearCart(HttpServletRequest req,int gid) {
        Cart cart = (Cart)req.getSession().getAttribute("cart");
        cartService.deleteDetail(cart.getId(),gid);
        //重定向
        return "redirect:getCart";
    }
    //新增购物车
    @RequestMapping("updateCartNum")
    @ResponseBody
    public int update(int gid,double price,int num,HttpServletRequest req) {
        //创建商品对象
        Goods gd = new Goods();
        gd.setId(gid);
        gd.setPrice(price);
        Cart cart = (Cart)req.getSession().getAttribute("cart");
        cartService.changeNum(cart.getId(), gd, num);
        return 1;
    }
}
```

5）编写页面 cart.jsp

cart.jsp 是用于显示购物车中的商品信息的页面，主要显示当前登录用户的购物车中的商品信息，并且可以修改商品数量和生成订单，主要代码如例 16-45 所示。

【例 16-45】 cart.jsp

```
<script type="text/javascript">
    //数量+1
    function pNum(gid,p,no){
        var nums = $("#num_count"+no).val();
        $.ajax({
            url:"updateCartNum?gid="+gid+"&num=1&price="+p,
            method:"get",
            success:function(){
                location.href = "getCart";
            },
            error:function(){
                alert("服务器异常");
            }
        })
    }
    //数量-1  如果删除为0
    function mNum(gid,p,no){
        var num = -1; //数量
```

```
            var nums =  $("#num_count" + no).val();
            //验证是否需要删除
            if(Number(nums)<=1){
                if(confirm("确认要删除吗?")){
                    /* num = 0; */
                    location.href = "clearCart?gid=" + gid;
                    return;
                }else{
                    return;
                }
            }
            //异步
            $.ajax({
                url:"updateCartNum?gid=" + gid + "&num=" + num + "&price=" + p,
                method:"get",
                success:function(){
                    location.href = "getCart";
                },
                error:function(){
                    alert("服务器异常");
                }
            })
        }
        function clearCart(gid){
            if(confirm("确认要删除吗")){
                location.href = "clearCart?gid=" + gid;
            }
        }
</script>
</head>
<body style="background-color: #f5f5f5">
<%@ include file="header.jsp" %>
<div class="container" style="background-color: white;">
    <div class="row" style="margin-left: 40px">
        <h3>我的购物车<small>温馨提示：产品是否购买成功,以最终下单为准哦,请尽快结算</small></h3>
    </div>
    <div class="row" style="margin-top: 40px;">
        <div class="col-md-10 col-md-offset-1">
            <table class="table table-bordered table-striped table-hover">
                <tr>
                    <th>序号</th>
                    <th>商品名称</th>
                    <th>价格</th>
                    <th>数量</th>
                    <th>小计</th>
                    <th>操作</th>
                </tr>
                <c:set value="0" var="sum"></c:set>
```

```
                    <c:forEach items = " ${carts}" var = "c" varStatus = "i">
                        <tr>
                            <th>${i.count}</th>
                            <th>${c.name}</th>
                            <th>${c.price}</th>
                            <th width = "100px">
                                <div class = "input - group">
                                    <span class = "input - group - btn">
                                    <!-- 数量 - 1   -->
                                        <button class = "btn btn - default"
                                            type = "button" onclick = "mNum( ${c.gid},
                                                ${c.price}, ${i.count})"> -
                                        </button>
                                    </span>
                                    <input type = "text" class = "form - control"
                                            id = "num_count ${i.count}" value = " ${c.num}"
                                            readonly = "readonly" style = "width:40px">
                                    <span class = "input - group - btn">
                                    <!--  数量 + 1 -->
                                        <button class = "btn btn - default"
                                            type = "button" onclick = "pNum( ${c.gid},
                                              ${c.price}, ${i.count})"> + </button>
                                    </span>
                                </div>
                            </th>
                            <th>¥  ${c.money }元</th>
                            <th>
                                <button type = "button" class = "btn btn - default" onclick =
                                        "clearCart( ${c.gid})">删除</button>
                            </th>
                        </tr>
                        <c:set var = "sum" value = " ${sum + c.money}"></c:set>
                    </c:forEach>
                </table>
            </div>
        </div>
        <hr>
        <div class = "row">
            <div class = "pull - right" style = "margin - right: 40px;">
                    <div>
                        <a id = "removeAllProduct" href = "javascript:clearCart(0)"
                            class = "btn btn - default btn - lg">清空购物车</a>
                          
                        <a
                            href = " ${pageContext.request.contextPath}/getOrderView"
                            class = "btn btn - danger btn - lg">结算</a>
                    </div>
                    <br/>
                    <br/>
                    <div>
```

```
                    商品金额总计:<span id="total" class="text-danger"><b>
                    ¥  ${sum}元</b></span>
                </div>
            </div>
        </div>
    </div>
    <!-- 底部 -->
    <%@ include file="footer.jsp" %>
    </body>
```

6）编写页面 cartSuccess.jsp

cartSuccess.jsp 用于展示添加购物车成功之后的跳转页面，可以选择查看购物车或者继续购物，主要代码如例 16-46 所示。

【例 16-46】 cartSuccess.jsp

```
<div class="container">
<hr>
    <div class="row" style="width: 30%;margin: 0 auto;padding-top: 20px">
        <div class="panel panel-success">
            <div class="panel-heading">
                <h3 class="panel-title">购物车添加成功提示</h3>
            </div>
            <div class="panel-body">
                <h3 class="text-default">
                <span class="glyphicon glyphicon-ok-sign"></span>
                    添加购物车成功!</h3>
                <hr/>
                <a href="${pageContext.request.contextPath}/getCart" class=
                "btn btn-primary">查看购物车</a>    
                <a href="index.jsp" class="btn btn-default">继续购物</a>
            </div>
        </div>
    </div>
</div>
```

16.8 订单模块

订单模块主要用于实现对整个网站的订单管理，包括后台模块的订单管理，如订单预览、生成订单、我的订单、订单详情等操作和对应的订单列表显示功能，接下来本节将对订单模块的编写步骤做详细讲解。

1）编写 JavaBean 类

首先，需要设计订单类 Order 和订单详情类 OrderDetail。在 src 目录的 com.qfedu.domain 包下新建类 Order，该类用于封装订单信息，其主要代码如例 16-47 所示。

【例 16-47】 **Order.java**

```
public class Order {
private String id;
private int uid;
private int uaid;
private String createtime;
private double money;
private int flag;
private String username;
private String address;    //不是数据库的字段记录收货地址 收货人 + 手机号 + 收货地址
  … …
}
```

以上代码块列举了 Order 类的属性，作为一个完整的 JavaBean，Order 类还需要有无参构造方法和 getter/setter 方法，这些需要读者在开发过程中自行编写。

在 src 目录下的 com.qfedu.domain 包下新建类 OrderDetail，该类用于封装订单信息，其主要代码如例 16-48 所示。

【例 16-48】 **OrderDetail.java**

```
public class OrderDetail {
private int id;
private String oid;
private int gid;
private int num;
private double money;
  … …
}
```

以上代码块列举了 OrderDetail 类的属性，作为一个完整的 JavaBean，OrderDetail 类还需要有无参构造方法和 getter/setter 方法，这些需要大家在开发过程中自行编写。

2）编写 dao 层

在 src 目录的 com.qfedu.dao 包下新建接口 OrderDao，该接口用于定义订单相关的持久层方法，具体代码如例 16-49 所示。

【例 16-49】 **OrderDao.java**

```
public interface OrderDao {
    //新增订单详情
    @Insert("insert into t_orderdetail(oid,gid,money,num)
    values(#{oid},#{gid},#{money},#{num})")
    public int insertDetail(OrderDetail detail);
    //新增订单
    @Insert("insert into t_order(id,uid,uaid,money,createtime,flag)
    values(#{id},#{uid},#{uaid},#{money},now(),1)")
    public int insert(Order order);
    //修改订单状态
```

```java
    @Update("update t_order set flag=#{flag} where id=#{id}")
    public int update(@Param("id")String oid,@Param("flag")int flag);
    //查询订单列表
    @Select("select o.*,CONCAT(ua.name,'-',ua.phone,'-',ua.detail)address from
    t_order o left join t_useraddress ua on o.uaid=ua.id where o.uid=#{uid}")
    @ResultType(Order.class)
    public List<Order> queryByUid(int uid);
    //查询订单
    @Select("SELECT * FROM t_order o LEFT JOIN t_useraddress ua ON
        o.uaid=ua.id WHERE o.id=#{oid}")
    @ResultType(Order.class)
    public ViewOrder queryOrder(String oid);
    //查询订单子项
    @Select("SELECT g.*,od.num,od.money FROM t_orderdetail od LEFT JOIN
        t_goods g ON od.gid=g.id WHERE od.oid=#{oid}")
    @ResultType(Order.class)
    public List<ViewOrderDetail> queryDetailList(String oid);
    //查询订单详情  较为复杂的 SQL
    @Select("SELECT g.*,od.num,od.money,o.createtime,o.money,ua.name,ua.
        phone,ua.detail FROM t_orderdetail od LEFT JOIN t_goods g ON od.gid=
        g.id LEFT JOIN t_order o ON od.oid=o.id LEFT JOIN t_useraddress ua ON
        o.uaid=ua.id WHERE od.oid=#{oid}")
    @ResultType(Order.class)
    public ViewOrder queryDetail(String oid);
    //查询全部订单
    @Select("SELECT o.*,u.username FROM t_order o LEFT JOIN t_user u ON
        o.uid=u.id")
    @ResultType(Order.class)
    public List<Order> queryAll();
    //根据用户姓名和订单的支付状态查询订单(admin)
    @Select("<script>" +
            "select o.*,u.username from t_order o LEFT JOIN t_user u ON
            o.uid=u.id" +
            "<where>" +
            "<if test='username != null'>" +
                "and u.username like concat('%', #{username}, '%')" +
            "</if>" +
            "<if test='flag != null'>" +
                "and o.flag = #{flag}" +
            "</if>" +
            "</where>" +
            "</script>")
    @ResultType(Order.class)
    public List<Order> selectByNameAndFlag(@Param("username")String
        username,@Param("flag")Integer flag);
    //删除订单(admin)
    @Delete("delete from t_order where id=#{id} ")
    int deleteById(@Param("id")String id); }
```

这里用的是 Mybatis 对数据库的操作，所以不用创建对应的实现类，用 MyBatis 的注解实现 SQL 语句的编写。注意：该接口中定义了操作订单表的方法和操作订单详情表的方法。为了页面更好地显示数据，需要根据订单页面的要求定义订单视图类和订单详情视图类，订单视图类 ViewOrder 代码如例 16-50 所示。

【例 16-50】 ViewOrdere. java

```
public class ViewOrder {
     private String id;
     private String createtime;
     private double money;
     private int flag;
     private String name;
     private String phone;
     private String detail;
     private List < ViewOrderDetail > list;    //订单子项
      … …
}
```

订单详情视图类 ViewOrderDetail 代码如例 16-51 所示。

【例 16-51】 ViewOrderDetail. java

```
public class ViewOrderDetail {
     private String name;
     private double price;
     private int num;
     private int star;
     private String  picture;
     private String pubdate;
     private double money;
      … …
}
```

3）编写 service 层

在 src 目录的 com. qfedu. service 包下新建接口 OrderService，该接口用于定义订单模块的业务层方法，具体代码如例 16-52 所示。

【例 16-52】 OrderService. java

```
public interface OrderService {
//下单
public boolean save(String oid, int uid, int uaid);
//直接下单
public boolean insertDirect(int uid,String oid,int uaid,CartDetail cd);
//修改订单状态
public boolean update(String oid, int flag);
//查询订单列表
public List < Order > queryByUid(int uid);
//查询详情
public ViewOrder queryOrderDetailById(String oid);
```

```
//查询全部订单
public List<Order> queryAll();
//根据用户姓名和订单的支付状态查询订单(admin)
List<Order> selectByNameAndFlag(String username,Integer flag);
//删除订单
int deleteById(String id); }
```

在 src 目录的 com.qfedu.service.impl 包下新建接口 OrderService 的实现类 OrderServiceImpl,该类用于实现订单模块的业务层方法,具体代码如例 16-53 所示。

【例 16-53】 OrderServiceImpl.java

```
@Service
public class OrderServiceImpl implements OrderService{
 @Autowired
 private OrderDao dao;
 @Autowired
 private CartDao cartDao;
 //新增订单
 @Override
 public boolean save(String oid, int uid, int uaid) {
     List<CartDetail> cds = cartDao.queryByDetail(uid);
     double sum = 0;
     //添加到订单详情表
     for (int i = 0; i < cds.size(); i++) {
         OrderDetail detail = new OrderDetail();
         detail.setGid(cds.get(i).getGid());
         detail.setOid(oid);
         detail.setNum(cds.get(i).getNum());
         detail.setMoney(cds.get(i).getMoney() * 100);
         dao.insertDetail(detail);
         sum += cds.get(i).getMoney();
     }
     Order order = new Order();
     order.setId(oid);
     order.setUaid(uaid);
     order.setUid(uid);
     order.setMoney(sum);
     dao.insert(order);
     cartDao.deleteDetailByCid(cds.get(0).getCid());
     cartDao.updateEmpty(cds.get(0).getCid());
     return true;
 }
 //更改订单状态
 @Override
 public boolean update(String oid, int flag) {
     // TODO Auto-generated method stub
     return dao.update(oid, flag)>0;
 }
 //查询用户的所有订单
```

```
@Override
public List<Order> queryByUid(int uid) {
    // TODO Auto-generated method stub
    return dao.queryByUid(uid);
}
//查询订单详情
@Override
public ViewOrder queryOrderDetailById(String oid) {
    // TODO Auto-generated method stub
    ViewOrder vo = dao.queryOrder(oid);
    vo.setList(dao.queryDetailList(oid));
    return vo;
}
//查询所有订单
@Override
public List<Order> queryAll() {
    // TODO Auto-generated method stub
    return dao.queryAll();
}
//直接下单
@Override
public boolean insertDirect(int uid, String oid, int uaid, CartDetail cd) {
    Order order = new Order();
    order.setId(oid);
    order.setUaid(uaid);
    order.setUid(uid);
    order.setMoney(cd.getMoney());
    dao.insert(order);
    OrderDetail detail = new OrderDetail();
    detail.setGid(cd.getGid());
    detail.setOid(oid);
    detail.setNum(cd.getNum());
    detail.setMoney(cd.getMoney());
    dao.insertDetail(detail);
    return true;
}
@Override
public List<Order> selectByNameAndFlag(String username, Integer flag) {
    if(username != null && "".equals(username)) {
        username = null;
    }
    if(flag != null && 0 == flag) {
        flag = null;
    }
    //根据用户姓名和订单的支付状态查询订单(admin)
    return dao.selectByNameAndFlag(username, flag);
}
@Override
public int deleteById(String id) {
    // 删除订单
```

```
    return dao.deleteById(id);
}
```

4）编写 controller 层

在 src 目录的 com. qfedu. controller 包下新建类 OrderController，该类主要用于实现订单模块的控制器层方法，具体代码如例 16-54 所示。

【例 16-54】 OrderController. java

```
@Controller
public class OrderController {
  @Autowired
  private OrderService service;
  @Autowired
  private UserAddressService uaSrervice;
  @Autowired
  private CartService cartService;
  //下单
  @RequestMapping("addOrder")
  public String add(int t, int aid, HttpSession session, Model model)
  {
      User user = (User) session.getAttribute("user");
      String oid = RandomUtils.createOrderId();
      boolean res = false;
      if (t == 1) {//直接下单
          res = service.insertDirect(user.getId(), oid, aid,
          (CartDetail) session.getAttribute("direct"));
      } else {//购物车下单
          res = service.save(oid, user.getId(), aid);
      }
      if (res) {
          //下单成功
          model.addAttribute("oid", oid);
          return "pay";
      } else {
          return "index";
      }
  }
  //列表
  @RequestMapping("getAllOrder")
  public String all(Integer t,Integer aid,Model model) {
      model.addAttribute("orderList", service.queryAll());
      return "admin/showAllOrder";
  }
  //直接下单
  @RequestMapping("getDirectOrder")
  public String direct(Goods gs, Model model, HttpSession session) {
      User user = (User) session.getAttribute("user");
      List<ViewCart> cds = new ArrayList<>();
```

```
    ViewCart cd = new ViewCart();
    cd.setGid(gs.getId());
    cd.setMoney(gs.getPrice());
    cd.setPrice(gs.getPrice());
    cd.setNum(1);
    cd.setName(gs.getName());
    cds.add(cd);
    CartDetail detail = new CartDetail();
    detail.setGid(gs.getId());
    detail.setMoney(gs.getPrice());
    detail.setNum(1);
    session.setAttribute("direct", detail);
    model.addAttribute("cartList", cds);
    model.addAttribute("addList", uaSrervice.queryByUid(user.getId()));
    model.addAttribute("type", 1);
    //转发
    return "order";
}
//查询用户的所有订单
@RequestMapping("getOrderList")
public String olist(HttpSession session, Model model) {
    User user = (User) session.getAttribute("user");
    model.addAttribute("orderList", service.queryByUid(user.getId()));
    return "orderList";
}
//订单预览
@RequestMapping("getOrderView")
public String viewlist(HttpServletRequest request, Model model) {
    Cart cart = (Cart) request.getSession().getAttribute("cart");
    User user = (User) request.getSession().getAttribute("user");
    model.addAttribute("type", 2);
    request.setAttribute("cartList", cartService.queryCart(cart.getId()));
    request.setAttribute("addList", uaSrervice.queryByUid(user.getId()));
    return "order";
}
//列表
@RequestMapping("getOrderDetail")
public String list(String oid,HttpSession session, Model model) {
    model.addAttribute("od", service.queryOrderDetailById(oid));
    return "orderDetail";
}
//根据用户姓名和订单的支付状态查询订单(admin)
@RequestMapping("selectOrderByNameAndFlag")
public String selectByNameAndFlag(String username,Integer status,
Model model) {
    model.addAttribute("orderList",
```

```
service.selectByNameAndFlag(username, status));
        return "admin/showAllOrder";
}
            //删除订单(user)
@RequestMapping("userDeleteOrder")
public String userDeleteOrder(String oid,HttpSession session, Model
model) {
        service.deleteById(oid);
        return "redirect:getOrderList";
        }
        //修改订单(admin)
        @RequestMapping("sendOrder")
        public String sendOrder(String oid, Model model) {
            service.update(oid,3);
            return "redirect:getAllOrder";
        }
}
```

5）编写页面 order.jsp

order.jsp 用于显示订单预览信息，它提供了订单信息的显示和生成订单，主要代码如例 16-55 所示。

【例 16-55】 order.jsp

```
<div class = "container" style = "background - color: white;">
 <div class = "row" style = "margin - left: 40px">
     <h3 >订单预览< small >温馨提示：请添加你要邮递到的地址</small ></h3 >
 </div >
 < form action = "addOrder" method = "post">
 < div class = "row" style = "margin - top: 40px;">
 < input type = "hidden" name = "t" value = " $ {type}">
     < div class = "col - md - 10 col - md - offset - 1">
         < table class = "table table - bordered table - striped table - hover">
                 < tr >
                     < th>序号</th>
                     < th>商品名称</th>
                     < th>价格</th>
                     < th>数量</th>
                     < th>小计</th>
                 </tr >
                 < c:set value = "0" var = "sum"></c:set >
                 < c:forEach items = " $ {cartList}" var = "c" varStatus = "i">
                 < tr >
                     < th > $ {i.count}</th >
                     < th > $ {c.name}</th >
                     < th > $ {c.price}</th >
                     < th > $ {c.num}</th >
                     < th > $ {c.money}</th >
```

```
                    </tr>
                    <c:set var="sum" value="${sum+c.money}"></c:set>
                    </c:forEach>
                    <tr>
                     <td colspan="5">
                         <h5>收货地址</h5>
                         <select name="aid" style="width:60%"
                              class="form-control">
                              <c:forEach items="${addList}" var="a"
                                   varStatus="ai">
<option value="${a.id}">${a.name}  
     ${a.phone}  ${a.detail}
</option>
</c:forEach>
     </select>
     <a href="self_info.jsp">添加收货地址</a>      </td>
  </tr>
</table>
</div>
</div>
<hr>
<div class="row">
<div style="margin-left: 40px;">
  <h4>商品金额总计:<span id="total" class="text-danger"><b>
¥  ${sum}</b></span></h4>
</div>
</div>
<div class="row pull-right" style="margin-right: 40px;">
   <div style="margin-bottom: 20px;">
   <button  id="btn_add" class="btn  btn-danger btn-lg" type="submit">提
          交订单</button>
  </div>
</div>
</form>
</div>
```

6）编写页面 orderList.jsp

orderList.jsp 用于展示用户的订单信息，主要代码如例 16-56 所示。

【例 16-56】 orderList.jsp

```
<script type="text/javascript">
function showOrder(orderId){
location.href="${pageContext.request.contextPath}/getOrderDetail?oid="+orderId;
}
function deleteOrder(orderId){
location.href="${pageContext.request.contextPath}/userDeleteOrder?oid="+orderId;

}
```

```
function changeStatus(orderId){
location.href = "${pageContext.request.contextPath}/changeStatus?oid=" + orderId;
}
</script></head>
<body style="background-color:#f5f5f5">
<%@ include file="header.jsp" %>
<div class="container" style="background-color: white;">
<div class="row" style="margin-left: 40px">
    <h3>我的订单列表  
    <small>温馨提示:<em>${loginUser.username}</em>有<font
color="red">${orderList.size()}</font>个订单</small></h3>
</div>
<div class="row" style="margin-top: 40px;">
    <div class="col-md-12">
        <table id="tb_list" class="table table-striped table-hover
            table-bordered table-condensed">
        <tr>
            <th>序号</th>
            <th>订单编号</th>
            <th>总金额</th>
            <th>订单状态</th>
            <th>订单时间</th>
            <th>收货地址</th>
            <th>操作</th>
        </tr>
        <c:forEach items="${orderList}" var="order" varStatus="i">
            <tr>
                <th>${i.count}</th>
                <th>${order.id}</th>
                <th>${order.money}</th>
                <th>
                    <font color="red">
                        <c:if test="${order.flag eq 1 }">
                            未支付
                        </c:if>
                        <c:if test="${order.flag eq 2 }">
                            已支付,待发货
                        </c:if>
                        <c:if test="${order.flag eq 3 }">
                            已发货,待收货
                        </c:if>
                        <c:if test="${order.flag eq 4 }">
                            已收货,未评分
                        </c:if>
                        <c:if test="${order.flag eq 5 }">
                            订单完成
                        </c:if>
                    </font>
                </th>
```

```
                    <th>${order.createtime}</th>
                    <th>${order.address}</th>
                    <th>
                        <button type="button" class="btn btn-danger
                            btn-sm" onclick="showOrder('${order.id}')">订单详情</button>
                        <button type="button" class="btn btn-danger
                            btn-sm" onclick="deleteOrder('${order.id}')">删除订单</button>
                        <c:if test="${order.flag eq 3 }">
                            <button type="button" class="btn
                                btn-warning btn-sm" onclick="changeStatus
                                ('${order.id}')">确认收货</button>
                        </c:if>
                    </th>                    </tr>
            </c:forEach>
        </table>
        </div>
</div>
</div>
```

7）编写页面 orderDetail.jsp

orderDetail.jsp 用于展示用户订单的详情信息，比如订单对应的收货人信息、订单内部的商品信息等，主要代码如例 16-57 所示。

【例 16-57】 orderDetail.jsp

```
<script type="text/javascript">
function pay(orderId,money){
    location.href="pay.jsp?oid="+orderId+"&omoney="+money;
}
function payWeiXin(orderId,money){
    location.href="payWeixin.jsp?oid="+orderId+"&omoney="+money;
}
</script>
</head>
<%@ include file="header.jsp" %>
<div class="panel panel-default" style="margin: 0 auto;width: 95%;">
<div class="panel-heading">
    <h3 class="panel-title"><span class="glyphicon
        glyphicon-equalizer"></span>  订单详情</h3>
</div>
<div class="panel-body">
<table cellpadding="0" cellspacing="0" align="center" width="100%"
        class="table table-striped table-bordered table-hover">
    <tr>
        <td>订单编号:</td>
        <td>${od.id}</td>
        <td>订单时间:</td>
        <td>${od.createtime}</td>
    </tr>
```

```
<tr>
    <td>收件人:</td>
    <td>${od.name}</td>
    <td>联系电话:</td>
    <td>${od.phone}</td>
</tr>
<tr>
    <td>送货地址:</td>
    <td>${od.detail}</td>
    <td>总价:</td>
    <td>${od.money}</td>
</tr>
<tr>
    <td align="center">商品列表:</td>
    <td colspan="3">
        <table align="center" cellpadding="0" cellspacing="0" width="100%"
              class="table table-striped table-bordered table-hover">
            <tr align="center" class="info">
                <th>序号</th>
                <th>商品封面</th>
                <th>商品名称</th>
                <th>商品评分</th>
                <th>商品日期</th>
                <th>商品单价</th>
                <th>购买数量</th>
                <th>小计</th>
            </tr>
            <c:forEach items="${od.list}" var="item"
                   varStatus="i">
                <tr align="center">
                    <th>${i.count}</th>
                    <th>
                        <img src="./fmwimages/${item.picture}"
                             width="50px" height="50px">
                    </th>
                    <th>${item.name}</th>
                    <th>${item.star}</th>
                    <th>${item.pubdate}</th>
                    <th>${item.price}</th>
                    <th>${item.num}</th>
                    <th>${item.money}</th>
                </tr>
            </c:forEach>
        </table>
    </td>
</tr>
<tr>
    <td align="right" colspan="4" style="margin-right: 40px;">
        <a href="${pageContext.request.contextPath }/getOrderList"
              class="btn btn-danger btn-sm">返回订单列表</a>
```

```
                  
                <c:if test = "${od.flag eq 1 }">
                    <button type = "button" onclick = "pay('${od.id}',
                    '${od.money}')" class = "btn btn-warning btn-sm">支付</button>
                </c:if>
            </td>
        </tr>
</table>
    </div>
</div>
```

8）编写页面 showAllOrder.jsp

showAllOrder.jsp 用于展示锋迷网的所有的订单信息，主要是在后台系统中进行显示，主要代码如例 16-58 所示。

【例 16-58】 showAllOrder.jsp

```
<script type = "text/javascript">
function sendOrder(id){
location.href = "${pageContext.request.contextPath}/sendOrder?oid = " + id;
}
 $(function(){
     $("#search").click(function(){
         var username = $("input[name = 'username']").val();
         var status = $("select[name = 'orderStatus']
option:selected").val();
location.href = "${pageContext.request.contextPath}/selectOrderByNameAndFlag?
username = " + username + "&status = " + status;
     })
})
</script>
</head>
<body>
<div class = "row" style = "width:98%;margin-left: 1%;margin-top: 5px;">
<div class = "col-xs-12 col-sm-12 col-md-12 col-lg-12">
    <div class = "panel panel-default">
        <div class = "panel-heading">
            订单管理
        </div>
        <div class = "panel-body">
            <div class = "row">
                <div class = "col-xs-5 col-sm-5 col-md-5 col-lg-5">
                    <div class = "form-group form-inline">
                        <span>用户姓名</span>
    <input type = "text" name = "username" class = "form-control">
                    </div>
                </div>
                <div class = "col-xs-5 col-sm-5 col-md-5 col-lg-5">
```

```
                <div class="form-group form-inline">
                    <span>订单状态</span>
        <select name="orderStatus" class="form-control">
                        <option value="0">-----------</option>
                        <option value="1">未支付</option>
                        <option value="2">已支付,待发货</option>
                        <option value="3">已发货,待收货</option>
                        <option value="4">已收货,未评价</option>
                        <option value="5">完成订单</option>
                    </select>
                </div>
            </div>
            <div class="col-xs-2 col-sm-2 col-md-2 col-lg-2">
            <button type="button" class="btn btn-primary" id="search">
            <span class="glyphicon glyphicon-search"></span></button>
            </div>
        </div>
        <table id="tb_list" class="table table-striped
            table-hover table-bordered table-condensed">
            <tr>
                <td>序号</td>
                <td>订单编号</td>
                <td>总金额</td>
                <td>订单状态</td>
                <td>订单时间</td>
                <td>用户姓名</td>
                <td>操作</td>
            </tr>
    <c:forEach items="${orderList}" var="order" varStatus="i">
            <tr>
                <td>${i.count}</td>
                <td>${order.id}</td>
                <td>${order.money}</td>
                <td>
                    <c:if test="${order.flag eq 1}">
                        未支付
                    </c:if>
                    <c:if test="${order.flag eq 2}">
                        已支付,待发货
                    </c:if>
                    <c:if test="${order.flag eq 3}">
                        已发货,待收货
                    </c:if>
                    <c:if test="${order.flag eq 4}">
                        已收货,未评价
                    </c:if>
                    <c:if test="${order.flag eq 5}">
                        订单完成
                    </c:if>
```

```
                    </td>
                    <td>${order.createtime}</td>
                    <td>${order.username}</td>
                    <td>
                        <c:if test="${order.flag eq 2}">
                            <button type="button" class="btn btn-danger btn-sm"
                            onclick="sendOrder('${order.id}')">发货</button>
                        </c:if>
                    </td>
                </tr>
                </c:forEach>
            </table>
        </div>
    </div>
</div>
</div>
</body>
```

16.9 收货地址模块

收货地址模块主要用于实现用户的收货地址的管理，用户需要在订单中添加自己的收货地址信息，如果从未添加过地址可以新增收货地址，也可以对已存在的地址进行编辑修改。接下来，本书将对收货地址模块的编写步骤做详细讲解。

1）编写 JavaBean 类

在 src 目录下新建 com.qfedu.domain 包，在该包下新建类 UserAddress，该类用于封装收货地址信息，其主要代码如例 16-59 所示。

【例 16-59】 UserAddress.java

```
public class UserAddress {
  private int id;
  private String name;
  private String phone;
  private String detail;
  private int uid;
  private int flag;
   ......
}
```

以上代码列举了 UserAddress 类的属性，作为一个完整的 JavaBean，UserAddress 类还需要有无参构造方法和 getter/setter 方法，这些需要读者在开发过程中自行编写。

2）编写 dao 层

在项目的 src 目录下新建 com.qfedu.dao 包，在该包下新建接口 UserAddressDao，该接口用于定义用户收货地址模块的持久层方法，具体代码如例 16-60 所示。

【例 16-60】 UserAddressDao. java

```
public interface UserAddressDao {
    //新增
    @Insert("insert into t_useraddress(name,phone,detail,uid,flag )
values( #{name}, #{phone}, #{detail}, #{uid},1)")
    public int insert(UserAddress ua);
    //修改
    @Update("update t_useraddress set name = #{name},phone = #{phone},
        detail = #{detail} where id = #{id}")
    public int update(UserAddress ua);
    //修改状态
    @Update("update t_useraddress set flag = #{flag} where id = #{id}")
    public int updateDea(@Param("id")int id,@Param("flag")int flag);
    //查询地址
    @Select("select * from t_useraddress where uid = #{uid} order by flag desc")
    @ResultType(UserAddress.class)
    public List<UserAddress> queryByUid(int uid);
    //删除地址信息
    @Delete("delete from t_useraddress where id = #{id} ")
    int deleteById(int id);
}
```

在上述代码中，主要是实现了用户收货地址模块的一些数据库表操作，这里采用的是 Mybatis 操作的数据库，而且这里使用的是 Mybatis 注解的方式操作数据库，其中的修改状态在后面设置收货地址默认状态时会用到。

3）编写 service 层

在 src 目录下新建 com. qfedu. service 包，在该包下新建接口 UserAddressService，该接口用于定义用户收货地址模块的业务层方法，具体代码如例 16-61 所示。

【例 16-61】 UserAddressService. java

```
public interface UserAddressService {
//新增
public boolean insert(UserAddress ua);
//修改
public boolean update(UserAddress ua);
//查询地址
public List<UserAddress> queryByUid(int uid);
//删除地址
public int deleteAddress(int id);
//修改状态
public int updateDea(int id,int uid);
}
```

在项目的 src 目录下新建 com. qfedu. service. impl 包，在该包下新建 UserAddressService 接口的实现类 UserServiceAddressImpl，该类用于实现用户模块的业务层方法，具体代码如例 16-62 所示。

【例 16-62】 UserAddressServiceImpl. java

```
@Service
public class UserAddressServiceImpl implements UserAddressService{
@Autowired
private UserAddressDao dao;
//新增收货地址
@Override
public boolean insert(UserAddress ua) {
    // TODO Auto - generated method stub
    return dao.insert(ua)> 0;
}
//查询收货地址
@Override
public List< UserAddress > queryByUid(int uid) {
    // TODO Auto - generated method stub
    return dao.queryByUid(uid);
}
//修改地址
@Override
public boolean update(UserAddress ua) {
    // TODO Auto - generated method stub
    return dao.update(ua)> 0;
}
//删除地址信息
@Override
public int deleteAddress(int id) {
    // TODO Auto - generated method stub
    return dao.deleteById(id);
}
//修改为默认
@Override
public int updateDea(int id,int uid) {
    List< UserAddress > queryByUid  =  queryByUid(uid);
    if(queryByUid != null && !queryByUid.isEmpty()) {
        for(UserAddress userAddress : queryByUid) {
            if(userAddress.getFlag() == 3) {
                //将以前的默认状态修改为普通状态
                dao.updateDea(userAddress.getId(),1); }
        }
    }
    return dao.updateDea(id,3);
}
}
```

4）编写 Web 层

在 src 目录下的 com. qfedu. controller 包下新建 Servlet 类 UserAddressController，该类主要用于实现用户收货地址模块的 Web 层方法，具体代码如例 16-63 所示。

【例 16-63】 UserAddressController. java

```
@Controller
public class UserAddressController {
@Autowired
private UserAddressService service;
@RequestMapping("userAddressDel")
public String userAddressDel(int id,HttpServletRequest req) {
    //删除地址信息
    service.deleteAddress(id);
    return "redirect:userAddressShow";
}
@RequestMapping("userAddressAdd")
public String add(UserAddress ua,HttpServletRequest req) {
    //获取登录信息
    ua.setUid(((User)req.getSession().getAttribute("user")).getId());
    //保存地址
    if(service.insert(ua)) {
        //保存成功,刷新地址列表
        req.setAttribute("addList", service.queryByUid(ua.getUid()));
    }
    return "self_info";
}
@RequestMapping("useraddressupdate")
public String update(UserAddress ua,HttpServletRequest req) {
    //保存地址
    if(service.update(ua)) {
        //保存成功,刷新地址列表
        req.setAttribute("addList", service.queryByUid(ua.getUid()));
    }
    return "self_info";
}
@RequestMapping("userAddressShow")
public String show(UserAddress ua,HttpServletRequest req) {
    req.setAttribute("addList", service.queryByUid(((User)req.getSession().
        getAttribute("user")).getId())
);
    return "self_info";

}
@RequestMapping("userAddressDef")
public String userAddressDef(Integer id,HttpServletRequest req) {
service.updateDea(id,((User)req.getSession().getAttribute("user")).getId());
    return "redirect:userAddressShow";
}
}
```

5）编写 self_info. jsp 页面

self_info. jsp 是本系统的收货地址页面，用户可以在该页面编辑收货地址相关信息，然后提交到 contoller，最后持久化操作。self_info. jsp 的具体代码如例 16-64 所示。

【例 16-64】 self_info.jsp

```
<script type="text/javascript">
  function deleteAddr(id){
      var res = confirm("是否删除");
      if(res == true){
          window.location.href = "userAddressDel?id = " + id;
      }
  }
  function defaultAddr(id){
    var res = confirm("是否设为默认");
      if(res == true){
          window.location.href = "userAddressDef?id = " + id;
      }
  }
  </script>
  </head>
  <body>
  <%@ include file = "header.jsp" %>
  <! -- 网站中间内容开始 -->
  <div id = "dingdanxiangqing_body">
    <div id = "dingdanxiangqing_body_big">
      <div id = "big_left">
          <p style = "font - size:18px;margin - top: 15px">订单中心</p>
          <a id = "big_left_a" href = "dingdanxiangqing.html">我的订单
              </a><br/>
          <a id = "big_left_a" href = "">意外保险</a><br/>
          <a id = "big_left_a" href = "">团购订单</a><br/>
          <a id = "big_left_a" href = "">评价晒单</a><br/>
          <p style = "font - size:18px">个人中心</p>
          <a id = "big_left_a" href = "self_info.html">我的个人中心</a><br/>
          <a id = "big_left_a" href = "">消息通知</a><br/>
          <a id = "big_left_a" href = "">优惠券</a><br/>
          <a id = "big_left_a" href = "">收货地址</a><br/>
      </div>
      <div id = "big_right" style = "height: 500px;overflow: scroll;">
          <div style = "margin:0 20px;">
           <h3>收货地址</h3>
           <hr>
           <table class = "table table - striped table - hover table - bordered">
               <tr>
                   <th>序号</th><th>收件人</th>
                   <th>手机号</th><th>地址</th>
                   <th>操作</th>
               </tr>
               <c:forEach var = "address" items = "${addList}" varStatus = "i" >
                   <tr>
                       <Td>${i.count}</Td>
                       <td>${address.name}</td>
                       <td>${address.phone}</td>
```

```
                    < td > $ {address. detail}</td >
                    < td >
                        < button onclick = "deleteAddr( $ {address. id})"
                            class = "btn btn - danger btn - sm">删除
                            </button >   
                        < button class = "btn btn - default btn - sm"
                            data - toggle = "modal" data - target =
                            " # myModal $ {address. id}">修改</button >   
                        <! -- 弹出模态框 -->
                        < div class = "modal fade" tabindex = " - 1"
                            role = "dialog" id = "myModal $ {address. id}">
                         < div class = "modal - dialog" role = "document">
                            < div class = "modal - content">
                                < div class = "modal - header">
                                    < button type = "button" class = "close"
                                        data - dismiss = "modal">
                                        < span > &times;</span >
                                    </button >
                                    < h4 class = "modal - title">修改地址</h4 >
                                </div >
                                < form action = "useraddressupdate" method = "post"
                                      class = "form - horizontal">
                                    < div class = "motal - body">
                                        < div class = "form - group">
                                            < label class = "col - sm - 2
                                              control - label">收件人</label >
                                            < div class = "col - sm - 10">
                                                < input type = "hidden" name = "id"
                                                value = " $ {address. id}">
                                                < input type = "text" name = "name" class =
                                                "form - control" value = " $ {address. name}">
                                            </div >
                                        </div >
                                        < div class = "form - group">
                                            < label class = "col - sm - 2
                                            control - label">电话</label >
                                            < div class = "col - sm - 10">
                                            < input type = "text"
                                            name = "phone" class = "form - control"
                                            value = " $ {address. phone}">
                                            </div >
                                        </div >
                                        < div class = "form - group">
                                            < label class = "col - sm - 2
                                            control - label">详细地址</label >
                                            < div class = "col - sm - 10">
                                                < input type = "text"
                                                name = "detail" class = "form -
                                                control" value = " $ {address. detail}">
                                            </div >
```

```
                                    </div>
                                </div>
                                <div class = "motal - footer">
                                    <button type = "submit"
                                    class = "btn btn - primary">修改</button>
                                    </div>
                                </form>
                            </div>
                        </div>
                        </div>
                        <button onclick = "defaultAddr( ${address.id})"
                            class = "btn btn - primary btn - sm">设为默认</button>
                        <c:if test = " ${address.flag == 3}">
                            <span class = "badge" style = "background -
                                color:red;">默认</span>
                        </c:if>
                        <c:if test = " ${address.flag == 1}">
                            <span class = "badge">普通</span>
                        </c:if>
                    </td>
                </tr>
            </c:forEach>
        </table>
    </div>
    <br>
    <div class = "container" style = "width:960px;">
        <form action = "userAddressAdd" method = "post"
                class = "form - horizontal">
            <div class = "form - group">
                <label class = "col - sm - 2 form - label">收件人</label>
                <div class = "col - sm - 3">
                  <input type = "text" class = "form - control"
                      name = "name"/>
                </div>
            </div>
            <div class = "form - group">
                <label class = "col - sm - 2 form - label">手机号</label>
                <div class = "col - sm - 3">
                  <input type = "text" class = "form - control"
                      name = "phone"/>
                </div>
            </div>
            <div class = "form - group">
                <label class = "form - label">详细地址</label>
                <textarea rows = "3" class = "form - control"
                      name = "detail" ></textarea>
            </div>
            <div class = "form - group col - md - 12">
                <input type = "submit" class = "btn btn - primary" value = "
```

```
                                             添加地址">
                                             <a href = "${pageContext.request.contextPath}/
                                             getCart" class = "btn btn - primary">返回购物车</a>
                                </div>
                        </form>
                    </div>
                </div>
        </div>
</div>
```

上述代码是收货地址模块的前端页面，通过 form 表单提交用户在前台输入框中填写的信息，并自定义函数，通过拿到的 address.id 删除地址或者将地址设置为默认状态，通过 flag 字段值的变化更改默认状态。通过 foreach 遍历展示出该用户已经添加的所有收货地址相关信息。最后通过超链接，返回至购物车页面，即可重新结算，提交订单，更改收货地址，完成购物。

16.10 本章小结

本章首先从开发背景、需求分析、开发环境、系统预览等层面对项目案例(锋迷网)进行概括性介绍，然后讲解了该系统数据库的设计与创建，接着讲解了开发环境的搭建，最后分模块讲解了项目的编码实现。通过本章知识的学习，大家需要掌握 Spring+Spring MVC+Mybatis 的框架整合以及框架技术在项目开发中的具体应用。

16.11 习题

思考题

(1) 简述本章项目案例中用户模块的功能。

(2) 简述本章项目案例中购物车模块的实现思路。

(3) 简述本章项目案例中订单模块的实现思路。

图书资源支持

感谢您一直以来对清华版图书的支持和爱护。为了配合本书的使用，本书提供配套的资源，有需求的读者请扫描下方的“书圈”微信公众号二维码，在图书专区下载，也可以拨打电话或发送电子邮件咨询。

如果您在使用本书的过程中遇到了什么问题，或者有相关图书出版计划，也请您发邮件告诉我们，以便我们更好地为您服务。

我们的联系方式：

地　　址：北京市海淀区双清路学研大厦 A 座 701

邮　　编：100084

电　　话：010-83470236　010-83470237

资源下载：http://www.tup.com.cn

客服邮箱：tupjsj@vip.163.com

QQ：2301891038（请写明您的单位和姓名）

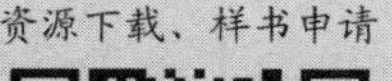

书圈

扫一扫，获取最新目录

课 程 直 播

用微信扫一扫右边的二维码，即可关注清华大学出版社公众号“书圈”。